THE SERENGETI LION
A Study of Predator-Prey Relations
George B. Schaller

"Predators are the best wildlife managers," writes George Schaller. They weed out the sick and old and keep herds healthy and alert. Yet the large predators of the world have been and are still being exterminated because they are thought to harm wildlife. Mr. Schaller's work, based on three years of study in the Serengeti National Park, describes the impact of the lion and other predators on the vast herds of wildebeest, zebra, and gazelle for which the area is famous.

The Serengeti Lion is the most intensive study ever made of the lion. Mr. Schaller's field observations provide abundant data on all aspects of lion behavior, including its social system, population dynamics, hunting behavior, predation patterns, and other factors which contribute to an understanding of this great cat as a member of the natural community. The book is thoroughly illustrated with maps, charts, sketches, and fifty photographs of lions, including several shots taken of the process of killing prey.

Among the questions Mr. Schaller explores are: Why are lions more sociable than other cats? Why are lions so aggressive at kills? Why are lions territorial but not cheetah? Mr. Schaller relates these questions to the basic evolutionary processes of adaptation and selection, and his conclusions reflect this broad biological perspective.

A thorough study of lions also entails knowledge of their competitors, and Mr. Schaller devotes individual chapters to the leopard, cheetah, wild dog, and others. The book also includes much information on the lion's prey species, including distribution, migration, and social organization. These data form the basis of Mr. Schaller's concluding chapter on the dynamics of predation, and he hopes that they may contribute to the preservation of the Serengeti and its wildlife.

GEORGE B. SCHALLER is a research associate with the New York Zoological Society and adjunct assistant professor at Rockefeller University. He is the author of *The Mountain Gorilla, The Year of the Gorilla,* and *The Deer and the Tiger,* and is editor of the *Wildlife Behavior and Ecology* series published by the University of Chicago Press.

ISBN: 0-226-73639-3
Printed in U.S.A.

THE SERENGETI LION

WILDLIFE BEHAVIOR AND ECOLOGY
George B. Schaller, Editor

THE SERENGETI LION

A Study of Predator-Prey Relations

George B. Schaller

Drawings by Richard Keane

THE UNIVERSITY OF CHICAGO PRESS
Chicago and London

599
Sch 15

International Standard Book Number: 0–226–73639–3
Library of Congress Catalog Card Number: 78–180043
The University of Chicago Press, Chicago 60637
The University of Chicago Press, Ltd., London
© *1972 by The University of Chicago*
All rights reserved. Published 1972
Printed in the United States of America

73.11-1001

HIGHLAND COMMUNITY
JUNIOR COLLEGE LIBRARY
HIGHLAND, KANSAS

Contents

To the memory of

Fairfield Osborn

President of the New York Zoological Society

His concern with the preservation of our natural heritage was a stimulus for this study and many others.

Acknowledgments

It is with pleasure that I acknowledge the help of several institutions and many individuals who made this study possible. The project was financed by the National Science Foundation and the New York Zoological Society, and it was sponsored by the Institute for Research in Animal Behavior of the New York Zoological Society and Rockefeller University. The late Fairfield Osborn, D. Griffin, W. Conway, P. Marler, and several other members of these institutions provided encouragement and assistance. A fellowship from the John Simon Guggenheim Memorial Foundation enabled me to finish writing this report.

J. Owen, then the director of the Tanzania National Parks, invited me to study the lions, and these few words are inadequate to fully express my gratitude to him for his interest in the project and for enabling me to spend three years in the Serengeti, years which were among the most enjoyable of my life. The government of Tanzania and the trustees of the Tanzania National Parks kindly permitted me to work in the Serengeti Park. The Serengeti Research Institute, established in 1966 as a result of J. Owen's vision and energy, provided excellent housing and other facilities, and its director, H. F. Lamprey, helped the project in many of its phases.[1]

Several members of the Tanzania National Parks provided me with assistance. M. Nawaz and his staff kept my Land Rover functioning, and Chief Park Warden P. A. G. Field supported my various endeavors. Deputy Chief Park Warden M. I. M. Turner, who knows the Serengeti better than anyone, was most generous in making his extensive field notes available to me, in taking time to report incidents of interest to me, and in permitting me to accompany him on an antipoaching patrol so that I could learn about this vitally important activity without which no lions would have been left for me to study. S. Makacha, a former

1. This is publication No. 86 of the Serengeti Research Institute.

park guide, carefully observed and recorded the behavior of lions in Lake Manyara National Park for two years. With initiative and enthusiasm he collected many valuable and unique data.

The National Geographic Society provided photographic assistance, and its advice and other help is gratefully acknowledged.

Several members of the Serengeti Research Institute assisted me considerably. H. Kruuk not only reported many interesting observations to me but also was most helpful with his comments and advice in matters concerning our mutual interest in predators. He also let me peruse his manuscript on hyenas prior to publication, a generous gesture which provided me with useful material with which to compare data on lions. R. Sachs did most autopsies and handled the parasitological material. I am particularly grateful to him for his contributions to this project. For other parasitological assistance I am indebted to A. S. Young, B. Schiemann, H. Hoogstraal, and A. Harthoorn. A. R. E. Sinclair aided me in many ways. I would like to thank him especially for aging my buffalo kills and providing me with much valuable unpublished data on wildebeest and buffalo. R. O. Skoog kindly shared some of his zebra records with me. H. Baldwin of the Sensory Systems Laboratory provided the project with radio telemetry equipment. I am grateful to him as well as to W. Holz and D. Baldwin for their efforts in tracking several lions and for their generosity in taking notes for me while doing so.

Many persons were helpful in reporting kills and in providing other forms of assistance. A. and J. Root photographed the project for the National Geographic Society; they also gathered much precise information on the animals they filmed, and I would like to thank them for many of the observations which appear in this report. S. and L. Trevor, H. and A. Braun, N. Myers, G. Dangerfield, P. Jarman, R. Campbell, R. Bell, S. T. Moore, and H. Hendrichs, among others, provided me with useful records. To follow lions throughout a night can be a tedious task. For help in this aspect of the work I am grateful to A. Laurie, S. Cobb, C. Craik, A. Cassels, S. Makacha, and A. Tengelin. G. Lowther was a pleasant field companion for a week while we attempted to lead the life of hominids. B. Bertram kindly gave me several observations of lions after I left the Serengeti. M. Lubin and M. Norton-Griffiths helped with the statistical aspects of the study.

C. A. W. Guggisberg shared his extensive knowledge of lions with me and permitted me to use his fine library. Through the generosity of the Elsa Wild Animal Appeal I was able to visit G. Adamson and his lions in Meru Park. Adamson knows lions as individuals better than any person, and I benefited greatly from my conversations with him. P. Joslin introduced me to the Asiatic lions in the Gir Sanctuary. L. Robinette kindly permitted me to use his preliminary aging criteria for gazelles. B. Mitchell of the Nature Conservancy sectioned several lion teeth for me. P. Leyhausen helped me with information on several aspects of cat behavior. U. de Pienaar generously made his study of predation available to me prior to its publication. To all these persons I wish to express my gratitude.

R. Keane drew the lion sketches from my photographs. The excellence of these sketches reflects in part his firsthand knowledge of the Serengeti lions. S. Torossian kindly prepared the figures.

H. Kruuk and P. Leyhausen read the whole manuscript and A. R. E. Sinclair checked chapters 7 and 13. I would like to thank them for their many helpful comments and criticisms.

My wife, Kay, contributed so much to this project that it actually was a joint venture. She not only managed our home and schooled our sons, Eric and Mark, but also spent many hours watching lion, cheetah, and wild dog. For one and a half months she carried on the field work entirely by herself. And, finally, she typed and criticized the various drafts of this report.

I THE STUDY

1 Introduction

Predation has a major influence on the dynamics of animal populations. Although there have been many studies concerning the effects of vertebrate predators on their prey, surprisingly few of them have documented the impact of large carnivores on the population level of hoofed animals. The large predators have either been exalted for their beauty or damned for the harm they supposedly do to those wildlife populations in which man has a vested interest. To observe a lion rush from concealment and in a flurry of violence bring down a zebra,[1] to observe a pack of wild dogs pursue and catch a gazelle are dramatic events, loathsome to some who watch them but beautiful to others who admire the precision of these actions, the unrestrained yet dispassionate vitality of the moment. But predation must be judged on the basis of its effect on populations, not on individuals. In recent years studies by Murie (1944), Mech (1970), Pimlott (1967), and others on the wolf and by Kruuk (1972) on the spotted hyena have shown that, in general, the young, old, and sick animals are killed, although a predator may affect its prey differently from area to area depending on such factors as the size and movements of the prey population. Both wolf and hyena capture their quarry by running it down, a technique different from the stalk and brief rush used by most large cats. It may be conjectured that stalking predators affect prey populations differently than coursing ones in areas where both types occur, as they do in much of Africa. With the exception of a study on puma by Hornocker (1970) and on tiger by Schaller (1967) few data exist on the effects of predation by large cats. This study on the lion was designed to add to the available information.

Exceeded in average length and weight only by the tiger, the lion is the second largest feline predator in the world. Within historic times it was distributed from Greece throughout the Near East to India and

1. Scientific names of animals mentioned in the text are given in Appendix A.

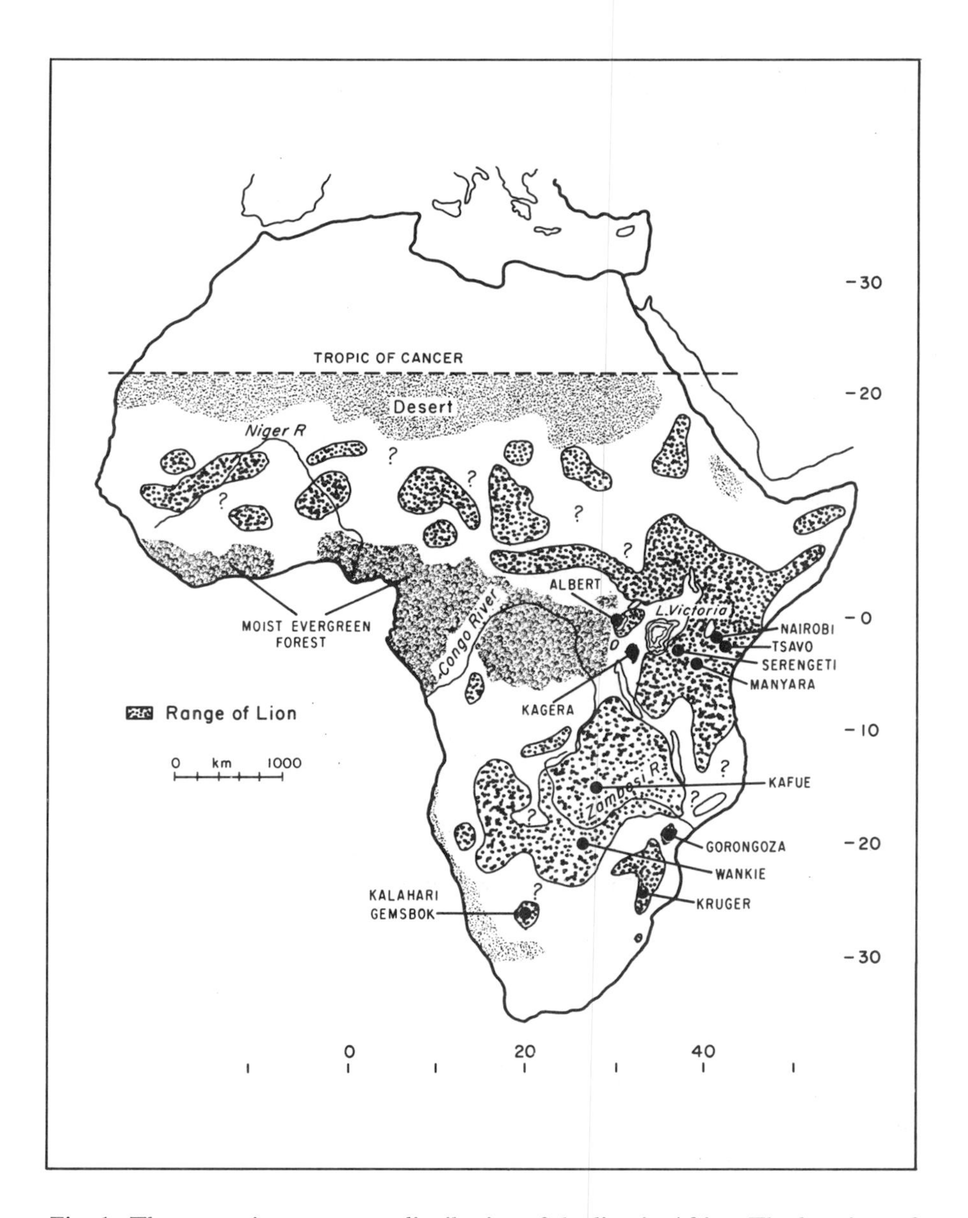

Fig. 1. The approximate present distribution of the lion in Africa. The locations of wildlife reserves mentioned in the text are also indicated. (Prepared with the assistance of C. A. W. Guggisberg.)

over much of Africa, except in the driest deserts and in the rain forests (fig. 1). Few animals have had more of an impact on the traditions and myths of the peoples it encountered. Because of its tremendous strength and majestic appearance, the lion was elevated to the position of king of beasts as long ago as the first century. It was a symbol of the sun, venerated as an animal god in the Near East, a "bold, brave, beautiful, and gallant animal," as Gesner called it in 1563. Lions became the companions of royalty. The Egyptian pharaoh Rameses II went into battle accompanied by a lion, and Assyrian kings bred them in captivity as early as 850 B.C. For centuries, too, man has pitted his strength against this cat as a means of gaining renown for courage. In 1370 B.C. the pharaoh Amenhotep II hunted lions with bow and arrow from a chariot, and during the seventh crusade, Saint Louis amused himself near Tunis by shooting lions from horseback with a crossbow. The Romans slaughtered lions in their arenas as public spectacles. "Julius Caesar celebrated the consecration of his forum with the slaughter of 400 lions, one giraffe and forty elephants. . ." (Guggisberg, 1961).

The lions of central Europe—whose profiles man scratched into the cave walls in France more than 15,000 years ago—died out before historic times. They were apparently not different from *Panthera leo* of today (Kurtén, 1968). European lions survived in remote parts of Greece into recent times. When Xerxes marched through Greece in 480 B.C. his baggage camels were attacked by lions. Herodotus considered them common at that time; by 300 B.C. Aristotle thought them rare, and by A.D. 100 they were gone. They survived in Palestine until the Crusades and well into this century in Syria, Iraq, and Iran. The only known population of the Asiatic lion, *P. l. persica* (Meyer, 1826), occurs now in the Gir Sanctuary of Gujurat State in India where about 175 animals represent the remnants of a population which 150 years ago was widely spread over the northern half of the country.

In Africa, on the other hand, lions continued to survive in the vast woodlands and plains, which for the most part saw little European penetration until the nineteenth century. Of the ten subspecies of African lion recognized by Haltenorth and Trense (1956), two—*P. l. leo* (L., 1758) the Berber lion of North Africa, and *P. l. melanochaita* (Hamilton Smith, 1846) the Cape lion of South Africa—have become extinct. Although in recent years agriculturalists have eliminated much lion habitat and pastoralists with their livestock have come into increasing conflict with the cats, huge tracts of land both inside and outside of reserves still harbor them, particularly in Kenya and Tanzania where *P. l. massaicus* (Neumann, 1900), the subject of this study, leads a natural life, roaming at will and preying largely on the local fauna.

Many books have been written about the African lion, most of them by hunters and game wardens who either shot the animals or made casual observations in the course of other duties (Selous, 1908; Roosevelt and Heller, 1922; Stevenson-Hamilton, 1954; Cowie, 1966). Lions have also been raised as pets and then returned to the wild (J. Adamson, 1960, 1961; G. Adamson, 1968; Carr, 1962). Many of these books contain interesting anecdotes and life-history notes, but none provides an accurate picture of, for example, the lion's social system and its predatory behavior. Guggisberg (1961) was the first scientist to study lions when he observed those in Nairobi National Park intermittently between 1953 and 1960. His book *Simba* is an excellent summary not only of his own work but also of the literature; it is, in effect, a companion volume to this one in that it treats the lion in history, in art, and in other aspects such as man-eating—information which would have been redundant in this report. A number of articles on the lion, particularly on food habits (Wright, 1960; Mitchell et al., 1967; Pienaar, 1969) and on territorial behavior (Schenkel, 1966a), have been published recently, but an intensive study of the lion had not been made when I began field work in 1966.

Mr. John Owen, director of the Tanzania National Parks,invited me to join the Serengeti Research Institute to study lions in the Serengeti National Park, especially to answer the question "What effect does lion predation have on prey populations?" The park administration was interested in maintaining the lion population while preserving the million or so hoofed animals that frequent the region, probably the largest such concentration in the world. As the major predator in the park, lions were thought to influence the dynamics of various prey species, and my task was to find out to what extent they did so.

The Serengeti has long been famous for its lions and for the huge herds of wildebeest, zebra, and gazelles which make their annual migration to the plains at the onset of the rainy season, then trek back to the woodlands during dry weather. In March, 1892, Baumann (1894) was the first European to visit the area, but he only noted laconically that "game was visible in large numbers." Jaeger (1911) crossed the plains in April, 1907, but saw no lions. It was in 1913, after a visit by White (1915) to the Bologonja and Mara rivers, that the Serengeti became known for its wildlife as a result of descriptions such as this: "Never have I seen anything like that game. It covered every hill, standing in the openings, strolling in and out among the groves, feeding on the bottom lands, singly, or in little groups. It did not matter in what direction I looked, there it was; as abundant one place as another."

He shot several lions that first trip, and in 1925 returned for three months to shoot fifty-one more around Seronera, the site of the future park headquarters (Turner, 1965). In 1920 the first car reached Seronera, and in the same year a Swedish cameraman, Oskar Olsson, obtained good lion photographs (Guggisberg, 1961). The Akeley-Eastman Safari camped at Seronera to collect specimens in 1926, and in 1928 and 1933 Martin and Osa Johnson (1929, 1935) photographed there for several months. Although these and other exploits made the Serengeti lions world famous, they provided little factual information by which past conditions could be compared with those of today.

The Serengeti National Park is the best area I have seen in East Africa for studying lions and other large predators. It is a huge, isolated area with open woodlands and plains that are ideal for finding and observing lions. The gently rolling terrain readily permits access by Land Rover to most places except during the rainy season when grass is high and rivers swollen. Lions within the park are rarely molested by man, although Waikoma and other tribesmen enter it to poach and sometimes catch lions in their snares. The Masai, who settled around the Serengeti plains during the first half of the nineteenth century (Fosbrooke, 1948), speared lions that raided their cattle. Even in 1954, Masai with about 25,000 head of cattle still lived in the Seronera area (Pearsall, 1957), but all were resettled outside of the park in 1959. As early as 1929 about 2,286 sq km (900 sq m) of land[2] around Seronera were designated as a game sanctuary, although the lions in it were shot until 1937 (Moore, 1938). In 1940 a total of 14,401 sq km from Lake Victoria in the west to the Crater Highlands in the east were gazetted as a park. The boundaries were realigned in 1951 and only 11,887 sq km included in the park. Increasing conflict between the demands of Masai for grazing lands and the requirements of the wildlife led to a drastic change in the park boundaries in 1959. The Crater Highlands and the eastern plains were excised from the park and a large area extending north to the Mara River was included instead. Two further small blocks were added in 1968, bringing the total park area to about 13,250 sq km (about 5,100 sq m). Throughout these vicissitudes the Seronera portion of the park remained protected. The number of lions there, and the tolerance they show to such disturbances as being approached closely in a car, is undoubtedly due in part to

2. 1 m = 3.2 feet; 1 km = .6 miles; 1 ha. = 2.5 acres; 2.6 sq km = 1 sq mile; 1 kg = 2.2 lbs; 1 cm = .39 inch.

their having had few or no adverse contacts with man in the past thirty years. It became my main study area.

The cooperation which an investigator receives from the Tanzania National Parks and the Serengeti Research Institute is another important incentive for working in the Serengeti area; housing and laboratory facilities are provided and airplanes are available. Some fifteen scientists work on ecological problems at the institute, and the mental stimulation and assistance they give is certain to make any research program more productive and enjoyable.

To determine the effects of lion predation was a complex endeavor. Among other problems, five species of large predator preyed on some twenty species of hoofed animals in the Serengeti region—in contrast to the North American predation studies in which only one carnivore subsists mainly on one or two kinds of prey. The project was, therefore, designed along several broad lines of inquiry.

1. With only a few published notes on Serengeti lions available (G. Adamson, 1964; Kühme, 1966; Kruuk and Turner, 1967), it was necessary to focus attention on all aspects of lion life-history, particularly on those having direct relevance to predator-prey relations. These include group structure and movements, general behavior within the group, population dynamics, and predation. Knowledge about each prey population is also essential if the effects of predation are to be evaluated. I censused various species in selected areas and collected data on mortality, on mating and birth seasons, and on other topics, but I was unable to gather detailed information on each species. Fortunately, the ecological surveys by Pearsall (1957) and Darling (1960) and the studies by Grzimek and Grzimek (1960), Talbot and Talbot (1963), Klingel (1967), Watson (1967), and others all contain useful information on certain prey. In addition valuable unpublished facts were given to me by several investigators, particularly by those working on zebra and buffalo.

2. The lion is only one member of the Serengeti predator community. To understand the role of the lion in the ecology of the area and in the dynamics of predation in general, the movements, social structure, food habits, and so forth of the other large predators must also be considered. Predators interact in various ways, indirectly by preying on the same species, and directly by scavenging each other's kills. Throughout the study I collected information on cheetah, leopard, and wild dog (Schaller, 1968, 1970) and I present these data to be better able to evaluate the lion as a predator. Observations on hyenas were given to

Kruuk (1972), who was conducting a major study on this species, and he incorporated them into his report, a brief summary of which is here included.

3. It was desirable to study another lion population for comparison with that in the Serengeti. In this I was fortunate to obtain the help of Stephen Makacha, who for two years observed the lions in the Lake Manyara National Park, an area of 91 sq km lying along the base of the Rift to the east of the Crater Highlands (Makacha and Schaller, 1969).

Field work began in June, 1966, and continued uninterruptedly until September, 1969, except for one and a half months when my wife, Kay, took notes on the Serengeti lions while I visited India and spent two weeks with the lions of the Gir Sanctuary. A study of only three and one-quarter years is not long enough to clarify all aspects of behavior in a long-lived predator like the lion. Weather conditions may change drastically from year to year, influencing the movements of prey which in turn affect the population dynamics of lions. Also, the Serengeti Park is the same size as the state of Connecticut, and the wildlife areas bordering it are equally as large, making it clearly impossible for one investigator to cover all parts of it thoroughly. Some of the conclusions about predator-prey relations in this report must, therefore, remain tentative, and I would like to emphasize that they apply only to the period of my study. Wildlife populations do not remain static; for example, a rinderpest epidemic as reported by Talbot and Talbot (1963) could change not only the food habits of lions considerably but also affect the existing balance between predator and prey. It is also important to realize that any conclusions about lion behavior and predator-prey relations refer only to the Serengeti National Park. Although my findings may have wider applications, only detailed studies in other areas can show whether this is so. The Serengeti, with its migrating herds, is a special situation, and the impact of lion predation on a smaller and more sedentary prey population may be different.

The main purpose of this report is to present facts about the large predators in the Serengeti National Park in the hope of contributing to an understanding of how these magnificent animals can best be preserved and possibly managed, and in order to add to our limited knowledge of the general biological problem of predator-prey relations. The pressures which the environment exerts on the structure of a society can be profound, as recent primate studies have shown (Rowell, 1967;

Gartlan, 1968), but this topic has so far received little attention from biologists. The large predators with their diverse social systems offer a good opportunity for studying the possible selective forces which act on their societies. Man, though phylogenetically a primate, has lived ecologically as a social carnivore for some two millions years, and possibly more can be learned about the evolution of his social system by studying the lion, hyena, and wild dog than by examining some vegetarian monkey (Schaller and Lowther, 1969). This book about predators mirrors man as much as any of the recent ones about monkeys and apes. The first twelve chapters present the factual information and the last two provide summaries, comparisons, and in general place the study in its biological perspective. Most readers may wish to read the last chapters first and then peruse some of the others.

Before continuing with the scientific aspects of the work, I would like to mention that the study gave me great pleasure. The Serengeti, with its flat-topped acacias and huge herds of animals, is the quintessence of Africa, a region of light and space which makes research there an aesthetically rewarding experience. To this scene the predators add tension, enhancing its vitality with an aura of impending violence. There is the lion, indolent, "radiating a kind of lazy, lordly power born of the carelessness of authority," as Edey (1968) phrased it. In contrast, the leopard seems a furtive creature of the moon and the cheetah an elegant aristocrat. In recent years the spotted hyena has metamorphosed in the public eye from a mobile garbage bin into a powerful and skilled hunter, gaining a measure of respect in spite of its "botched and skulking form" (Moorehead, 1967). There, too, is the wild dog whose mere existence arouses abhorrence and intemperance even in such a conservationist as Carl Akeley who said, after shooting down a whole pack, "That's the first shooting on this whole trip that I've enjoyed" (Akeley and Akeley, 1932).

These predators, each so different in the superficial impression they convey, differ also in their habits, thereby raising the innumerable biological questions without which even the most interesting research would become routine. As I began to recognize many of the animals individually, the study moved from an abstract plane with an emphasis on collecting quantitative data to a more emotional and hence more satisfying one. Knowing the history of many animals, I had empathy with their problems, I anticipated their future. Male No. 134, for example, is to me not merely a nomadic lion who gained and lost a territory but a living entity, a part of the vastness of the plains and the kopjes that jut from it; he represents memories of the immense silence at night when even his footsteps were noisy interruptions, of the heat waves at noon transforming distant granite boulders into visions of castles and zebra into lean Giacometti sculptures. These facets of the study are difficult to transmit yet they are a part of it and they might be kept in mind by anyone reading the dry facts in this report.

2 The Serengeti Region

The Serengeti National Park is not an ecological entity in that migrating herds travel freely back and forth across its boundaries. The eastern plains near the Crater Highlands are used for several months each year, and the herds also move into the Maswa Game Reserve and into the Ikorongo, Endulen, and Loliondo Game Controlled Areas where the killing of animals except on license is forbidden but other human rights are unrestricted. Some, too, enter the Masai Mara Game Reserve in Kenya for brief periods. The total number of square kilometers actually used by the migratory species is about 25,500, and, following Watson (1967), I consider this area the Serengeti ecological unit (figs. 2 and 3).

Physiography

The Serengeti region is a vast, austere area of plains and hills lying in northern Tanzania between latitudes 1° and 3°30′S and longitudes 30°50′ and 36°E. The geographical boundaries of the unit include in the east the western wall of the Rift Valley and the Crater Highlands, which rise to an altitude of 3,587 meters, with Ngorongoro Crater as their most spectacular feature. Settlements border the unit in the south, the Speke Gulf of Lake Victoria in the west, and the Isuria Escarpment and the Masai Mara Game Reserve of Kenya in the north. The whole area is a high plateau broken by hills. The plateau slopes gently and irregularly from an altitude of about 1,900 m in the eastern part to 1,170 m near the shores of Lake Victoria. Drainage is westward, but the principal rivers, such as the Grumeti, Mbalageti, and Duma, are seasonal, retaining nothing but pools during the summer; only the Mara and Bologonja rivers in the north part are perennial.

Stretching westward from the Crater Highlands and the Rift are the Serengeti plains, covering some 5,200 sq km. "And all this a sea of grass, grass, grass, grass, and grass. One looks around and sees only grass and sky," wrote Jaeger (1911) after his visit in 1907. The plains, when verdant and covered with herds of wildebeest, zebra, and gazelle from

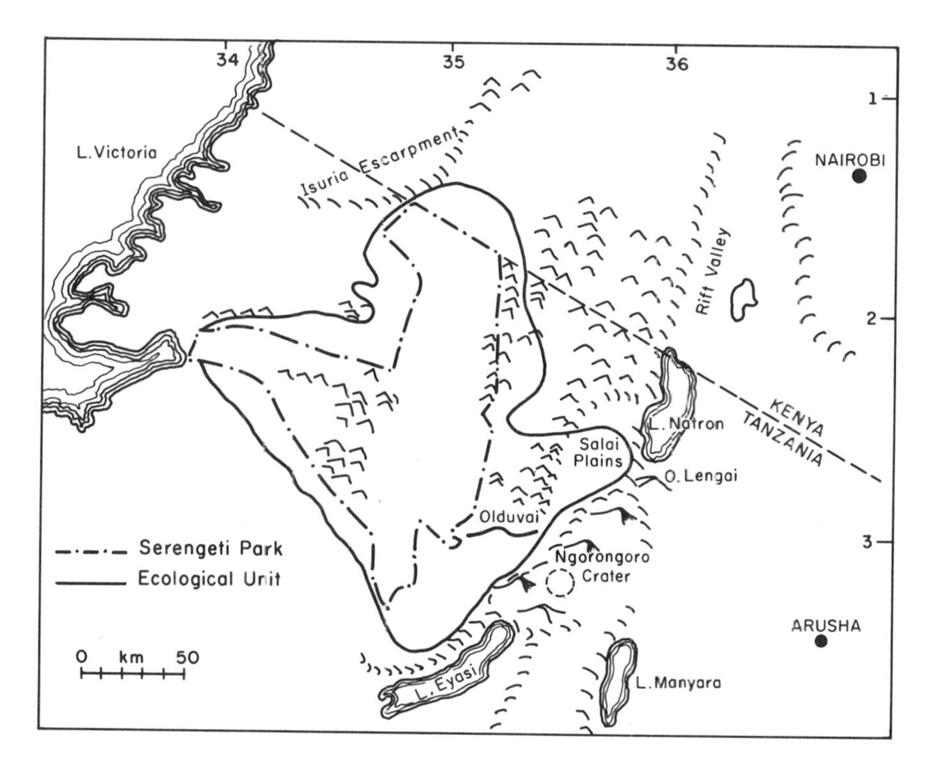

Fig. 2. The physical features of the Serengeti ecological unit and surrounding areas.

horizon to horizon, surely present one of the loveliest sights on earth. Inselbergs or kopjes composed of weathered granite and gneiss stud the plains, but the main topographical feature is the Oldoinyo Gol Mountains—hills of quartzite and gneiss covered with scrubby *Acacia* and *Commiphora* trees. Olduvai Gorge, famous as an early hominid site, cuts through the plains to the south of these mountains and seasonally drains the alkaline Lake Lagaja into the Olbalbal depression at the base of the Crater Highlands.

At the western boundary of the plains the Itonjo and Nyaraboro hills rise abruptly to a height of 300 m above the level of the Duma and Mbalageti rivers. To the north, between the Mbalageti and Grumeti rivers and forming the drainage between them, are other hill ranges—the Nyamuma and Simiti, to name only two—which run east and west down the Corridor, as this area of the park is called. After the grass has burned each year, these hills are dry and barren, with quartzite, shales, meta-cherts, and other rocks exposed on the steep slopes. The northern part of the park, the Northern Extension, is gently rolling,

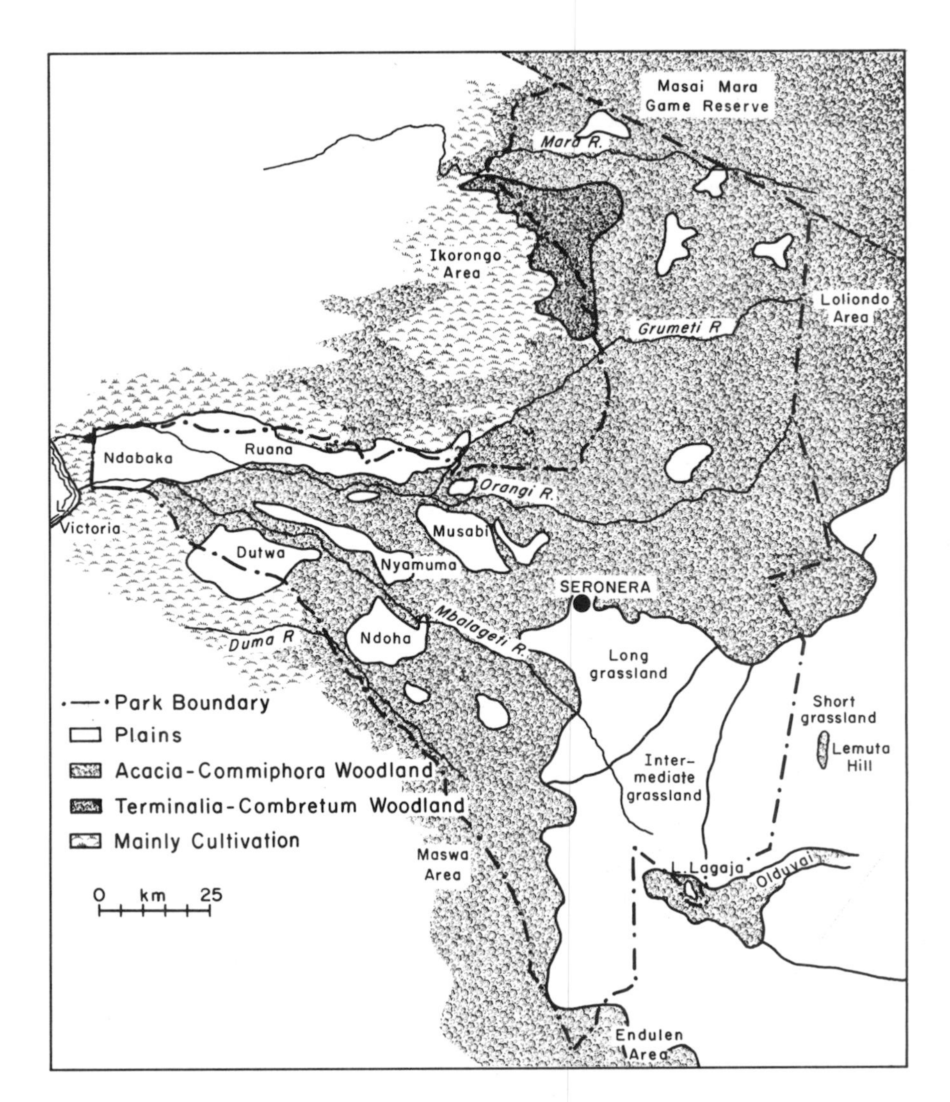

Fig. 3. The major vegetation types of the Serengeti ecological unit.

becoming somewhat steeper toward the north, and cut by numerous streams revealing the granites and gneisses that underlie this region. Occasional kopjes and isolated hills break the vista, but, except for the Kuka and Lobo hills along the eastern part of the ecological unit, the area lacks the prominent topographic features found in the Corridor.

Vegetation

The ecological unit supports two main vegetation types: grassy plains on which trees are either absent or limited to the banks of streams, and open woodlands or, more precisely, wooded grasslands, with trees widely scattered, giving the area a pleasing park-like aspect (fig. 3). A comprehensive classification of vegetation types has not been published, but the descriptions in Pearsall (1957), Anderson and Talbot (1965), and Watson (1967) are useful. For purposes of this report, a brief general outline of the major habitats, particularly of those that have an effect on the behavior of the predators and their prey, is sufficient.

The Plains

Most of the plains consist of Precambrian basement rocks of gneisses, quartzites, and schists, overlain by a calcareous tuff formed from ash deposited during the eruption of Kerimasi some 150,000 years ago (Pickering, 1968). The distribution of the grass types has been strongly affected by the climate and kinds of soil, as well as by grazing and burning. The grass growth in the eastern part of the ecological unit has been influenced by the volcanic activity of Oldoinyo Lengai, an active volcano standing in the valley of the Rift. This volcano has erupted at least fourteen times since 1880, and its ash has been carried westward by the prevailing wind. The Salai plains, flat and dusty, about twenty-five miles to the west of the volcano, are characterized by mobile or stabile sand dunes covered sparsely with *Sporobolus marginatus*, *Digitaria macroblephara*, and other grasses, most of them 15 cm or less high and with a basal cover of 10 to 20%. To the west of this dune country, extending into the park, is another area of short grasslands. The soil there is a sandy loam with a hard calcareous pan at a depth of .6 to .9 m that prevents the vegetation from reaching the water table. The grass is 10 to 15 cm high with a basal cover of about 30%, cropped down to ground level by the numerous ungulates which use this area of the plains more heavily than any other (Watson and Kerfoot, 1964). Common species include the sedge *Kyllinga nervosus* and grasses of the genera *Sporobolus* and *Eragrostis*. Forbs are locally abundant, particularly *Solanum incanum*, *Hypoestes forskali*, and *Indigofera basiflora*. These reach a height of .3 m. or more and provide lions with cover for stalking prey in this otherwise open terrain. Where the soil has been broken by the hooves of animals and other causes, wind erosion and trampling have created bare depressions and terraces. The short grasslands are never burned.

To the south and west of the short grasslands is a broad transitional zone with grasses .6 to .9 m tall and a basal cover of 30 to 50%. The calcareous pan lies at a depth of 1.3 or more meters. The species composition of the grass varies somewhat from area to area; *Pennisetum mezianum*, *Sporobolus pellucidus*, *Themeda triandra*, *Cynodon dactylon*, and *Andropogon greenwayi* are some of the common grasses. This transitional zone is so heavily grazed by the wild ungulates during the rainy season, particularly in the more easterly portion, that it often has the appearance of short grassland. The area is sometimes burned if enough vegetation remains after grazing.

In the west of the plains, and along their northern edge, is a zone of long grasslands growing on fairly well-drained clay soils and lacking the calcareous pan. *Themeda triandra*, *Pennisetum mezianum*, *Andropogon greenwayi*, and *Digitaria macroblephara* are the dominant and most abundant species, reaching a height of 1.0 to 1.3 m and with a basal cover of 50 to 60%. It is possible that the lack of trees on this grassland is the result of yearly fires that sweep across most of it. Except at the beginning and at the end of the wet season, and during extremely dry years, the long grasslands are only lightly grazed, probably because *Themeda* and other grasses grow tall and coarse so rapidly that most ungulates find them rather unpalatable.

The Woodlands

An open woodland covers much of the northern and western sectors of the park and surrounding areas, as well as the hills rising from the plains. Most trees are thorny and rarely exceed 9 to 12 m in height; their canopy coverage averages only about 15 to 20%, with as little as 5 to 10% in the north where trees are confined largely to the valleys and ridge tops. The dominant trees are acacias (*A. tortilis*, *A. clavigera*, *A. gerrardii*, *A. hockii*, *A. senegal*, and others), some species growing scattered, others in almost monotypic stands. For example, *A. drepanolobium*, the whistling thorn acacia, is characteristic of sites with impeded drainage which it covers in dense brushy stands sometimes many hectares in extent. *Commiphora trothae*, looking like gnarled apple trees, are abundant, as are *Balanites aegyptiaca* and *Albizia harveyi*. In the north is a small area of broadleafed woodland in which *Terminalia trichopoda* and *Combretum molle* are the dominant trees. Shrubs are not a conspicuous feature of the woodlands, although *Grewia* spp., *Ormocarpum trichocarpum*, and others are found throughout. The ground is covered mainly with grasses about 1 m high of the same species as those found in the long grasslands of the plain.

Thickets occur predominantly in parts of the Corridor and in the northern part of the ecological unit. Those in the Corridor are small, often 30 m or less in diameter, and very dense with *Salvadora persica*, *Acacia mellifera*, *Acalypha fruticosa*, and other shrubs and trees, as well as an aloe (*Aloe volkensii*) and the wild hemp plant (*Sanseviera robusta*), growing in profusion. The thickets in the north may be several hectares in extent, particularly those in valleys. The trees, largely evergreen and up to 8 m high, include *Croton dichogamus*, *Euclea divinorum*, *Tecklia nobilis*, and *Acacia brevispicata*. Although seemingly impenetrable, each thicket is riddled with a network of elephant and buffalo paths.

A distinctive floral community grows on kopjes. The candelabra euphorbia (*E. candalabrum*) and many other succulents occur there. Common shrubs and trees include *Lannea fulva*, *Rhus natalensis*, *Cordia ovalis*, *Capparis fascicularis*, and *Ficus* spp., to name just a few. Because of their dense shrubby vegetation and rocky retreats, kopjes are often used by the large cats to hide their newborn young. The kopjes in the plains are also one of the few spots where shade can be found, although most have been denuded by pastoralists.

Confined to a narrow strip along both banks of most streams is riverine forest. The canopy ranges from 10 to 25 m in height and is often dense with trees, saplings, shrubs, creepers, and forbs growing in profusion to ground level. *Acacia kirkii* grows there commonly, as do stands of the fever tree (*Acacia xanthophloea*) with its lemon-colored bark. Other riverine trees include the sausage tree (*Kigelia aethiopica*), on whose broad branches leopards like to rest, several kinds of fig (*Ficus* spp.) and clumps of the wild date palm (*Phoenix reclinata*). Fires burn through most of the woodlands during the dry season, leaving the terrain open with little cover behind which predators can stalk their prey. The dense vegetation along the rivers persists, however, and this the cats use when approaching animals that have descended to the stream beds to drink.

The woodlands in the Corridor are broken here and there by extensive plains—notably the Musabi, Nyamuma, Ndoha, Dutwa, Ruana, and Ndabaka plains—which total almost 2,500 sq km in area. Some of the plains, like the Ndoha, are found on tops of granite shields. Grasses .4 m or so high consist predominantly of *Cynodon dactylon* and *Sporobolus marginatus*. Other grasslands, like the Musabi plain, are found in areas of impeded drainage on black cotton soil, which, when wet, makes travel by car exceedingly difficult. *Pennisetum mezianum*, 1.3 m high, is dominant. The Ndabaka area is an alluvial flood plain near Lake Victoria, its *Themeda*, *Pennisetum*, *Setaria*, *Erichloa*, and other grasses heavily grazed by resident wildlife.

SEASONS

The average year is divided into a dry season from June to October, followed by a period of rain in November and December. January and February tend to be dry with only occasional showers. About one-third to nearly one-half of the total annual precipitation falls from March to May. The average annual rainfall between 1937 and 1959 at Banagi was 772.2 mm with a variation from 466.4 mm to 1,074.4 mm, and the number of days with rain in the year varied from 60 to 113 with an average of 82 (Grzimek and Grzimek, 1960). Between 1963 and 1969 an average of 814.5 mm (620–1,031 mm) of rain fell at Seronera (table 1). These figures are fairly typical of the woodlands area, but the eastern plains receive at least one-third less rain. For instance, Seronera received 902 mm of rain in 1967, but the Gol kopjes only 613 mm and the Barafu kopjes 579 mm. Temperatures are pleasant (fig. 4), without extremes, rarely exceeding 32°C (90°F) or dropping below 10°C (50°F). Relative humidity ranges from a mean monthly minimum of about 15% during the height of the dry season to about 40% during the rains; the mean monthly maximum is about 85% with the daily peak in the evening during the wet season and early morning during the dry one. The prevailing winds come from the east and reach their highest velocity during the dry season. In 1968 the average monthly windspeed at Seronera during the dry season was about 6.5 to 8 km per hour, as measured by a Casella cup counter anemometer, but it probably averaged twice as much on the open plains; from December to April average windspeed at Seronera was only 4 to 5 km per hour (H. Braun, pers. comm.).

The seasons vary considerably each year from this average pattern primarily as a result of the erratic and unreliable appearance of the rains. In 1966, for example, the rains during November and December failed and the migratory herds did not move to the plains until March, 1967. The following year the main rains started a month earlier than usual, in February, but in 1969 there was almost no precipitation on the plains from March to May. This variability in the amount and distribution of the rain affects not only the movements of the prey but also has a profound influence on the food habits and other aspects of behavior of lions.

At the onset of the dry season in late May the grasses in the woodlands are still somewhat green. A few of the major rivers flow and water is found widely, as pools in the tributaries and in wallows and other depressions. The plains dry out more rapidly than the woodlands, and

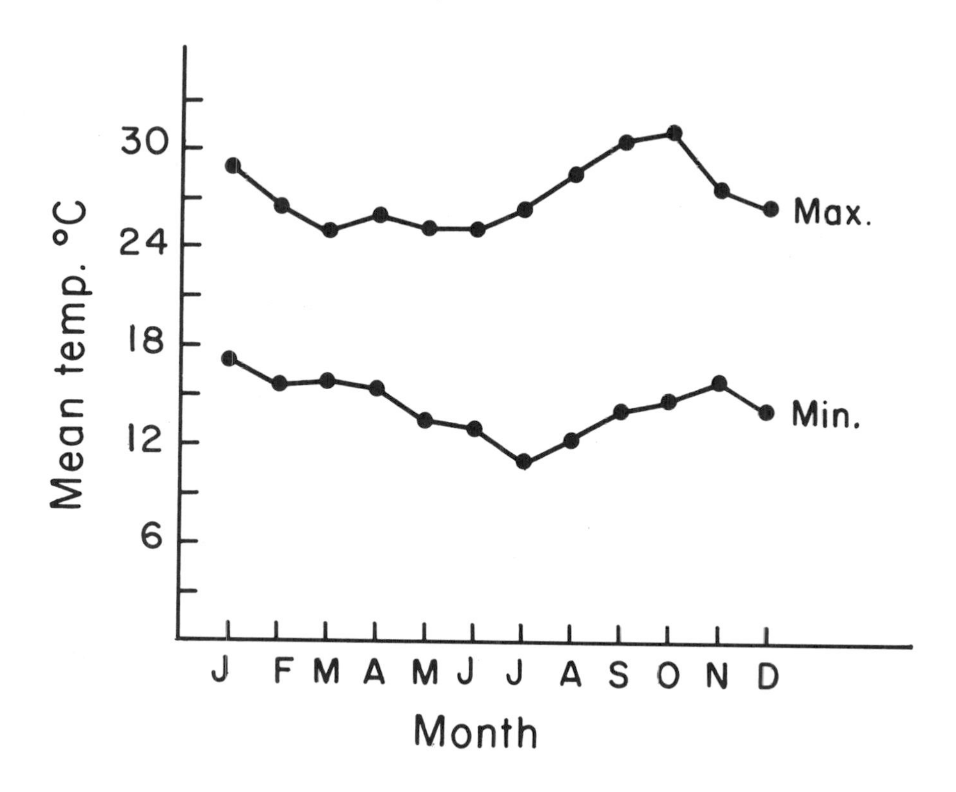

Fig. 4. Mean monthly minimum and maximum temperatures at Seronera, 1968. (Data collected by H. Braun.)

the grasses there have either been eaten down to a stubble or, farther west, are tall and dry. Water in the erosion pans disappears rapidly, and a month or two later the main water sources consist of Lakes Lagaja and Magadi and a few alkaline pools. With water and forage scarce, the migratory herds leave the plains, the zebra and wildebeest first, followed by Thomson's gazelle. Mainly ostriches, Grant's gazelle, and a few Thomson's gazelle remain.

During the height of the dry season the plains present a bleak appearance. The long grasslands have been burned, leaving the ground black and bare except for occasional scorched tufts. In the eastern plains the dry stubble crackles underfoot. Wind whips over the rises, and sand-colored dust devils spin along. Now and then an ostrich appears vibrating in the heat haze on the gray horizon. The Crater Highlands are almost lost in the smoke haze from the fires in the surrounding country. The woodlands dry up progressively. Fires set by man, both inside and outside the park, burn over three-fourths of the woodlands between June and October, eliminating the dry grass, killing saplings, and leaving dead trees as ashy skeletons on the ground. The migrating prey move ceaselessly, first west, then north, feeding on the dry grasses that remain or concentrating in an area where a local shower has stimulated a flush of green.

Towering thunderclouds balanced on black columns of rain herald the onset of the wet season in late October or November. Showers are scattered and local at first, so that one spot may be green and another, a kilometer or two away, still dry. But with heavy rains in late November and December, the woodlands and plains are suddenly transformed from the predominating colors of black and gray to an intense green as the grasses and the leaves of trees sprout anew. Completely rainy days are rare, however, although on some of them the clouds hang below the tops of the hills. The migratory species leave behind the good forage in the woodlands and flood back onto the plains. By the end of December the grasses in the woodlands have again reached their full height. The rivers are full, at times impassable, for much of the rain is torrential and lost through runoff. With the variable weather from January to May, the plains may be either wet or dry. The herds there eddy back and forth, moving as far east as the Salai plains, but during a dry spell retreating westward and southward into the woodlands, only to surge out into the open once more with renewed rain. This pattern continues until the onset of the dry season.

3 Study Methods

The main purpose of this study was to collect information which would lead to an understanding of lion predation in the Serengeti National Park. To accomplish this task, various data on movements, mortality, population densities, and so forth were collected for both the predator and prey species. Some of the specific study methods are described later under the relevant headings, but a few of the general ones are mentioned here to serve as a background to the study.

General

The park is so large that I was unable to study all parts of it with equal intensity, and, although most portions were visited at one time or another, nearly all observations were made in a block of land 3,800 sq km (1,500 sq m) in size, including about 2,280 sq km of plains, ranging from short to long grasslands, and 1,520 sq km of woodlands (fig. 5). The study area was roughly rectangular, about 95 × 40 km in extent, with Seronera, where my family and I lived, at its approximate center. The amount of time that was spent in various parts of this area changed with the seasons. From December to May, when travel was difficult in the woodlands, the lions in the plains received most of my attention; during the dry season, after the long grasses had been burned, I worked primarily in and around the woodlands. However, the area was too large and some of it too difficult of access to provide much detailed information from individually known animals on such aspects as births and deaths and changes in group composition. To study these aspects, as well as various social interactions throughout the year, an area of about 250 sq km, containing two large lion prides, was selected around Seronera. Through frequent contact with visitors these animals had become so accustomed to cars that they permitted approach to within a few meters and indeed sometimes rested in the shade of a vehicle on hot days.[1]

1. Tourists occasionally disturbed the lions by such inconsiderate behavior as driving too fast or interfering with a hunt, but most cars remained in the vicinity of the lions for only a few minutes, mostly in the morning and evening. The average number of minutes spent by 100 consecutive cars with lions was 10 (range 1–76).

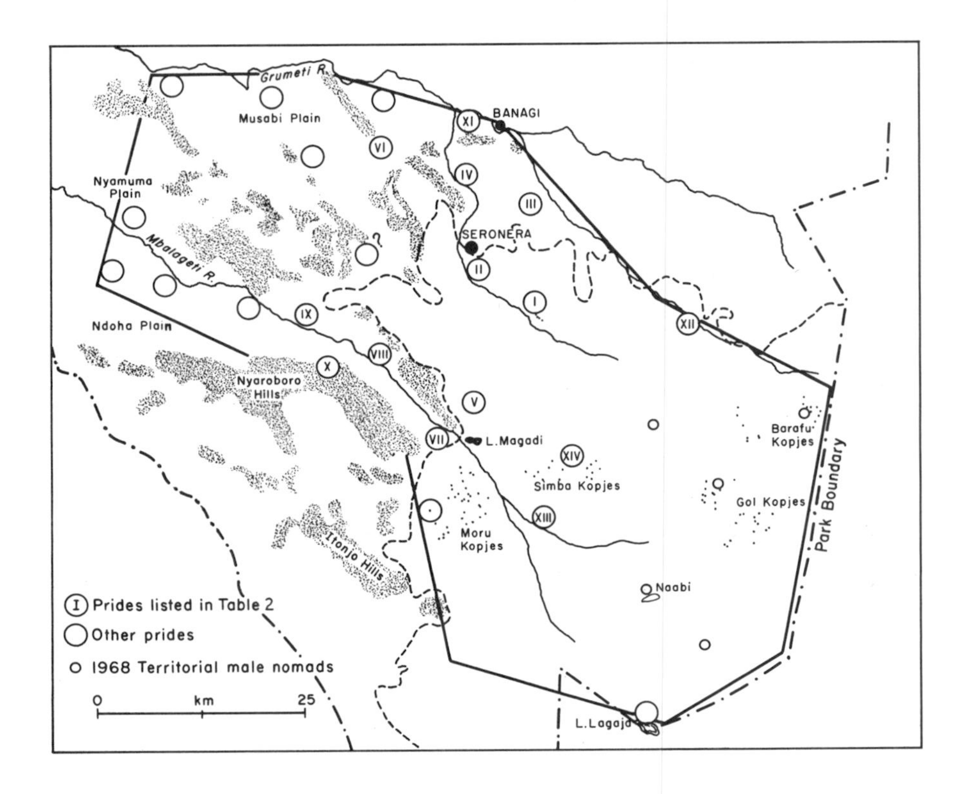

Fig. 5. The main study area in the Serengeti Park, showing the distribution of lion prides and of territorial nomadic males in the plains during the rainy season of 1968.

All observations were made from a Land Rover. As early as 1933 Johnson (1935) noted that lions "pay very little attention to a motor car . . .," whereas they usually flee from a person on foot. This is also true for most lions today. In addition, the large size of the study area and the potential danger involved in watching lions closely on foot made it desirable to use a car, particularly at night. Most lions permitted a car to approach to within 25 m without moving away from it and some failed to respond even at 5 to 10 m. Sometimes animals walked or trotted into a thicket, hid behind a termite mound, or sought other cover at the sight of a vehicle. They then often crouched and, if approached, either fled again or growled, lunged toward the car, or in other ways behaved aggressively. A few lions, especially at the periphery of the park, flee immediately on perceiving a car, presumably because they have been harassed by motorized hunters. On one occasion a subadult male attacked my car from a distance of 25 m and bit two holes in the fender.

Most lions do not respond to persons in and around their cars, even if these sit on the roof, hang out of windows, stand by the open door, or indulge in other acrobatics, as some tourists are wont to do. But lions usually flee as soon as they perceive a person away from a car. On the dozen or so occasions when I met lions while hiking either alone or with another person, they bolted at distances of 30 or more meters. However, one lioness with cubs merely growled at my approach.

My daily routine consisted of a dawn search for lions and their kills. Since all lions in a 75 sq km area may be lying under one tree, the search often took several hours. In the course of the study, these searches and other research activities required 149,000 km of travel. After spotting lions, I drove in most instances to within 50 m of them and watched from a few minutes to twenty-four or more consecutive hours, using the unaided eye, 7 × 50 binoculars, and when necessary a 20 × scope. The composition of the group was noted, and if I recognized one or more members their location was indicated on a map. If the lions were hunting, I stayed at least 50 m behind them in order not to disturb the prey. If, as was usually the case, the lions were inactive, I searched for others, sometimes returning later in the day to observe the evening social activity or hunting after the onset of darkness. Lions were watched for about 2,900 hours, a figure which includes neither the many hours which I spent inattentively near them at night nor the times when my wife or assistants observed them for me.

Lions are active primarily at night. To observe their behavior at that time, 104 whole nights and numerous partial ones were spent with them. Certain individuals or groups were watched continuously for two or more days. Some lions behaved abnormally when a car followed them throughout the night, leaving only certain ones suitable for that purpose. At night even those animals which usually ignored the car seemed on occasion to make attempts to elude it by suddenly descending into a ravine or otherwise disappearing. To keep a moving lion in sight at night was not always easy, and in fact Kühme (1966) considered it impossible. No lights could be used, for this would disturb any prey in the vicinity, and an infrared device I had proved unsatisfactory. As a result we observed lions primarily on bright moonlit nights except in a few instances when a radio could be used for tracking. The moon often provided enough light to enable an observer to see the animal for 100 m or more; it would be lost from sight only when it crossed or recrossed stream beds, entered thickets, or when I inadvertently fell asleep.

Tagging

To study the life history of lions adequately, it is essential to recognize many animals individually. Resident lions are easy to identify by notches in their ears, scars, and other minor markings. But animals which roam widely and are rarely encountered need to be permanently tagged to eliminate all chances of error in identification. A few, of course, are unmistakable, for they lack an ear, the tail, or have some obvious deformity. I recognized about sixty lions individually by their natural markings, including the thirty-eight resident adults of three prides around Seronera. In addition, 156 lions (79 males and 77 females), one year old and older, were marked between September, 1966, and April, 1967, by placing colored and numbered metal cattle tags in one or both of their ears. Five small cubs were similarly tagged after they were caught by hand. Attached to each tag was an inch-long plastic tab, which, however, tore off within six months in most instances. One or both ears were also notched to make certain that the animal remained recognizable as having been marked if the tags were lost. The park authorities rightly thought it inadvisable to use more conspicuous and durable methods of marking because lions are the main tourist attraction. Eighty-six percent of the marked lions were resighted and identified at least once. The 14% which were not seen again either lost their tags, moved outside of the study area, died, or had been marked in places that were seldom visited. Of the lions that were observed again, 58% were encountered one to ten times each and 42% were met eleven or more times. Resightings totalled 1,750. Most lions retained their tags for over a year and a number of them for over two years. An unknown percentage lost both tags, some probably torn out by other lions during fights at kills or during grooming sessions.

Before tagging, an animal was shot in its flank with a drug-filled syringe fired from a carbon-dioxide-powered Cap-Chur gun (Palmer Chemical and Equipment Co.). Succinylcholine chloride at a concentration of 50 mg per cc was the drug used. A total of 166 lions were injected and handled, several two or more times, and of these one (0.6%) died. The breathing of three other animals faltered, but artificial respiration, applied by alternately lifting and pressing down on the foreleg with the animal lying on its side, revived them after a few minutes. I determined the correct dosage by injecting increasing amounts into different animals until one permitted itself to be handled. The average dose for an adult lioness weighing an estimated 114 to 136 kg (250–300 lbs) was 65–70 mg, or about 1 mg per 1.86 kg; for an adult male lion, weighing 159 to 182 kg (350–400 lbs) it was 80–85 mg or about 1 mg per 2.09 kg. Large lionesses were given as much as 75 mg, males up to 90 mg, and small individuals correspondingly less. After a large meal animals needed about 5 mg more than usual to become fully relaxed. A normal dose of the drug failed to take effect if the lion had been running. After three such unsuccessful tagging attempts, the animals were shot only while resting or feeding. At least six animals failed to be more than lightly affected by an average dose even though the drug seemed to have been injected properly. Given an accurate dose, the lion reclined on its side and became approachable after 4 minutes (1.5–8 min); it raised its head again, seemingly alert but often still unable to stand, after about 25 minutes (13–36 min).

I drove slowly to within about 20 to 25 m of a lion and shot it while it was standing or lying; shy animals could be approached more closely at night than during the day. With the impact of the syringe, the lion usually grunted, looked around, occasionally fled some 50 m, then resumed its former activity. Sometimes it lunged at a neighbor as if the latter had been responsible for the sudden pain, and twice it glanced up and circled the tree under which it had been lying as if suspecting that something had fallen. Two animals climbed trees after having been darted, and two charged aggressively to within 10 m of the car. Frequently a lion pulled the syringe out, puncturing it with its teeth in the process, and on five occasions a lion other than the one shot did so. After a minute or two the animal showed the effect of the drug by licking its lips and sometimes by eating grass or biting at twigs. Then its legs became unsteady and finally it collapsed and rolled on its side, a typical resting position of lions which usually elicited no response from others. Driving along behind the back of the animal, I quietly stepped from the car. If the dose was too small, a tarp was placed over the animal's legs and while it was clawing that I clamped a tag into the ear. When possible, blood samples and ectoparasites were also collected. Other data, such as weight and tooth casts, could have been obtained, but it was felt that the additional disturbance of the immobile but seemingly conscious animal, as well as of the other group members, was not justified. In most instances the tagging went so smoothly that other lions sometimes watched the operation from a distance of only 10 to 15 m without fleeing. On several occasions other lions approached the immobile one and sniffed it, and once an animal pawed at it playfully. When I covered an estrous lioness with a tarp, her male consort dragged the tarp 60 m, returned to her, nudged her back, then reclined by her. I was twice charged while tagging, once by a courting male whose female was being handled, another time by a lioness as I gave her daughter artificial respiration. Twice, too, other lions attacked the tranquilized animal and had to be driven off with the vehicle. To prevent such occurrences and also to make certain that the breathing of a drugged animal remained normal, I stayed with a newly tagged lion until it had fully recovered. Most lions did not associate the tagging with the presence of the car; at least they were not noticeably shyer afterwards. A few, however, were affected. For instance, one male, which had to be given artificial respiration, subsequently retreated from the car, whereas his four male companions showed no difference in response. It required a year of frequent contacts with this animal before he accepted the car as indifferently as he had done prior to tagging.

Radio Telemetry

The Sensory Systems Laboratory, Tucson, Arizona, was associated for two years with the Serengeti Research Institute to provide telemetry equipment suitable for studying the movements of predators. Six radio transmitters were made available. Each was mounted on a collar which was then fastened around the neck of the lion. The animal ignored the collar except for a shake of its head or for scratching it once or twice, but its companions attempted to pull or bite it on several occasions. One lion was followed for two days, another for four days, and a third for nine days before the radios ceased to transmit effectively; a fourth lion was tracked for twenty-one days but we had to replace the collar twice because the equipment did not function properly. Since the transmitters had a maximum range of less than a mile on the plains, it was necessary to remain with the animal continuously, and we usually did so in twenty-four-hour shifts. At night the position of the antenna on the car and the intensity of the signal indicated the direction and approximate distance of the lion. In addition, the temperature of three free-ranging lions was monitored continuously by radio. A probe was placed into the fascia of the neck, and temperature fluctuations were recorded for three periods of 1, 1½, and 2½ days.

Food Habits

To discover precisely what lions eat was one of the most important aspects of the study yet also one of the most difficult, because biases were so readily introduced into the data. The analysis of feces, which has provided useful information in many carnivore studies, was an unsatisfactory method in the case of the Serengeti lion. Dung beetles bury the droppings during the rains, and vultures, hyenas, and jackals eat them, probably because of their high protein content (Neseni and Heidler, 1966). Lions, unlike tigers and hyenas, do not defecate at a certain site, and a large sample of feces is difficult to find in the short time necessary to give an accurate impression of food habits in an area where the available prey changes constantly. In twenty hours of walking along the Mbalageti River at a time when lions were abundant there but the grass fairly high, I found only eight droppings; this was clearly not a productive method. Consequently, I obtained most information by finding lions actually on a kill. This technique biases the data in favor of large prey or the adults of a species because a gazelle fawn, for example, may be eaten in ten minutes whereas an adult may last for several hours. In the plains, where lions feed almost exclusively on large prey, the effect of this bias is too small to be significant, but in the woodlands it may be appreciable. Also difficult to assess is the percentage that each prey species contributes to the lions' yearly diet. Lions and their kills are difficult to locate in the woodlands when the grass is high, especially since vultures more often than not fail to indicate the site by their presence. Furthermore, lions in various parts of the park have a somewhat different diet, being dependent upon the erratic availability of prey. To combine kill records uncritically for a huge area, much less for those at several parks as Wright (1960) has done, may lead to errors.

Either the lower jaw or the whole head of a kill was collected, whenever possible, for aging the animal according to the eruption and wear of the teeth. The sex of the adults of most species was easy to check, but that of the young was difficult without chasing the predator from its kill. This was clearly undesirable, and therefore most small young remained unsexed. The bone marrow of some species was checked visually for the presence or absence of fat as a crude measure of the prey's physical condition. Data were collected on 883 lion kills, as well as on 256 cheetah kills, 164 leopard kills, and 198 wild dog kills—a total of 1,501.

Given the propensity of lions to scavenge, it was particularly important to find out if they were eating on an animal they had killed by themselves, if they had appropriated it from another predator, or if they had found it dead. Signs indicating a probable lion kill included tooth marks on the throat and scratches on the shoulder and rump.

The presence of only a partial carcass, or lacerations on the hindlegs of the prey, often meant that the lions had scavenged the meat from hyenas. And vulture feathers and feces on and around the body while lions were still feeding on it indicated a possible disease death. While these and other signs were indicative, the cause of death remained equivocal at times, especially if only the bones were left. On one occasion, for example, a pack of wild dogs killed a wildebeest. The dogs were driven off by several hyenas, and a short while later two lions came and took the remains. On another occasion, a cheetah killed a Thomson's gazelle. A hyena appropriated the carcass but was driven off in turn by a pack of wild dogs. Such complex situations were not uncommon, and care had to be taken before attributing a kill to a particular predator.

Age Classes

Lions were placed for convenience into one of four age classes. In many instances I knew the birth month of a cub and watched the animal grow into a subadult; this provided me with a rough scale by which to estimate the ages of other young lions. Due to occasional periods of starvation, Serengeti cubs are stunted in growth when compared to captives of similar age. One small cub which I found dying at Seronera and which was then raised in captivity was at the age of ten months nearly twice as large as free-living cubs of similar age. However, adult lions show no obvious aftereffect from this lack of food.

Small Cubs (0–1 yr)

Cubs usually remain hidden until they are mobile at about 4–6 weeks of age. The color of their eyes is gray-blue at that age but changes gradually to amber when they are between 2 and 3 months old. Their natal coat is soft, woolly, and grayish-yellow, often with many dark spots on the forehead. This pelage is replaced slowly, starting when the cubs are nearly 3 months old, until at the age of about 5 to 5½ months they have a sleek, short-haired, tawny coat similar to that of adults. The legs and sides are often faintly spotted, a pattern that may persist into adulthood. The conspicuous tail tuft is absent at birth, except for some black hairs, but it becomes evident when the cubs are about 5½ months old and prominent at 7 months.

The sexes of small cubs are easy to distinguish because the scrotum of the male is visible. In addition, male cubs are stockier and seem to have a broader head than females even at the age of only a few weeks. By the age of 6 months some males have a slight throat ruff.

Large Cubs (1–2 yrs)

The best criterion for distinguishing older small cubs from large cubs is tooth replacement. Between the ages of about 13½ and 15 months, the deciduous canines are replaced by permanent ones. G. Adamson (1968) noted that "at the age of thirteen months, the three females Suki, Sally and Shaitani started to shed their milk teeth and grow the second teeth. It was not until six weeks later that the male Susua followed suit. I had noticed the same among Elsa's cubs and have no doubt that it is normal for females to develop their teeth earlier than males." When 12 to 15 months old most cubs are only the size of a leopard, weighing perhaps 45 kg, but they have a growth spurt soon after that, possibly because, with their permanent canines, they are better able to compete for meat. One female cub, sick but in good physical condition, weighed 68 kg (150 lbs) at the age of 22 months. At the age of 2 years, female cubs are about two-thirds the size of an adult lioness and male cubs are somewhat larger, weighing perhaps 90 kg to 100 kg. The males have the massive shoulders and head that presage the adult animal, though their mane consists only of a crest on the top of the head and nape, tufts on the cheeks, and scraggly patches on the neck and chest.

Subadults (2–3½ to 4 yrs)

With their more slender build and shorter muzzle, subadult females can be told from adult females until they are around 3½ years old, even though their height and length may be similar. Their inconspicuous nipples and taut abdomen indicate that they have not had cubs. Most have their first litter around the age of 4 years, and by then too their face has lengthened and their body has become bulkier, making it difficult to distinguish them from adults. Any female with cubs was classified as an adult regardless of her age.

The males begin to grow rapidly when they are 3¼ to 3½ years old and are transformed within 6 months from rather husky youngsters to almost their adult size and weight. The growth rate of the mane varies with the individual, with some having but a short ruff at the age of 4 years and others a heavy mane.

Adult (4+ yrs)

Both females and males continue to grow until they are about 6 years old, mainly becoming more massive. The mane of a male becomes heavier with age, growing backward and downward until it covers his head, except the face, his neck, chest, and shoulders; tufts sometimes sprout from the elbows as well. The mane of subadults is sometimes light blond in color, but the majority of adults have brownish ones often tinged with rust, yellow, and black hairs. Old males tend to have rather short, scruffy manes though a few of them retain luxuriant ones. Adults can be divided into age classes based on the relative amount of wear on the teeth (see table 5).

Five lionesses from Kenya, weighed by Meinertzhagen (1938), ranged from 122 to 182 kg (mean 151 kg) in weight and 14 males from 150 to 189 kg (mean 172 kg). The total length of these females varied from 241 to 269 cm (95–106 in) and that of the males from 246 to 284 cm (97–112 in). Two lionesses from Rhodesia weighed 137 and 145 kg without their stomach contents; four males weighed 167, 171, 176, and 184 kg, respectively (Wilson, 1968). One pregnant adult lioness with 9 kg of buffalo meat in her stomach weighed 126 kg (267 lbs) when she died as a result of having been tranquilized. Her total length was 240 cm, including 86 cm of tail (Sachs, pers. comm.). An adult male which died in a fight weighed 168 kg (369 lbs), including 12.7 kg of zebra meat in his stomach. One male that was shot near the park boundary weighed 196 kg, including 2.8 kg of stomach contents, and his total length was 272.5 cm of which 91.0 cm was tail (Sachs, pers. comm.).

II THE LION

4 Group Structure and Movements

Serengeti lions are of two basic types: *residents*, which remain a year or more or, in some cases, their whole life, within a limited area; *nomads*, which wander widely, often following the movements of the migratory herds. These categories are not mutually exclusive—a nomad may become a resident and vice versa—but there is, nevertheless, a dichotomy between the two types of life. In popular usage any aggregation of two or more lions is a pride. A *pride* in this report denotes specifically any resident lionesses with their cubs, as well as the attending males, which share a pride area and which interact peacefully. Any aggregation of pride members or nomads is termed a *group*. The region occupied by a pride is a *pride area*, that by a nomad a *range*.

The Pride

Most cats tend to be solitary except when courting or when a female has young. This, however, does not imply that they are unsocial, as Leyhausen (1965a) and Schaller (1967) have pointed out, and transitory groups of several adults and young have been reported, for example, among house cats and tigers. Nevertheless the lion is unique among cats in the extent of its social life.

Size and Composition

Naturalists have been rather confused by the social organization of lions, judging by the descriptions in works such as those by Percival (1924), Carr (1962), and Cowie (1966). For instance, Stevenson-Hamilton (1954) wrote: "Exactly how a pride forms and breaks up is not perfectly clear, and the probability is that there is no definite habit or custom in the matter at all." Selous (1908), after observing a group of three males, several lionesses and their cubs, came to the conclusion "that such large assemblages . . . are in all probability only of a very temporary nature, the chance meeting and fraternisation of several families." On the other hand, Guggisberg (1961) and Schenkel (1966a) noted that lions form fairly cohesive social units. After I had observed lions around Seronera for several weeks and learned to recognize all adults individually, it became apparent that the size and composition of groups changed from day to day, and that only certain animals consistently associated whereas others never did. Those that interacted peacefully belonged to the same pride. A lion pride is not a cohesive social unit in the sense of all members being together all the time; instead, the members, or groups of them, may be widely scattered. Therefore the only criterion that can be used to find out whether some lions belong to a certain pride is to observe their contacts with the other lions in the area, a task that may require several months (in a group of 20 lions there are 190 possible relationships). For example, on one occasion I located several lionesses and cubs of the Masai pride. Eleven kilometers away were several others belonging to the same pride, and 5 km further on were the rest. The casual observer would have no intimation that these animals belonged in fact to the same pride. Not once in over three years of observation did I see all the members of the Seronera pride together. Between January and August, 1967, for instance, this pride contained 13 adults. The average number of these adults together at any one time was 3.6, with a variation of 1 to 11 based on 129 encounters.

Table 2 presents 14 prides whose size was determined accurately at some point during the study and which were known to have retained their identity for at least one year. The average number of animals in a pride was 15 with a variation of 4 to 37. Ten other prides were not studied in as great detail but were of similar size: one at Musabi plains numbered at least 22 individuals, another at the Kilimafeza mine at least 26. It is difficult to compare these figures with published information because previous observers failed to note that a pride may be split into several groups. Guggisberg (1961) mentioned a group of 40 on the Kapiti plains in Kenya and another of 32 in the Amboseli Reserve. The largest group seen by Stevenson-Hamilton (1954) in Kruger National Park numbered 35. These large groups contained, presumably, most pride members. A. Jacobs, an anthropologist, told me of seeing 57 lions spread over about .4 km in Ngorongoro Crater in 1957, but he was not certain that they formed one pride.

Most groups are quite small. De Pienaar (1969) noted that average group size in Kruger National Park was 3 to 4. Wright (1960) tallied 347 groups in East Africa and derived an average of 6, not including solitary individuals. In the Kafue National Park, Zambia, Mitchell et al. (1965) found that the average group size was 4 to 5 with a maximum of 15. To compare these figures with some from the Serengeti, I added all lions seen in the woodlands, except those belonging to the four study prides around Seronera, regardless of whether the lions were nomad or resident. Among 3,123 lions tallied, in 778 groups and singly, were 177 solitary individuals, 203 groups with 2 individuals, 103 groups with 3, 63 groups with 4, 42 groups with 5, 30 groups with 6, and so forth down to one group with 25. Average group size was 4. The Seronera pride averaged 3.1 individuals per group in 1968 and the Masai pride 5.4 individuals. These figures are similar to those reported for other areas.

Each pride consists of 2 to 4 adult males, several adult females, and a number of subadults and cubs (table 2). Prides change from year to year, of course, as some adults die and cubs grow up, and the date with each pride in table 1 corresponds to the time when the composition was first known accurately. If the compositions had been given for late 1968, rather than for 1966 and 1967, there would have been fewer cubs and more subadults. All of the Serengeti prides I studied had at least two adult males. In the Nairobi National Park, on the other hand, prides sometimes had but one male for several years. The few subadult males in the 14 prides were partly due to chance—several large male cubs in the population would have entered the subadult class a few months later—and partly due to the fact that the males usually left the pride by the age of 3 to 3½ years.

Prides XIII and XIV differed from the others in that they shared the same two males even though the lionesses and cubs in each pride were distinct, and, as far as I knew, never associated. The males spent most of their time with pride XIII in 1966 and 1967 but then remained largely with pride XIV in 1968. A similar situation was noted in the Lake Manyara National Park (Makacha and Schaller, 1969), except that there the two males divided their time about equally between the two prides involved.

Although each pride member readily associates with every other member, some are together more often than others. Cubs, of course, tend to be with their mother, but, in addition, lions often form a companionship with others of their sex and, in the case of subadults, with others of their own age. Pride structure can be understood only if the adult males are considered separately from the females, as will be done here from now on. Table 3 shows the degree of association between each of the females in the Seronera pride in 1967 and 1968.

Since a female sometimes belonged to several groups in the course of a day, her association was recorded only once after she had settled down in the morning. The degree of association was measured by the formula

$$a = \frac{2N}{n_1 + n_2}$$

where $n_1 + n_2$ represent the number of times each pair of lionesses was seen and N the number of times they were together. No two animals remained together all the time to attain a value of 1.00, but some showed values of .40 or more, and these were considered to be companions. In 1967, females A, B, E, and F were together frequently. The first two had cubs and the last two were old and without cubs. Another grouping consisted of C, an elderly lioness, and G and H, both $3\frac{1}{2}$ to $4\frac{1}{2}$ years old. All three had small cubs born within a few weeks of each other. Females J and K were both young adults which remained peripheral to the pride though they belonged to it. Females D and I formed no individual attachments. The companionships had changed somewhat by 1968. Female E died, and both B and F became rather solitary, breaking their tie with A that had lasted throughout the previous year. Instead, female A associated closely with H and I. Females C, G, and H were still together, often joined by I, and females J and K also continued their contacts. Further changes occurred in 1969. Female F died, and J and K split up when the former had cubs. The remaining seven females all had cubs late in 1968 and subsequently formed one fairly cohesive group. Between January and June, 1969, these lionesses were seen on 66 days, and on 55 (83%) all seven were together.

The Masai pride females showed a grouping pattern similar to that of the Seronera pride. The lionesses were encountered on 97 days in 1967 and on 55 (56%) of these days all seven were together during the midday rest. Four of the females had cubs of the previous year and a fifth gave birth early in 1967. In 1968 the lionesses tended to split up (table 4), but the general degree of association was still greater than that among the Seronera females in 1967 and 1968. Females L and Q gave birth in January, 1968, and became almost inseparable companions, an association that lasted to the end of the study, even though L lost her cubs in October, 1968. In general, the lionesses of the Masai pride formed a more cohesive social unit than did those of the Seronera pride.

Both centrifugal and centripetal forces affect the formation and maintenance of companionships. Estrous females tend to separate from the others, as do those having newborn young. Then, too, prey size has an effect on the optimum number of animals that can use a kill. For instance, when the seven lionesses in the Seronera pride fed mainly on such large prey as wildebeest between January and May, 1969, an average of 6.4 associated whenever I encountered them. But from June to August, when Thomson's gazelle constituted their principal food, an average of only 3.6 were together. Then, in September, after zebra moved into the area, all seven tended to form a group again.

Although some individuals become peripheral to the pride and even leave it entirely for unknown reasons, other factors draw the lionesses together. The most potent of these factors is the presence of small cubs. Several females may give birth during the same month and these tend to become companions, an association that may persist long after the cubs have died. Cubs that have been raised together often remain companions as subadults and adults. The association of females J and K of the Seronera pride is an example of this. Possibly some lionesses become friends, to use a somewhat anthropomorphic term, for no reason other than that they find each other's company congenial.

However, companionships have no influence on pride composition. No matter how widely females are scattered or how frequently they meet each of the other members, they still constitute a closed social unit which strange lionesses are not permitted to join. On one occasion an old and seemingly sick lioness of the Magadi pride switched to the Plains pride for a few days, the only such instance noted. Consequently the composition of the lionesses in a pride remains constant from year to year, except for deaths and emigration of young. All pride lionesses are directly related and consist of daughters, mothers, grandmothers, and perhaps yet another generation. Table 5 gives the approximate age of lionesses in four prides based mainly on the relative wear on the teeth as judged from a distance with binoculars. The Loliondo pride contained mostly young and prime animals, whereas the Seronera pride had only one prime lioness and several elderly ones. The difference probably reflects a variable success in raising young in the past.

Pride males may be alone or together, with females or without them. For example, both Black Mane and Brown Mane of the Masai pride were alone on 11% of the days when I encountered them between August, 1966, and April, 1969, and the figure for Limp was 7%; all three were together on 48% of the days, sometimes with lionesses, sometimes not. Each of the males was with a lioness, either of his own pride, the Seronera pride, or a nomad, on 70 to 80% of my encounters with them. Similarly, the two Seronera pride males were seen on 115 and 120 days, respectively, between June, 1966, and August, 1967. They were together on 67% of the occasions and one or both were with lionesses of their pride on 90% of the occasions. Although pride males clearly are companions, often traveling, feeding, and resting together, they do not form such close social bonds with lionesses; their contacts with them remain rather casual except during courtship.

In contrast to the variable ages of pride females, most males are of similar age. In many instances they are probably brothers or at least pride mates which in their youth became companions. The majority of pride males are young or prime adults; only in the Kamarishe pride was one of the males old, as was one of the two nomadic Cub Valley males that took over the Masai pride just at the termination of my study.

Subadults in the pride often associate closely with each other until they either become integrated into the pride as adults, in the case of females, or leave the pride to become nomads. Subadults frequently form a companionship which may persist for years. For example, two surviving female cubs of lionesses A and B of the Seronera pride were raised together and finally left the pride together as subadults.

Changes in Pride Composition

In March or April, 1963, at least 4 lionesses of the Seronera pride gave birth to a total of 12 cubs, according to the park wardens. In June, 1966, at the beginning of this study, the cubs were just over 3 years old. There were 10 females and 2 males, all still with the pride. The 6 adult lionesses raised the total number of females to 16. Between June and October, 1966, five subadult lionesses became peripheral to the pride and by December had left it, not as a group but singly and in pairs. One returned the following June, briefly, but after that I never saw those animals again or at least did not recognize them. The two subadult males also left late in 1966. Two other subadult females (J and K) became peripheral to the pride at that time too. In 1967 they had occasional contacts, but these were rare in 1968 and 1969 (table 3). Once I saw them together 20 km south of the pride area, and another time female J was alone at the Gol kopjes about 40 km east of her usual haunts. But both females retained tenuous social ties with the pride throughout the study, and female J even had cubs twice within the pride area. Thus, of the 10 female cubs born in 1963, 5 left the pride, 2 became casual visitors, and only 3 remained as full members. No cubs were born in 1964 and 1965, or at least none survived, but two female cubs born to lionesses A and B in 1966 grew up. They associated with the pride through 1968, but in January, 1969, when 2¾ years old, they became peripheral and almost ceased to have contact with the others. No cubs that were born in 1967 and the first half of 1968 survived, but 7 out of 18 born late in 1968 were still alive at the end of 1969. Two lionesses died of old age, one in 1967 and the other in 1968. In the course of seven years the number of subadult and adult lionesses in the pride could have increased from about 6 in 1963 to 18 in 1969, but death and emigration reduced the number so that by the end of the study there were only seven permanent members, several of them with small cubs. These changes are particularly interesting because they suggest how the size of a pride is limited and how oscillations occur around a basic number of lionesses which may well represent the optimum for that pride.

The Masai pride consisted of six adult lionesses and one subadult in June, 1966. Of litters born in March and August, 1966, seven cubs (one male, six females) survived. All were still pride members at the age of 3–3½ years when the study terminated. One female cub, born in January, 1968, was alive late in 1969 and will enter the subadult class in January, 1970. One old lioness disappeared in 1969 and was presumably dead. In nearly four years the number of subadult and adult lionesses increased from 7 to 13. No changes among the 13 lionesses of the Loliondo pride were noted in 1967 and 1968, while in 1969 I did not see all the animals often enough to be certain that no changes occurred. Of 22 cubs in 8 litters born between September, 1966, and April, 1967, 14 (7 males, 7 females) survived and all were still in the pride area when 2½–3 years old. In the Masai and Loliondo prides, and in several other prides as well, the number of lionesses increased considerably as cubs grew up to become subadults, but the study did not last long enough for me to find out how many of these young animals ultimately emigrated.

It is worth reiterating that the adult lionesses of a pride constitute a remarkably stable social unit, the composition of which is mainly affected by deaths and the acceptance of some subadults that have grown up in the pride. Such adherence of a lioness to her pride is well illustrated by Blondie from Nairobi Park. When Guggisberg (1961) first knew her in 1953 she was already an adult. When she died in

February, 1967, at the age of at least 17 years and possibly as much as 22 years (Foster and Kearney, 1967) she was still in the same area.

Adult males exhibit a pride tenure pattern strikingly different from that of the lionesses, as shown well by data collected in Nairobi Park. When Guggisberg (1961) began his observations there in March, 1953, one male, Leo, had been "for a long time master of the Park." He was still there in February, 1954, but late that year or early in 1955 two males called Hildebrand and Hadubrand took the area from Leo. Late in 1961, after having been in the park for six years, two new adult males replaced Hildebrand and Hadubrand (Guggisberg, pers. comm.). One of these males was extremely aggressive toward strange lions he encountered, killing several of them, with the result that the park authorities castrated him in July, 1962. This had little effect on his behavior and he was shot in May, 1964 (Schenkel, 1966b). His companion, Spiv, which, according to the park rangers, was born in the park in 1955 or 1956, stayed on in the area. After he had been pride male for about six years, he was replaced in November, 1967, by a new male, which Guggisberg (pers. comm.) saw running around the park roaring at that time. This male was still there in September, 1969. Over a period of seventeen years, four different males or pairs of males successively owned, so to speak, that one area and associated with the lionesses there.

The pride males in the Serengeti were replaced more frequently than those in the Nairobi Park. Of 12 prides I observed for two or more years only 3 (25%) retained the same males throughout the period. These were the Mbalageti, Mukoma, and East Moru prides. The nature of the change from one set of males to the next differed somewhat from pride to pride although in only a few was it possible to witness the details. The lionesses of the Plains pride were accompanied by two adult males in mid-1966. These males were last seen with the pride in late February, 1968, and soon after that they apparently left on their own volition for no new males took over the pride in 1968 and 1969 although the lionesses remained in the area and solitary nomads were seen with them twice. The two males of the Nyaraswiga pride disappeared about November, 1967, possibly having been expelled by the Loliondo males, but the lionesses remained in the area without males until at least September, 1968. The four lionesses of the Simba pride lacked pride males of their own in 1966 except for intermittent contact with a pair of nomads. However, in April, 1967, the two males of the E. Moru pride were first seen with the Simba pride. The association was occasional that year but frequent the next and continued into 1969.

Two adult males occupied the Loliondo pride area in June, 1966. On September 28, 1966, two new males accompanied the lionesses. Both were at most 4 to 4½ years old, nearly fully grown in body but still with a sparse mane, the only instance of such young animals establishing themselves as pride males. They were still with the pride in August, 1969. The Kamarishe lionesses associated with two males in December, 1966. Later a nomadic pair was once seen with them, and by July, 1967, two new males, a prime animal and an old one, were with the pride and remained with it most of the time until the end of the study. The males of the Nanyuki pride were also replaced by new ones.

I was fortunate to observe the ouster of the Magadi pride males by several nomads. In March, 1967, a group of five nomadic males, all about 3½ to 3¾ years old, frequented the Moru kopjes. During the rains in April and May they roamed around the Gol kopjes and other parts of the plains and by the following dry season had retreated to the Moru kopjes again (fig. 6). They returned to the plains from December, 1967, to May, 1968, where they established a temporary territory around Naabi Hill. I did not see them between May and August. On

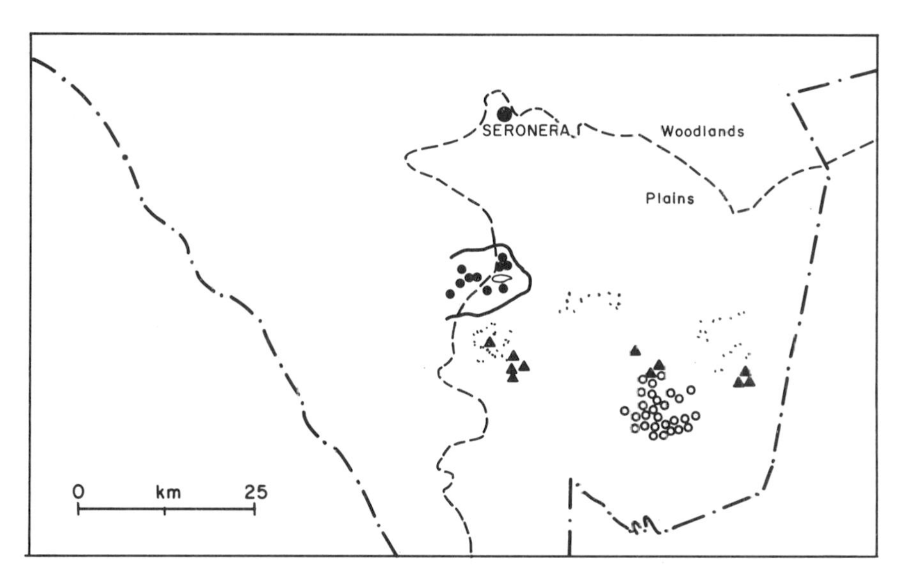

Fig. 6. Movement pattern of a group of five males showing their transition from complete nomadism (triangles—March to October, 1967) through temporary residency on the plains (open circles—December, 1967, to May, 1968) to complete residency after they took over the Magadi pride (closed circles—from September, 1968, onward). Each symbol represents a sighting.

September 10, 1968, the Magadi pride was waiting at the edge of a pool into which a bull buffalo had retreated. On that day the group consisted of two out of the three adult males that had been with the pride at least $2\frac{1}{4}$ years, six adult lionesses, and seven large cubs. The five nomadic males, now about 5 years old, traveled up the valley at 0925 hours. When they saw the pride about 150 m away, they advanced at a trot, roaring. The cubs fled immediately, and the two pride males also turned to retreat. However, a lioness attacked the nomads and chased one of them for 30 m. The pride males and also several lionesses then acted as if to charge, but they soon fled. The nomads chased one or another until all were out of my sight. These five males returned 45 minutes later, killed the buffalo, and remained by it all day. A lion roared in the distance at 1900. The nomads became nervous and two of them trotted 100 m away as if to flee. Then all five roared and moved into the direction from which the sound had come, only to return to the kill at 2320. A solitary nomad joined them at the kill and after considerable antagonism was permitted to eat. The five nomads stayed with the buffalo until the morning of September 12. I did not meet these animals again until November, when two out of the five nomadic males were seen with the pride lionesses. At least one was still with the pride the following June, but I could not find out if all five remained together. I saw one of the former pride males as a solitary nomad in January and again in February, 1969, first 20 km southeast then 40 km east of his former pride. The two other pride males were not encountered again.

A highly complex situation, described later in detail, arose in the Seronera pride. This pride had two males in 1966, but one was killed in September, 1967, by the Masai pride males and the other was driven from the area in December, probably by the Kamarishe pride males. No new males established themselves with the Seronera lionesses, which were instead shared by both the Masai and Kamarishe pride males though not at the same time. After May, 1969, the Masai males spent progressively more time with the Seronera pride until by August they had switched completely to that pride. One of the three males died in August but the other two were still there in December, 1969, according to Bertram (pers. comm.). The Masai pride lionesses, abandoned by their males, were joined in August by the two nomadic Cub Valley group males.

These descriptions show that the males are rather transitory members of prides in contrast to the permanent social units of lionesses. Changes among males are of three types: some males seem to leave the pride on their own volition and are not immediately replaced by others; most are actively expelled by nomads which then appropriate the area and the lionesses it contains; and a few pride males replace their neighbors while retaining jurisdiction over their own area, so that they associate with two distinct prides of lionesses. The longest tenures of males with a pride are the two instances of six years from Nairobi Park. The frequent changes in the Serengeti are surprising, for many of the replaced males seem to be vigorous animals. "It has often been said that a lion over 10 years old has passed his peak of usefulness as an exhibition animal and is on the downgrade. Severe though this restriction may seem, it is certainly true that somewhere between that age and 15, signs of senility are bound to appear" (Crandall, 1964). If the same applies to free-living animals, then males which take over a pride at the age of 5 years have at most ten years during which they can be considered in their prime and during which they can retain a pride. However, the number of males challenging those in a pride also influences the outcome regardless of age. The single Seronera male fled before the two Kamarishe males, and the three prime Magadi males gave up their pride to five young nomads.

The changes that occurred between June, 1967, and September, 1969, in two prides in the northern half of the Lake Manyara Park were in some ways similar to but in others different from those observed in the Serengeti lions. According to S. Makacha (pers. comm.), the same two adult males associated with both prides until November, 1968, when one of them disappeared, apparently having been shot after he wandered outside of the park. In March, 1969, four months after the death of his companion, the remaining male, Chongo, was still in his usual haunts but for the first time a new adult male ventured into his area, and this intrusion was followed by a brief one of three males from the southern portion of the park.

The Chemchem pride in the center of the park consisted of 3 adult lionesses, 3 subadult ones, 1 subadult male, and 2 small cubs in June, 1967 (Makacha and Schaller, 1969). One lioness died in September, 1967, apparently of old age, and her cubs disappeared shortly afterwards. The subadult male was repeatedly chased and finally driven from the pride by the adult males in December, 1967, when about 2½ years old. A new lioness was seen with the pride for the first time in June, 1968, and she may once have been a member. She remained with the pride and had cubs in December, 1968. One adult lioness disappeared in January, 1968, with her two cubs, but suddenly reappeared in July without cubs. The following January she left the pride once more but returned in July with a new litter.

The Mahali pa Nyati pride consisted in 1967 of 2 adult lionesses, 2 large female cubs, and a large male cub. In March, 1968, one of the adult females left the pride and was not seen again until she suddenly reappeared in February, 1969, with a cub. She stayed with the pride only until May and then moved to the north out of the park into the vicinity of a village where she injured a person. The young male was chased on several occasions by the adult males in 1968 but remained with the pride. In mid-1969, Chongo, the remaining adult male, began to associate with the young male, now about 3½ years old. This was my only record of a pride male tolerating and even sharing the pride with a young male, possibly his son. The young male[1] mated with one of his sisters in June and Chongo did so with the other young female, most likely his daughter.

Three of the six adult lionesses were only sporadic members of their prides, disappearing suddenly from the area only to join the others again a few months later. They were not peripheral animals in the same sense as females J and K of the Seronera pride. Similar behavior was not observed in the Serengeti, although a lioness, particularly F of the Seronera pride, was sometimes not seen with the others for several weeks.

Emigration of Subadults

Some subadult lionesses leave the pride to become nomadic. It would be of interest to discover the factors that determine such behavior. Schenkel (1966b) felt that "as the cubs lose their baby-characteristics, increasing intolerance of the older generation towards them can be observed, and with 1½ to 2 years the cubs are finally treated as trespassers." The situation in the Serengeti was more complex than that. In the Seronera pride, for instance, of 12 subadult females whose fate was known, 7 left the pride between the ages of 2⅔ and 3½ years, 2 became peripheral, and 3 stayed.

In the Seronera pride I observed eight instances of aggression by lionesses and pride males toward young females, 3 to 4 years old. Only the males and females C and D were involved in the attacks. The subadult females J and K, which later became peripheral to the pride, were chased three times, female G twice, and females H and I once

1. This male was shot in April, 1970, while feeding on a person (Makacha, pers. comm.).

each. The attacks were in some instances not precipitated by any obvious behavior on the part of the subadults. On one occasion, for example, females J, K, and D approached a male and females C and H through high grass. Suddenly the male rushed the newcomers and chased female K nearly one kilometer. Female C joined in the pursuit, as did D, which had just been with K. Then all three chased female J almost .5 km, and finally pursued K once more. On another occasion, females D, G, and H were walking along a riverbank when D ran into a thicket and flushed females J and K from it. D chased J about $\frac{1}{3}$ km before the latter lay down, but then greeted her, as did G and H. K fled from the area.

Aggression may in some instances be due to the fact that a lioness has a litter of small cubs, as Schenkel (1966b) pointed out. The seven lionesses of the Seronera pride all had litters between August and November, 1968. It is probably not coincidental that as the lionesses brought their cubs to the group, the two subadult females became peripheral, sometimes lying 100 or more meters from the adults even though no actual aggression was seen. Their transition from pride member to nomad was rapid and occurred when they were about 32 months old. They were still with the group on December 25, 1968. On January 27, 1969, they rested beneath a tree when several adults approached. They fled at a trot and females C and D, which had also been so antagonistic toward the previous generation, chased them for 1.5 km. They never joined the pride again, although on July 12 they were seen to meet the others. On that occasion, A. Root watched them sit 50 m from three adult lionesses of their former pride. After 20 minutes, a lioness walked over and greeted one of the subadults, and then all rubbed heads mutually for $1\frac{1}{2}$ minutes. But suddenly the subadults walked away, looking back at the others. They were attacked immediately and chased 500 m.

At times a subadult female elicits attack by behaving as if unsure of herself while with members of the pride. Once, for instance, female G greeted females C and I but seemed nervous as she glanced over her shoulder. Suddenly female C attacked her. A pride male ran up and bit her in the lower back; she slapped him, then retreated and lay down. Similarly, female I reclined 45 m from peripheral female K without responding to her. But when female K walked off rapidly, she followed 100 m with her head held low in threat. Another time, on August 29, 1969, female H walked slowly to within 5 m of peripheral female J. The latter approached and greeted rather uneasily and was immediately slapped. On the other hand, when these peripheral lionesses behaved as if they belonged in the area, the others did not attack them. Once two lionesses saw female J sitting at 40 m and ignored her; another time female K casually approached female A on a kill to within 10 m without being driven away.

It is likely that lions, which have few distinctive markings by which individuals can recognize each other at a distance, often use behavioral cues to ascertain whether or not an approaching animal belongs to the pride. A pride member joins others unhesitatingly, often running toward them, whereas a stranger typically crouches, advances a few steps, then turns as if to flee, and in general behaves as if uncertain of its reception. Since pride members attack strangers, any lion which behaves like one will be treated accordingly until its identity has been established. A subadult lioness which for some reason has been chased several times probably will approach others in increasingly hesitant fashion, which in turn elicits further attacks. It was my impression, for example, that females J and K of the Seronera pride remained peripheral mainly because their behavior usually caused aggression when they met a former pride member. These speculations fail, however, to explain why some subadults leave the pride and others do not. Perhaps individual character traits, such as a propensity toward nervousness, are of importance in determining this.

Subadult males leave the pride at about the same age as do females. The Loliondo, Kamarishe, and Magadi prides contained subadult males 2 to 3 years old which still associated closely with adults. Two subadult males of the Seronera pride became nomads at the age of about $3\frac{1}{3}$ years, one male in the Lake Manyara Park was chased from the pride when about $2\frac{1}{2}$ years old, and seven subadult males of the Mbalageti pride left as a group at the age of about $2\frac{2}{3}$ years. One male broke away from his nomadic mother at $2\frac{1}{2}$ years. Judging by the estimated ages of nomadic males and the absence of subadults more than $3\frac{1}{2}$ years old in prides, the majority become independent between the ages of $2\frac{1}{2}$ and $3\frac{1}{2}$ years. Schenkel (1966b) stated that cubs are treated as trespassers by the adults when $1\frac{1}{2}$ to 2 years old. A $1\frac{1}{2}$-year-old cub would probably not be able to survive by itself in the Serengeti. On the infrequent occasions when youngsters between the ages of $1\frac{3}{4}$ and $2\frac{1}{4}$ years old were encountered away from the pride they were usually in a group of subadults older than themselves.

I observed the responses of the Masai pride adults to a subadult male for over a year. He was born in August, 1966, and grew into a rather runty individual with a short, scraggly mane. In June, 1968, when he was 22 months old, he frequently rested some 10 to 40 m from the rest of the group solely because two lionesses (L and Q) harassed him. Both had small cubs. Later, after she also had a litter, female M sometimes joined in the fray. Between June, 1968, and January, 1969, these lionesses were seen to attack the young male on twelve occasions, joined once by a male as well. However, their response to him was erratic. On some days he was with the group, on others a low growl or a step in his direction sent him to the periphery, but on six occasions one or more lionesses chased him some 30 to 80 m and swatted at him. The other pride members, including the males, ignored the commotion. As the cubs grew larger, aggression toward the male first lessened, then ceased entirely. By May, 1969, he was once again a full member of the pride, a position he still retained at the age of 3 years 1 month when the study terminated.

Responses of Prides toward Other Prides and toward Nomads

Although both friendly and violent interactions between lions have occasionally been described (see Guggisberg, 1961), it was in most instances not known if the participants were nomads, members of the same pride, or members of different prides. In Nairobi Park the same two males killed two lionesses and a subadult male in 1961 and 1962; two more lionesses were killed in 1963. "Furthermore, I have observed three more instances of serious attack by one or both males of the central pride with, in two of these cases, the participation of the females. In every case mentioned the individuals attacked were foreign lionesses and all attacks occurred in areas which were frequently visited by the central pride" (Schenkel, 1966b). Such frequent unrestrained violence is, however, not typical of lions. In over two years of observation on lions in Manyara Park, which has a higher lion density than Nairobi Park, no serious fights were seen, and, in fact, no interactions between the neighboring Mahali pa Nyati and Chemchem prides were observed except in one instance when one litter of cubs switched to the other pride after their mother died.

There is a considerable amount of overlap between pride areas (see fig. 7), but direct confrontations are remarkably infrequent and actual combat is rare because lions avoid members of other prides whenever they become aware of them. On one occasion, for example, the two lionesses of the Nyaraswiga pride rambled along a river's edge when 200 m ahead of them a zebra herd bolted from a waterhole pursued by eight lionesses of the Seronera pride. One lioness pulled a lagging zebra down and all then ate. The Nyaraswiga lionesses trotted to within 150 m of the kill and watched it without approaching further. On another occasion, seven lionesses of the Seronera pride roared at 0520 as they moved through some terrain usually occupied by the Masai pride. Several Masai pride lionesses answered immediately from less than .5 km away, and the intruding animals turned abruptly and moved at a fast walk for five minutes toward their own area.

Interactions between Prides

I watched meetings between the Seronera and Masai prides a number of times and describe these in detail as an indication of the complexity which such interactions can attain. No direct contacts between the prides were seen between June, 1966, and June, 1967. On July 9, 1967, J. Root told me that she found the three Masai pride males resting 120 m from four lionesses of the Seronera pride, neither group aware of the other. One female spotted the males and walked slowly to within 80 m of them. She sat with her head held low and ears retracted; the males first strutted then trotted toward her. She fled, and the males chased her 1.5 km before stopping and roaring. On July 30, the males suddenly became alert and ran 1 km to a thicket where two Seronera pride lionesses had just killed a Thomson's gazelle. The first male to reach the scene chased one lioness 10 m, then pursued the other, which attempted to drag the gazelle away. She dropped the kill, and, while this male ate, the other two continued the chase. One almost caught a lioness, but she whirled around and slapped at him, then continued her flight. The males escorted her another .6 km without pressing the attack and finally stood and roared.

On September 1, the three Masai pride males were 1.5 km from one of the males and three of the lionesses of the Seronera pride on a zebra kill near the center of the Seronera pride area. The following morning at 0635 the Seronera male was lying in the same place, covered with blood. Tatters of his yellow mane were strewn over an area of 3 × 10 m. His right eye was closed and a deep gash angled across his left brow. His left flank was ripped to the bone, and a hole 8 cm wide penetrated his chest. Other deep punctures and cuts covered his body, particularly his rump. A bite in the top of his head had broken the sagittal crest. He breathed heavily. At 0655 female A walked up to him and seemingly sniffed his mane. Suddenly she growled and fled. Standing 40 m away was Black Mane of the Masai pride. He advanced slowly to within

1.5 m and faced the wounded male which, with his chin resting on his forepaws, gave a brief, low growl. Black Mane scraped the ground with his hind paws and walked to the remains of the zebra in a nearby thicket.

Three times the Seronera male raised his head in the next half hour, but the last time he lowered it with a groan. At 0830 he placed his head sideways on his paws and his breathing grew erratic as violent spasms racked his chest. At 0834 his breathing became steady again but feeble. His pupils enlarged somewhat, his right foreleg twitched, his bladder emptied. At 0835 a muscle in his right flank quivered and then his pupils became very large, his last movement.

The other two Masai pride males were nearby and only one showed evidence of a fight—a cut on his paw. One male mated with Seronera female I, which had been on the kill the previous day. This female, as well as female A, singly or together associated frequently with the Masai pride males until late November. On September 25, female H, with four cubs, approached one of the males as he rested with females A and I. She came closer slowly, head lowered and ears laid back, hissing and growling. But the male ignored her and she joined the group.

At noon on November 13, female C killed a zebra. When at 1515 hours two Masai males arrived, she first retreated 50 m and finally moved from sight. However, she returned with females H and I at 2300 and without hesitation joined the males at the kill. The following night the two males and female I were still by the carcass. At 2005 the males suddenly became alert, then fled while roaring. Coming from the opposite direction were the two Nyaraswiga pride males, which to my knowledge had never before penetrated this far into the Seronera pride area. When they heard the Masai pride males roar, they promptly fled, so that both pairs of trespassers ran in opposite directions. However, the Masai males then turned and chased the others at a fast trot. These silently crossed a river while their pursuers, failing to note this, continued roaring down a road for 1.5 km.

After the one Seronera pride male had been killed, the other seemed to lose his assurance. When he heard the roars of other males in his area he frequently walked silently in the opposite direction, looking furtively back over his shoulder instead of answering or investigating. He seemed nervous, often glancing around and even jerking to attention at the approach of a vehicle, something he had not done in the past. Males from three prides—Masai, Nyaraswiga, and Kamarishe—now penetrated far into his pride area, and he was clearly unable to prevent this without the help of his companion.

The death of the male also affected the pride in other ways. On November 1, three cubs were bitten to death by a lioness, judging from the tracks. As I was watching the lionesses of the Kamarishe pride at 0630 on December 8, the Seronera male ran past pursued at a distance by the two Kamarishe pride males. These stopped after 200 m and returned to a fallen tree under which a Seronera pride lioness had left her three small cubs. They were dead, and the spoor in the dew-soaked grass showed that the Kamarishe pride males had killed them earlier. One male picked up a cub, carried it 10 m, then ate its viscera. The other male carried a second cub away with him (plate 15), stopping occasionally to lick it, and, later, to nestle it between his paws. He had still not eaten it at 1645. At 1900 the mother of the cubs returned to the fallen tree. She sniffed at her remaining cub, licked it briefly—then ate it, starting at a small hole the vultures had torn into the abdomen. By 1915 only the head and forepaws remained uneaten. The Seronera pride male was never seen again, having probably left the area or possibly been killed.

From December, 1967, to May, 1968, the Masai pride frequented the eastern part of its area and the Kamarishe pride returned to its usual haunts with the result that no males associated with the Seronera pride lionesses for several months. Both sets of males returned in June and associated freely with the females which often were seen on one day with the males of one pride and on the following day with those of the other. Female J, for instance, courted with a Kamarishe pride male on November 30 and with a Masai pride male the next day. I saw only one direct interaction between the males of the two prides. On that occasion, a Masai pride male followed female B who was in estrus. She walked rapidly, the male at her heels, about 1 km to where female C, also in heat, reclined in high grass with a Kamarishe male. The female raised her head above the grass, making herself visible, but the male did not do so until the Masai pride male was 20 m away. Each stood motionless with bodies broadside for several seconds, but then the Kamarishe male lunged forward with a growl and the Masai pride male, though younger and larger, fled for over 1.5 km in the direction of his pride area.

Both sets of pride males associated occasionally with the Seronera pride lionesses early in 1969. Social contact was not, however, as close as that with the lionesses of their own pride. For example, they rarely were in bodily contact with the Seronera pride lionesses, except when courting or at a kill, and usually rested 30 to 50 m from the group. A change occurred in June, when the Kamarishe pride males returned to their own pride. Once, on June 29, they saw the Seronera lionesses 200 m away but did not join them. On the other hand, the frequency of the visits by the Masai pride males increased to such an extent that they seldom were with the lionesses of their own pride. In August, almost two years after the first friendly contacts, the Masai pride males abandoned their pride and became permanent members of the Seronera pride. Socialization in this instance was obviously a slow process, but the persistence of the males, coupled in the beginning with the tolerance some estrous lionesses showed toward them, ultimately ended in their acceptance. Only Limp was sometimes not tolerated. Possibly because of his poor physical condition and because of his complete parasitism at a time when food was in short supply, the lionesses attacked him and drove him from their vicinity on several occasions. The Cub Valley males had moved into the general Masai pride area in April, 1969, and in August they joined that pride in a rapid and seemingly smooth transition.

In contrast to the males, the lionesses of the Masai and Kamarishe prides never associated with those of the Seronera pride, and, in fact, avoided each other so assiduously that contacts were rare. On one occasion a lioness of the Seronera pride, followed by a Masai pride male, inadvertently advanced to within 200 m of several Masai pride females. These happened to roar, and the Seronera pride lioness fled at a trot for .5 km before stopping and staring back in the direction from which she had come. She roared twice, scraped the ground with her hindpaws, and continued on.

At 0655, on December 27, 1968, five Masai pride lionesses walked through the central part of their area obviously searching for the rest of the pride as one or another roared, then stopped, as if waiting for an answer. Filing across the skyline, barely visible in the high grass .6 km away, were seven lionesses and six cubs of the Seronera pride. When the Masai pride females spotted them, they trotted closer, apparently thinking that these were members of their own pride. The Seronera pride lionesses became aware of the others at 60 m; the cubs scattered and several lionesses growled, yet the Masai pride females still did not discover their mistake. One ran up to a Seronera pride female as if to rub heads in greeting. But suddenly she stopped, with noses almost touching, snarled and slapped at the other. For 15 seconds the lionesses milled around growling and snarling and roaring and several tussled briefly. One snarling lioness chased a Masai pride female and clawed her deeply in the rump, an injury which caused the thigh to wither in the following months until she could only hobble on three legs. Two Masai pride lionesses fled at the beginning of the melee and two more did so at the end. Only one remained. She walked for nearly a minute among the other lionesses with her canines exposed, snarling, then followed unharmed those that had fled. Although the Seronera pride females had driven four of the five others away, they now moved toward their own pride area, traveling 5 km in 65 minutes.

On March 2, 1969, at 1715, a Seronera pride lioness fed alone in the Masai pride area on a zebra that had died from disease. After she left the carcass, apparently to fetch her cubs, vultures descended. These were seen by three lionesses and four cubs of the Masai pride which appropriated the meat. The Seronera pride lioness reappeared with her cubs at 1905, stopped 200 m from the other lions and roared softly several times. She received no answer. Instead, two Masai pride lionesses approached her silently. She retreated and went to her pride, where somehow she induced six lionesses to follow her; all trotted toward the carcass at 1955, chased the Masai pride animals 50 m, and then ate the remaining meat.

Interactions between Prides and Nomads

Pride lionesses seldom interact with nomadic ones, a reflection on the efficiency of the communicatory system which enables lions to avoid each other. Once several males and females of the Masai pride trotted and walked rapidly 4 km to where several hyenas hooted around three nomadic lionesses on a zebra kill. The nomads retreated when the others approached and then sat 60 m away and watched the latter eat. After 5 minutes two males and a female attacked the nomads and chased them several hundred meters. On another occasion, two Masai pride males and a female rested all day within 50 m of an elderly nomadic lioness without responding to her. Their behavior was so unusual that I suspected that the nomad may have been a pride member at one time.

The response of lionesses toward nomadic males was generally antagonistic, especially when the lionesses had cubs. On June 4, 1968, for example, females L and Q of the Masai pride rested by a kopje as two young adult males walked into sight 60 m away. Attacking immediately, the lionesses slapped at the smaller of the two males until he fled with a bloody foot and tail. Both then rushed the other male, which retreated 30 m. One approached him snarling, reclined at his feet and lightly batted at him with a paw. He retreated once more, but suddenly both lionesses attacked again, chasing him 50 m. After they had returned to their cubs, the male approached again, this time to within 25 m, but the lionesses faced him growling until he left. On another occasion, four Masai pride lionesses hunted gazelle in high grass. One had just chased and missed a fawn and on glancing up saw a nomadic male standing 50 m away, watching. She took a step in his direction but suddenly seemed to realize that he was a stranger. She fled 100 m in a semicrouched posture. Two other lionesses approached him hissing, their heads low in threat, then chased him for 130 m. Another time, two Seronera pride lionesses crouched, growled, and twice lunged several meters at a nomadic subadult male when he slowly approached to within 10 m of them. The Kamarishe pride lionesses responded similarly toward two nomadic adults. On yet another occasion, four Masai pride lionesses chased a subadult male .4 km when he approached their kill, while the three pride males merely continued to eat. In all these instances the males retreated after being attacked.

On two other occasions the lionesses fled into another part of the pride area after meeting nomads. One evening at 1910 five Masai pride lionesses and the subadult male fled for 1.5 km when two nomads appeared from the grass 35 m away and followed them at a fast walk for several hundred meters. Another time, a nomadic male appeared out of a stream-bed 35 m from four Seronera pride lionesses. When these growled at him, he moved 50 m to a thicket where he met his companion, and then both walked toward the females which scattered and fled. The nomads pursued two of the lionesses for .8 km. Twice a lioness wheeled around to face the pursuers with mouth wide open, and the males did not press the attack.

Pride males may chase nomadic lionesses, as several of the above examples illustrate, and their response toward nomadic males, too, is antagonistic. Once a Seronera pride male and three lionesses spotted a pair of nomadic males. The lionesses pursued them about 180 m, but the male continued for 2 km. Similarly, awakened by much roaring near our house at 0430, I found the two Kamarishe pride males chasing two nomads.

The general antagonism between pride members and nomads may be modified in several ways. A pride lioness in estrus tolerates strange males, and a nomadic one in estrus is accepted by pride males. For instance, female B of the Seronera pride once mated with nomadic male No. 107. On October 29, 1967, nomadic lioness No. 78 was with a Masai pride male beside a zebra kill where they copulated several times during the night. At 0515 the following morning, the female suddenly fled and a few seconds later four lionesses of the Masai pride arrived and joined the male on the kill.

Male No. 61, a nomadic subadult about $2\frac{2}{3}$ years old, was frequently seen in the Masai pride area in 1966, either alone or in the company of other nomads, including female No. 78. He interacted with the Masai pride several times and I describe these meetings here to show how a male might become acquainted with a pride without becoming a member of it. On October 22 and 24, 1967, male No. 61 and a somewhat older male companion were in the area once again. On October 29, at 1600, a Masai pride female caught a Thomson's gazelle whose dying bleat attracted a male, two females, and several cubs of the pride. These rushed to the kill. Just as they arrived, male No. 61 stepped out of a thicket 5 m away. The lionesses halted and stared at him, but the Masai male, seemingly unaware of the stranger, attacked the lioness with the gazelle and relieved her of it. When a small cub hurrying in the direction of the kill passed the nomad, he pounced on it, bit it, and shook it. A lioness—not the cub's mother—growled and charged the male. He dropped the cub and ambled away, passing within 10 m of the eating Masai pride male which remained oblivious to the commotion. One lioness attacked the nomad again, but he turned 180° in midair, slapped at her, and walked rapidly away. The injured cub, bleeding from its side and leg, sat in the grass and miaowed, but none of the lions responded.

The following day at 1540 vultures descended on the remains of a Grant's gazelle. Attracted by them, male No. 61 ran over and fed on the scraps. Two minutes later three Masai pride lionesses arrived from a different direction. The one in the lead saw the male, returned to the others, and all lay down. Ten minutes later the male approached the lionesses, walking with head held low, not looking at them, and finally reclined 50 m away. At 1605 he ventured 10 m closer, but the lionesses crouched and faced him hissing. At 1705 he moved to within 17 m of them, and all still held their positions at 1905. On November 6, male No. 61 sat 50 m from a Masai pride male. The latter scraped the ground with his hindpaws and reclined, as if indifferent to the presence of the other.

On April 19, 1969, I saw male No. 61, now a young adult, with the Masai pride again. At 1045 he was 50 m from eight lionesses and the subadult male. By 1150 he had moved to within 12 m of the group. At 1130 a female and the subadult male rose and hissed at him, but the others continued their rest until at 1930 all except one lioness moved 200 m. Male No. 61 sniffed at the places where they had been lying, seemingly ignoring the remaining lioness, which watched him, chin on her paws. Slowly he drifted to within 5 m of her, still sniffing around. She growled and jerked her head at him when he was 3 m from her, but he, ignoring the gesture, walked past her and followed the group, nose to the ground, and reclined 30 m from it. At 2040 he was still there.

The interactions between prides and between prides and nomads raise several points and provide some generalizations. Neighboring prides, as well as certain nomads and prides, meet each other a number of times, and it is likely that some animals know each other individually as a result of such contacts. An observer seldom knows the history of an animal he sees, yet its behavior and the responses of others toward it are often influenced by what has happened in the past. On the whole, prides tend to be antagonistic toward members of other prides and toward nomads, but several factors temper the severity of interaction. For instance, any lioness in estrus accepts and is accepted by both pride males and nomads. Males, such as male No. 61 described above, reduce the tendency of lionesses to flee or to attack strangers by approaching slowly and halting at intervals—several hours at times—until their presence is accepted or at least ignored. Persistence by males in doing this could ultimately lead to their becoming fully acquainted with the lionesses. Even though interactions are at times seemingly violent, with the animals growling, slapping, and so forth, injuries, if any, consist of minor cuts. In fact, the combatants give the impression that they avoid physical contact. When a lion pursues a stranger it usually maintains a certain distance, at least 10 m, adjusting its speed to that of the intruder; even if it catches the other, actual contact is usually limited to a slap or two. Serious fights do occur but are rare. The various interactions permit one generalization. A resident lion will ultimately accept a nomad or a member from another pride of the opposite sex but not one of its own. Males and females both help to maintain the integrity of their pride by selectively excluding others of their own sex.

The Pride Area

Several authors have commented on land use by lions. Jearey (1936) wrote that lions "had well-defined territories in which they hunt and have their being." More recently Carr (1962) noted "that lions have an elaborate territorial system and a strict set of rules governing trespassing on one another's preserves." In the Nairobi National Park, Guggisberg (1961) found that "for the lioness, the territory means simply a hunting ground, while for the lion it is a region containing a certain number of lionesses which from time to time are ready to mate with him." Studying in the same area, Schenkel (1966a), using the terminology as defined by Burt (1943), found that a pride maintains a territory surrounded by a home range. However, the land tenure system of the Serengeti lions is more complex than these statements indicate.

Each pride confines itself to a definite area in which its members spend several years or, in the case of some lionesses, their whole life. The main requisites for the existence of a pride area are a water source and sufficient prey throughout the year, conditions existing in the woodlands and along their edges but, for the most part, not on the plains. Figure 5 shows that pride areas are distributed fairly evenly throughout the woodlands; additional prides also frequent the study area around the periphery, but since these prides center their activity outside the boundaries they have not been indicated. Possibly a small pride or two was overlooked in the hills north of the Mbalageti River. Two prides (I, XIII) used mainly the tree-fringed watercourses that meander into the plains, and one (XIV) remained entirely in the plains around the Simba kopjes. The rest of the plains within the study area—some 1,300 sq km of terrain—had no prides.

Figure 7 shows the size of pride areas around Seronera. The line enclosing each area is based on the peripheral locations at which female members of the pride were seen. Pride males occasionally used a larger area than the females. The Masai pride males, for instance, extended their wanderings to the hills west of the pride area (I), and the same two males encompassed the area of both prides XIII and XIV and, in addition, roamed west into the Moru kopjes. Because of the large area and difficult terrain frequented by most prides, I was unable to delineate all boundaries; for some, such as IX and X, only one end of the area was known to me. A few prides have not been indicated on the map. Pride areas III, XI, and VI, for example, were visited by lions from the north about which I knew little.

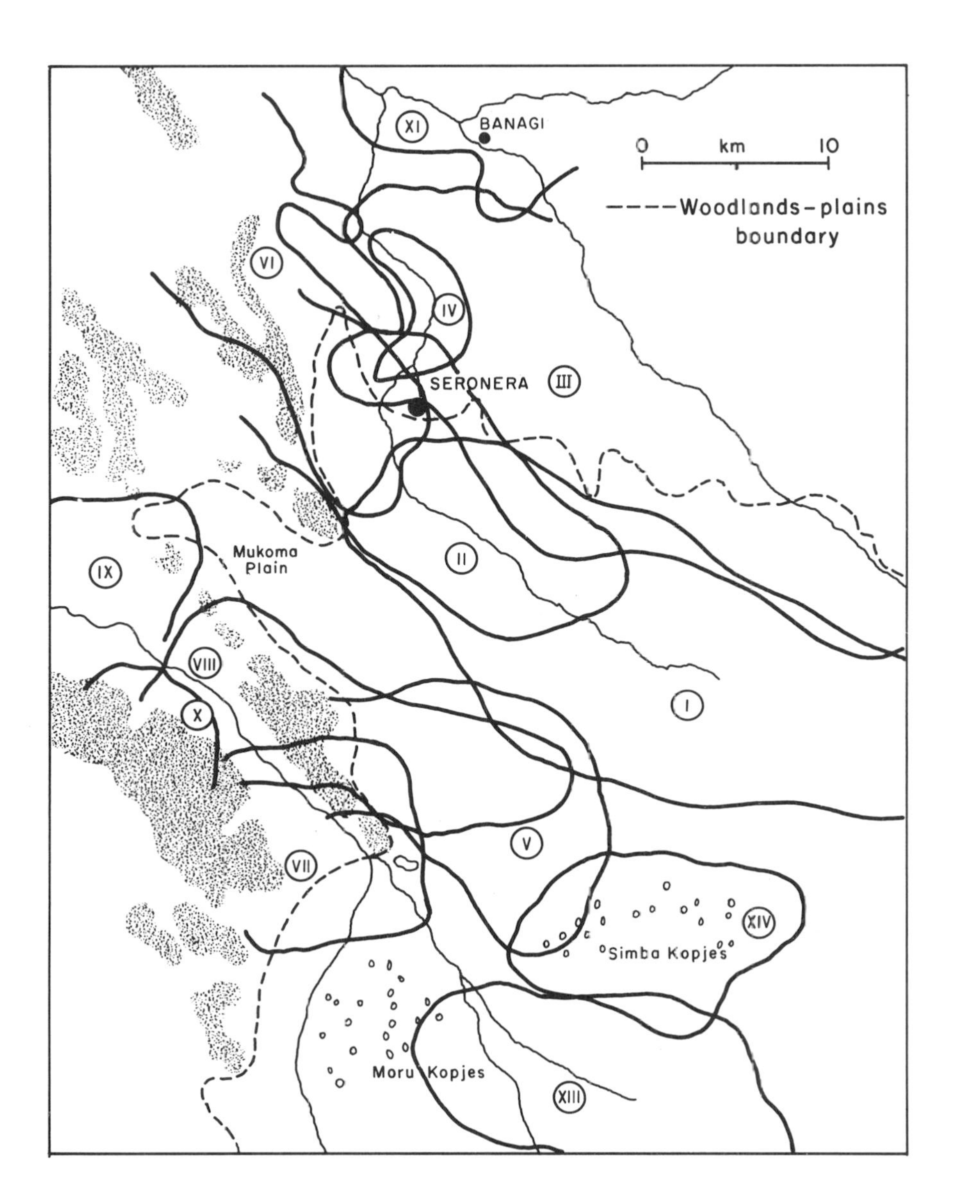

Fig. 7. Distribution of pride areas around Seronera illustrating their overlap. The numbers correspond to those in table 2.

Figures 5 and 7 show that the woodlands and the edge of the plains are wholly occupied by prides, that essentially no areas remain unused. A few small pieces of land, such as the Mukoma plain, are not claimed by a pride, probably because of the lack of prey for much of the year. With all suitable land already taken by prides, new ones can establish themselves only by squeezing in among the others as apparently was done by pride IV.

Both the Masai (I) and Loliondo (III) prides ranged over about 400 sq km each, the Seronera (II) pride over at least 210 sq km, the Kamarishe (VI) pride over at least 175 sq km, the Plains (V) pride over at least 150 sq km. The Nyaraswiga pride (IV) area was about 30 sq km in extent. The small East Moru (XIII) and Simba (IX) prides used 140 sq km and 120 sq km of terrain, respectively, and the males which they shared roamed over about 275 sq km. Males seldom associated with two prides at the same time in the Serengeti possibly because the huge pride areas made it physically difficult for them to keep more than one under their jurisdiction. In Manyara Park, two prides each occupied overlapping areas of about 20 sq km, and these were visited by the same two males which ranged over a total of 39 sq km. In Nairobi Park two lionesses with eight cubs remained within an area of 34 sq km, but the males that associated with them wandered over an area of 140 sq km (Guggisberg, 1961).

In general, size of pride area appeared to be related to the size of the pride itself, with, for example, the large Loliondo and Masai prides using more square kilometers of terrain than the small East Moru, Simba, and Nyaraswiga prides. Precise comparisons are of little value because the size of the pride area is in the final analysis related to the amount of prey available, and this varies considerably. Fourteen adult and subadult lions in Manyara Park remained within an area of 39 sq km, whereas in the Serengeti a pride with that number of members would use at least three times as much land.

Pride areas overlapped extensively, a fact illustrated best by the prides at Seronera, whose movements were studied more intensively than those in other parts of the park (fig. 7). Discrete pride area boundaries, such as those Schenkel (1966a) drew for prides in the Nairobi National Park, were rare in the Serengeti. The amount of overlap varied from pride to pride. Prides whose area encompassed parts of the plains shared little with others in the eastern portion because no prides existed in that direction. The Loliondo pride (III) occupied a considerable piece of terrain by itself; the Seronera pride (II) shared almost its whole area with neighbors, totalling four prides. The small Nyaraswiga pride (IV) was unique in that it occupied an area almost wholly within the one used by the Loliondo pride.

A map such as figure 7 is misleading to some extent because it gives the impression that lions wander rather haphazardly within a certain area. One lioness (female L) of the Masai pride was encountered on 305 days during the study, and figure 8 shows the places where she was

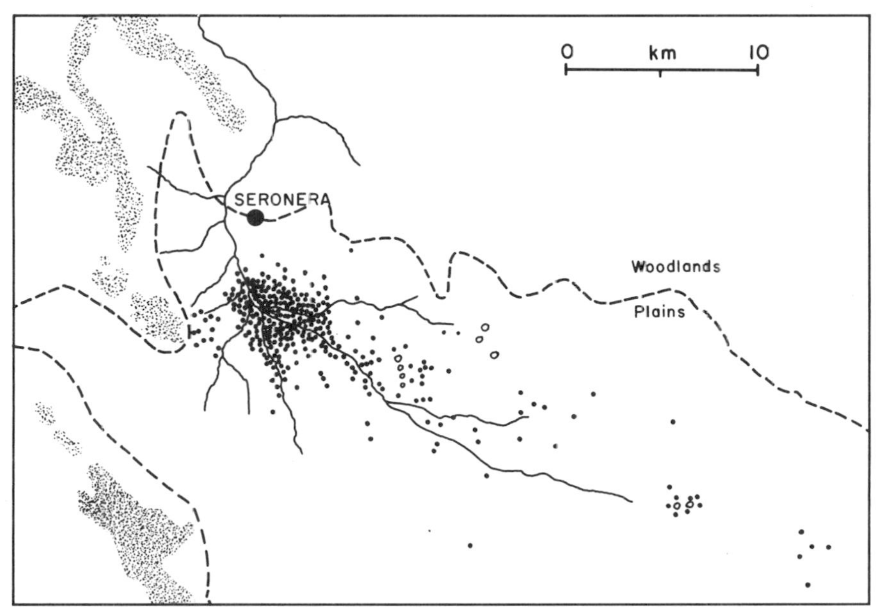

Fig. 8. The locations at which female L of the Masai pride was encountered, showing how she focused her activity in the western portion of the pride area.

seen, usually accompanied by other pride members. Although sightings are slightly biased in favor of the western portion of the pride area, where the lions were easier to find than in the eastern part, the lioness obviously used some parts of the pride area more frequently than others. The pride as a whole remained in the western 130 sq km from June, 1966, to March, 1967. With the onset of the rains in April and until July, a period when the western part was almost devoid of prey, the pride moved eastward into the central third of its area. It used the western third again from July to November. The rains in late 1967 and early 1968 were prolonged and heavy, and the pride stayed in the central and eastern portion until May, 1968. In June, 1968, it returned to the western third and remained until the end of the study. In 40 months, the eastern third was used only for about 3 months (7.5%), the central third 7 months (17.5%), and the western third 30 months (75%).

An eastward movement during the rains, when prey was scarce in the woodlands but abundant in the plains, was shown by most prides along the woodlands-plains border, including prides II, III, V, VI, VIII, and XIII. Most prides, except I, III, and XIII, failed to move far enough to the east to reach the prey concentrations on the intermediate grasslands. Prides such as the Seronera and Kamarishe, seemed forced to search for prey over a wider area than usual during the rains, and, with the prides east of them gone from their usual haunts, nothing hindered their movement in that direction. The number of prides that shifted eastward depended on the length of the rainy season. In 1967 only the Plains pride moved into the grasslands, but in 1968 all the prides bordering the woodlands did so, to mention only the effect of the shortest and longest rainy seasons. Such a seasonal shift was typical of prides primarily along the edge of the woodlands. Those in the woodlands also roamed more widely within the pride area during the rains, but, in general, the animals concentrated wherever the erratic prey happened to be.

During the dry season, prides moved less than during the rains. They often concentrated their activity for months along a certain stretch of river as long as prey was available there. For example, three prides each used a portion of the Mbalageti River from Lake Magadi downstream for 25 km, and the Seronera and Masai prides hunted mainly along their respective sections of the Seronera River. The most heavily used part within a pride area is here termed the *focus of activity*, following the definition of Jewell (1966). Whereas pride areas overlapped extensively, the foci of activity were discrete or almost so (fig. 9), a land tenure pattern superficially similar to that described for baboons in Nairobi Park by DeVore and Hall (1965) and for coatis on Barro Colorado Island by Kaufmann (1962). However, core areas, as these authors called them, were never entered by neighboring groups, whereas the foci of activity of lions may be visited by other prides. The Seronera pride area was completely overlapped by other prides, but the pride retained the exclusive use of its focus of activity for much of the year. The foci of activity of the Nyaraswiga and Loliondo prides overlapped somewhat. Usually the Nyaraswiga pride shifted its position when the other entered the area, but the fact that the two males disappeared after some 1½ years of tenure suggests that they were evicted by the males of the Loliondo pride. The focus of activity of the

Masai pride was about 65 sq km in size and that of the Seronera pride about 45 sq km. With animals spending much of their time within their focus of activity and with incursions into areas occupied by neighboring prides occurring mainly when those animals are somewhere else, frequent confrontations are prevented.

The concepts of territory and home range as defined by Burt (1943) have been used widely in vertebrate studies: "Every kind of mammal may be said to have a home range, stationary or shifting. Only those that protect some part of the home range, by fighting or aggressive gestures, from others of their kind, during some phase of their lives,

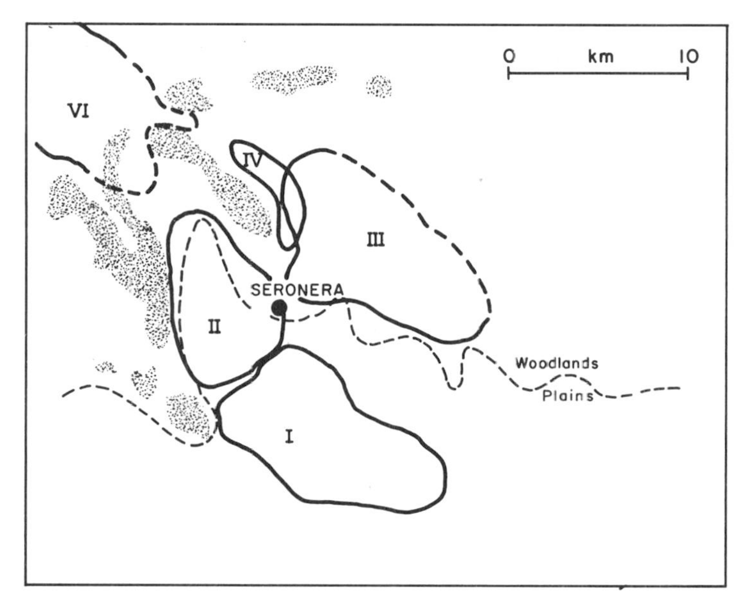

Fig. 9. The foci of activity of five prides.

may be said to have territories." Pitelka (1959) broadened the definition of territory to include any area of exclusive use, emphasizing the ecological partitioning of an area rather than the mechanisms employed to maintain it. Because the definition of *territory* has been extended so much that it has become ambiguous, I preferred to use the innocuous term *pride area* until I had presented all information about land tenure in lions. Now it is of interest to interpret the data on the basis of these definitions.

Pitelka's definition of a territory as an area of exclusive use could apply to those parts of a pride area which do not overlap with other prides. On this basis some pride areas could be considered territories, particularly those in Manyara Park, whereas others, such as the Seronera pride, could make no territorial claims and still others would have to restrict their claims to a small portion of the total area occupied by them. I found Pitelka's concept of territory to be of little use in helping to clarify land tenure in lions. Central to Burt's idea of territory is some form of aggressive behavior by the resident toward the intruder. Pride members certainly are aggressive toward their neighbors, both resident and nomad, but it may be argued that such antagonism does not indicate defense of a particular area but merely represents a general animosity toward any stranger. However, when the Seronera and Masai pride lionesses fought in an area usually occupied by the latter, the Seronera animals retreated toward their focus of activity even though they could be said to have won. When one of the Seronera pride males was killed, males from three other prides promptly penetrated deeply into the pride area, but when these prides encountered each other they fled toward their usual haunts. Nomads tended to be tolerant of strangers, in contrast to pride members.

These and several other observations suggest that central to the disputes was an incursion on other animals' land. Intruders were not chased to the border of the pride area, yet the attack induced them either to leave or shift to a part not occupied by the pride, producing the same effect as if they had been fully evicted. Nomads sometimes lived for weeks or months within a pride area, and prides wandered with impunity around areas usually occupied by others. However, the size of a pride area is important when evaluating how effectively intruders are kept out. Some pride areas are so large that it is physically impossible for a few lions to prevent others from entering. A search for nomads within the area of the Masai pride in April 21, 1969, revealed seven males, seven females, and twelve cubs singly and in groups, but all of them were in a part of the pride area not used by pride members at the time. It is of interest to note that in Manyara Park, where pride areas are small and easily defended, such incursions did not occur until after one of the two pride males was shot. Pride males may associate with nomadic lionesses, especially if these are in estrus, and pride lionesses sometimes tolerate visiting males, but such special relationships between members of the opposite sex in no way mitigate the antagonism of males and females toward strangers of their own sex.

Although the evidence favors the idea that lions defend a piece of land, it is not at all easy to delineate those parts of the pride area which the animals consider their territory, mainly because so few interactions are seen. A fairly definite boundary existed along the Seronera River beyond which the Masai pride lionesses never ventured westward (see fig. 8). However, the Seronera pride lionesses readily moved east across this boundary but retreated when they became aware of the Masai pride. This suggests that the Seronera pride lionesses considered the area west of the line their territory and the area east of it an undefended home range, used mainly in times of food scarcity. The Masai pride, on the other hand, seemed to consider its whole pride area as a territory—at least it chased intruders in all parts—but the animals visited the eastern portion so seldom that any territorial claims there were highly sporadic and actually unnecessary; it contained so little prey that permanent occupancy was not possible. The large size of pride areas obviously results in uneven defense of an area, most of the defense being carried out in the focus of activity where the pride spends most of its time. In a system such as this, the strict maintenance of boundaries is of no consequence, for not only is the pride usually in the part of its territory which provides the necessities of life most readily, but it can also reclaim any seldom-used portion by evicting intruders. Movements out of the territory and into the home range seem to be mostly the result of a food shortage.

The pride males had a tenure pattern much like those of the lionesses, and, in fact, their use of the pride area was often identical. The Masai pride males often ventured west into the focus of activity of the Seronera pride, an area never entered by the Masai pride lionesses. But the tendency of the Masai males to flee eastward when encountering males there indicated that they did not claim this part as their territory too; it was merely a home range a few square kilometers in extent. However, it became part of their territory toward the end of the study. Males with two prides under their jurisdiction had territories about twice as large as those of the lionesses with whom they associated.

The land tenure pattern of some females and all males changes in the course of their lives, because they become nomadic as subadults. Two types of ranges may exist: a circumscribed pride range, all or part of which may be defended as a territory, and a huge nomadic range. Some males often become territorial again as adults in an area different from that in which they were born, only to become nomadic once more when replaced by other males. In addition, some nomads establish temporary territories in the plains. The land tenure system of males is thus in a continuous state of flux. Although the concepts of territory and home range are applicable to lions, they are useful only if the complexities of the male and female systems and the variation between prides are taken into account.

The Nomad

Groups of nomadic lions are indistinguishable from resident groups but basic differences nevertheless exist. Whereas a pride is a relatively stable and closed social system in which the lionesses are all genetically related, nomadic groups are open, with animals joining and parting at intervals. Because I found it difficult to distinguish a nomad from a resident in the woodlands, unless all residents were known individually, as in the Seronera area, I studied nomads mainly in the plains where prides were scarce.

Group Size and Composition

A total of 3,111 nomads in 1,099 groups, including solitary animals, were classified in the plains between June, 1966, and May, 1969. Of these, 330 (10%) were alone. The others were in groups varying in size from 2 to 13 animals; groups with more than 6 individuals comprised only 7% of the sightings (fig. 10). Average group size was 2.8.

The composition of groups varies greatly, and, as with prides, sometimes changes from day to day. Some animals tend to remain solitary, others associate casually for varying periods, and still others form companionships that persist for years, as in the case of two young lionesses which I saw on the plains each rainy season between 1967 and 1969. Such companionships were so close that if I saw only one of a pair the

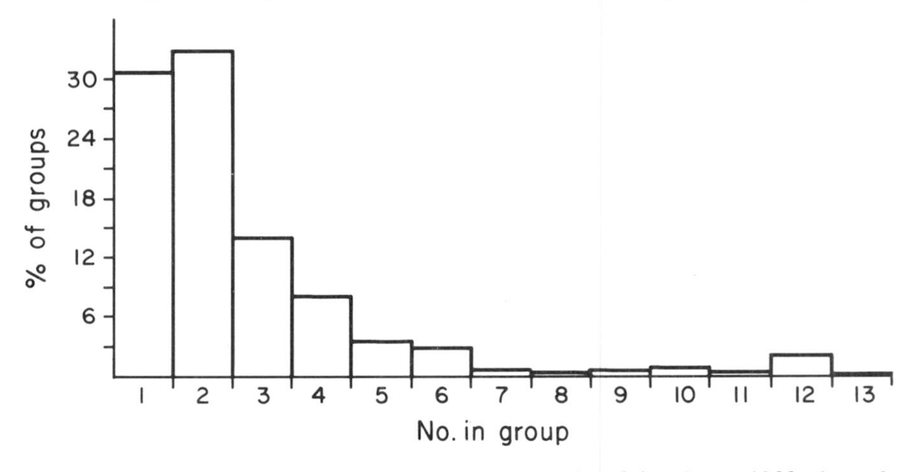

Fig. 10. The size of 1,099 nomadic groups seen on the plains, June, 1966, through May, 1969.

other was usually in the vicinity and the animals were together again on subsequent days. Some large groups may persist in one locality for months and give the appearance of a pride before suddenly splitting. The compositions of three such large groups were: 2 males, 7 females, and 2 cubs; 5 males and 7 females; 2 males, 3 females, and 8 cubs. Large groups may also consist of subadults which probably have left the pride together. One such was composed of 5 subadult males and 3 subadult females, another of 4 subadult males and 6 subadult females.

Of 330 lions seen alone, 47% were adult males, 22% were subadult males, 27% were adult females, 3% were subadult females, and 1% were large cubs. When compared to the population composition (see table 26), these figures show that adult males are alone nearly twice as often as expected, and subadult females much less frequently so. Most solitary individuals, of course, join other nomads occasionally.

Of 362 nomadic pairs, about half comprised two males, but, as table 6 shows, an adult and a subadult seldom traveled together; males chose as companions mainly males of their own age. Many such male companions probably were brothers or pride mates, but some were of different ages and their contacts were known to have been, at most, casual before they joined. About 29% of the pairs consisted of a male and female, both usually adult and sometimes courting; pairs of females were seen on only 14% of the encounters.

Although Percival (1924) stated that "occasionally two, but never more, big males are seen together," Figure 11 shows that as many as five males, either adult or subadult, may associate, but that single animals or groups of two are most common. In the woodlands, I once saw a group of seven subadult males which had left their pride together, and on other occasions groups of 7 and 8 subadult males, whose history was not known to me. Not included in figure 11 are instances when subadult and adult males were in the same group. Of 368 groups containing more than one male, only 9% contained both age classes. This suggests some form of avoidance on the part of young animals even though overt aggression is infrequent among males. Fifty-six percent of the adult males were solitary or in groups containing only males, and 44% were with one or more females (as compared to the 70–90% for some pride males); for subadults the corresponding numbers were 50% and 40%. Solitary males have little opportunity to obtain, much less maintain, a pride, and thus roughly half of the solitary males on the plains are likely to remain nomadic unless they form a companionship.

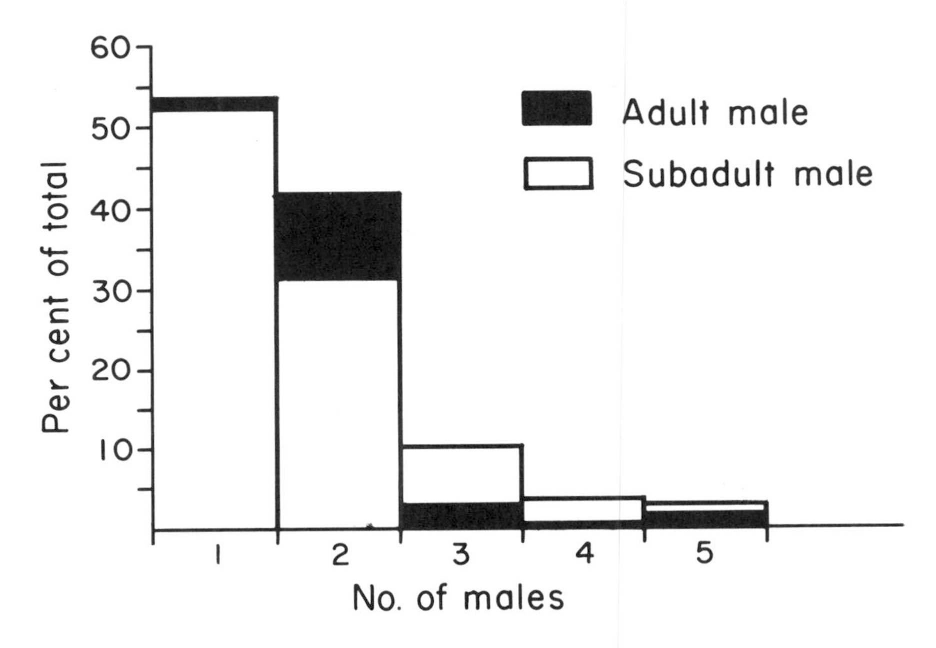

Fig. 11. The number of nomadic subadult and adult males associating in groups on the plains. Groups containing males of both age classes of males have not been included in the computations.

Changes in Group Composition

The social life of many nomads exists as a series of casual encounters, lasting from a few hours to a few days, with other nomads. Tailless, an adult male seen intermittently over a period of three years, provides a typical example. He had an adult male companion until June, 1967, but Cutlip, as I named him, disappeared. During my 44 encounters with Tailless, he was alone 14 times (32%) and with other lions 30 times, 24 of these with Cutlip. The largest group of which he was a member numbered 6 lions. He was known to have associated with 21 other nomads—9 males, 8 females, and 4 cubs. However, he was with male No. 61 five times, once each in 1966, 1968, and 1969, and twice in 1967. Since I encountered Tailless on only about 3% of the days in the study, he must have met many other nomads during his wanderings. His contacts with male No. 61 illustrate not only the tenuous nature of the social bonds but also that nomads are often old acquaintances, an important point when considering interactions between them.

Another typical nomad was male No. 57, tagged as a subadult on the Musabi plain in November, 1966 and seen 25 times after that. He was alone when tagged, but he associated briefly with another subadult male a month later. The following May he spent several days with a lioness. On June 9, 1967, at 1330, he was with a group consisting of subadult male No. 134, another subadult male, an adult male, and two adult females. But by 2100 he was alone again and remained so the following day. At about 0400 on June 11, he joined the subadult male and two lionesses which he had met two days previously on the remains of a wildebeest, ate briefly, and departed by 0630. The following midnight he met male No. 134 again and both ventured to within 50 m of, but did not join, two other subadult males on a wildebeest carcass. Male No. 134 departed alone at 0645. Male No. 57 rested all day and at 2125 joined yet another subadult male, and this one became his companion. Thus in the course of three days, male No. 57 met six males and two females and parted from all except one of them on his own volition. In subsequent months I saw him 13 more times, always with the male he joined in June, 1967, and three times also with an additional male. On November 9, 1968, he was shot by a hunter outside of the park.

Female No. 78 was seen 46 times, 11 of them alone, between July, 1966, and June, 1968. She was the focus of an unusual group. When I first saw her she was accompanied by a male cub just over a year old and subadult male No. 55, nearly 3 years old. In October, 1966, I observed for the first time a nomadic lioness (No. 53) with two subadult females and one subadult male (No. 61), about $2\frac{1}{4}$ years old, all presumably her offspring. On November 1 at 2200, as I watched female No. 78 and her males on a kill, male No. 61 from the other group suddenly appeared around the periphery of the feeding lions and later that night was accepted by them. He severed all ties with his presumed mother and sisters, and roamed between Seronera and the Simba kopjes in the company of female No. 78, and the two other males. During that time they met Tailless and Cutlip at least twice; the subsequent contacts between Tailless and male No. 61 are mentioned above. On January 5, 1967, male No. 122, another subadult, joined the group for several days. On May 31, female No. 78 was seen alone for the first time, and between June 18 and July 27 she was encountered alone 8 times. Her cub, now 2 years old, had joined the older males. On September 25, October 28, and November 4 she courted with Masai pride males. Late in November and December she joined her former companions, Males No. 55 and 61, but her cub was not there. She was around Naabi Hill with five adult males, including

Nos. 149 and 150, in January, 1968. I last saw her in June, 1968, with yet another male. During two years she was known to have had contact with fifteen different males and three females. Her offspring disappeared, but her companions, males No. 55 and 61, were often together until March, 1968. After that date I saw only male No. 61, several times with the Masai pride.

Nothing prevents nomads from remaining together, yet most of their contacts remain transitory. Male No. 57 associated with many subadults before he suddenly established a companionship with one of them. Obviously two animals have to find each other congenial to form such a close social bond, but the conditions that make one lion an acceptable companion in contrast to all others are not clear. Generally a companion of the same age and sex is preferred, and a solitary individual possibly is accepted by another lone one more readily than by a group in which social ties already exist. In addition, past contacts or lack of them undoubtedly influence the outcome of meetings, as do individual idiosyncrasies. Some males, for example, form no companionships. Among these was a young adult male, No. 128, which I saw twelve times between January, 1967, and March, 1969. He was alone eight times and with other males briefly three times.

The social contacts of male No. 159 during a period of nineteen full days in 1967 give an idea of the frequency and duration of interactions between nomads. Male No. 159 was an adult past his prime with a rather scruffy mane who, with a male companion, had established a temporary territory around Naabi Hill. During the 19 days he associated with 9 other lions, excluding 2 small cubs, but for varying lengths of time. Most social contacts were with his male companion, with whom he spent at least several minutes on 18 of the 19 days for an average of 72% of the day (fig. 12). He met a subadult male on seven days and their association was limited in time to 12% of the day; in fact, he chased the young male from his vicinity on three occasions. Lionesses roamed singly or together in various combinations around the area frequented by male No. 159. His contact with these females was variable. One he met on only two days, another on twelve days, and the time he spent with them ranged from five minutes or less in casual meetings during his wanderings to several hours when at a kill.

In contrast to the transitory contacts of these animals, another group of nomads, called the Cub Valley group, formed a permanent social unit which was similar to a pride and in fact might have become one had not the lack of prey on the plains during the dry season forced it to maintain a nomadic existence. Female No. 69, a nomad with two small cubs, was tagged in November, 1966, near Seronera. The following February her cubs were gone and she courted near the Simba kopjes with two males I had not seen before. Another lioness, No. 102, was alone east of Seronera in December, 1966. By February she was at the Barafu kopjes 55 km to the east with two other females. Sometime during late February, females Nos. 69 and 102 and a third lioness joined and centered their activity around a shallow valley just west of the Gol kopjes. These three lionesses became permanent companions. Except for a brief meeting with female No. 135 and on another occasion with an untagged one, they had few contacts with others.

The males that associated with the Cub Valley group had their independent history. Males No. 54 and 110, companions somewhat past their prime, were around the Simba and Gol kopjes and along the edge of the woodlands between October, 1966, and March, 1967 (see fig. 14). Male No. 58 was the companion of male No. 59 in November, 1966. By February male No. 58, a prime animal, had established another bond, this time with a large, elderly male whose hindquarters were emaciated. These two males joined the three females in March soon after the latter had settled into the valley. I did not visit the group in April, and by May a change had occurred; male No. 54 had replaced

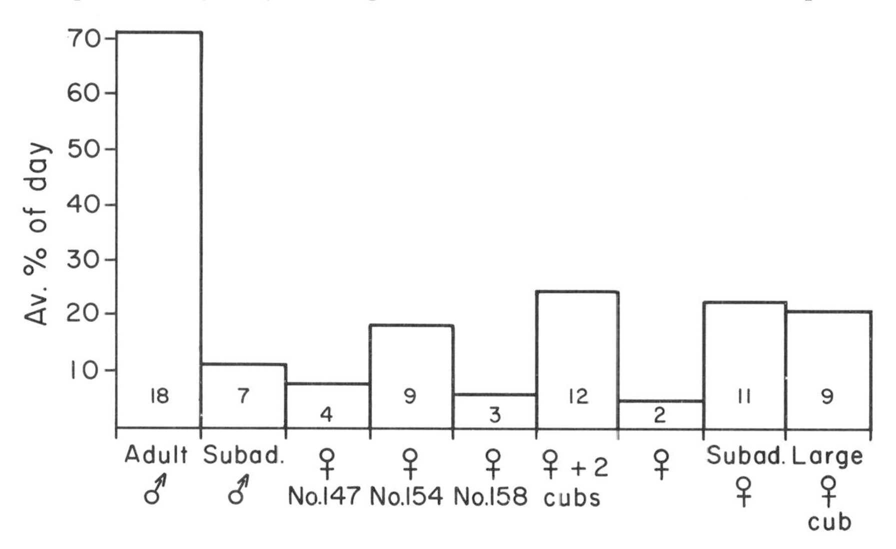

Fig. 12. The average percentage of day that each of 9 individuals associated with nomadic male No. 159 during 19 full days of observation, June–July, 1967. The figure within each bar represents the number of days that the animal was seen with male No. 159.

male No. 58 in the group and both males Nos. 58 and 110, now without companions, became solitary nomads which ranged widely to, among other places, the Barafu kopjes. The new male companions were together at least three months. By November, the original pairs—Nos. 54 and 110, and Nos. 58 and the old one—were together again, illustrating well the changeable nature of some companionships.

Male No. 58 and the old one remained closely attached to the three females and their cubs until 1969. During the dry season of 1968, all moved from the valley to the Mukoma plains and probably beyond. They were back in their usual valley during the 1969 rains. The dry season found them along the edge of the woodlands in the Masai pride area. In August, when the Masai pride males switched their allegiance to the Seronera pride lionesses, the two Cub Valley males abandoned their female companions of nearly two years and became the pride males of the Masai pride. During the study, the status of these two adult males thus changed from complete nomadism to localized nomadism within a temporary territory and finally to residence in a pride.

The Nomadic Range

The Serengeti ecological unit is so vast that I found it impossible to delineate the entire nomadic range of lions without adequate telemetry equipment. Some nomads, such as the Cub Valley group, appeared to remain within the study area throughout the year, spending the rainy season on the plains then moving to the woodland border for the dry season. Most nomads, however, drifted widely over the plains and through the woodlands, both inside and outside of the park, and for me to find such animals was mainly a matter of luck. While searching for nomads along 70 km of the Mbalageti River in August and September, 1967, at least 135 different lions were seen, not one of which was a tagged nomad. Yet about 75 nomads had been tagged.

A few nomads were seen often enough to provide a broad outline of their movement pattern. Many nomads followed the migratory herds from the woodlands to the plains, where the nomads were widely scattered as long as prey was available. When the migratory herds retreated to the woodlands at the onset of the dry season, the lions, too, trickled off the plains, some traveling deeply into the woodlands, others staying near the periphery of it. Kühme (1966) felt that the number of lions on the plains remains constant, regardless of prey abundance, but my observations did not confirm this conjecture. After the wildebeest and zebra moved to the plains, a week or more elapsed before many nomadic lions appeared. For example, in 1967 the herds arrived on the eastern plains on November 25. I searched the area on November 29 without finding a lion. By December 2 a few lions were near the herds and by December 10 there were many. But more nomads continued to arrive in the next few weeks. During the 1967–68 season, 43 tagged lions frequented the plains. Of these 25% had been encountered by the end of December, 70% by the end of January, and 97% by the end of March, an indication of a gradual increase in numbers. If prey became scarce, as it did in February, 1968, when many wildebeest and zebra moved northward into the woodlands, some lions, too, shifted their position. The retreat of the lions to the woodlands at the onset of the dry season in May and June was gradual, with some animals staying on the plains and preying on stragglers until July and August. During the height of the dry season, the short and intermediate grasslands were almost devoid of lions (fig. 13). On an average day of fieldwork during the rainy seasons in April–May, 1967, and January–May, 1968, an average of 8 to 10 lions were found, whereas from August to October it was unusual to locate one in the same area.

Many lions made the same woodlands-plains-woodlands circuit each year, but the number that did so was affected by the length of the rainy season. During a short season, as in 1969, few lions went to the plains. During the 1966–67 rainy season a total of 62 tagged nomads were on the plains. The following dry season I found 23 (37%) of these animals in the woodlands. During the 1967–68 rains, 40 (64%) of the 62 nomads were back on the plains. Nine lions there had lost their tags and could not be identified individually. Thus, probably about $\frac{3}{4}$ of the lions of

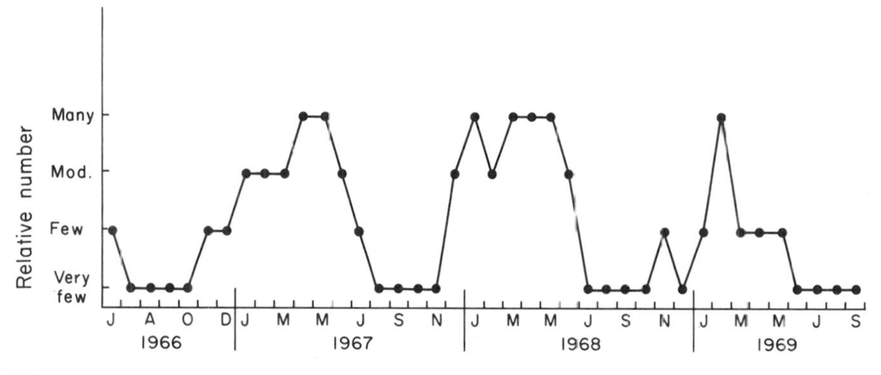

Fig. 13. Relative abundance of nomadic lions on the intermediate and short-grass plains.

the previous season returned to the plains. Three additional tagged ones had not been seen there the previous year. Conversely, two lions that were on the plains in 1966–67 remained in the woodlands during the following rains. Eleven (25%) of the 43 tagged lions were encountered in the woodlands in the subsequent dry season. The 1968–69 rains almost failed and many nomads came to the plains only in February. That year 15 (24%) of the original 62 lions were recognized, having made three consecutive seasonal circuits. Twelve untagged but individually recognizable lions provided similar figures: 5 were seen on the plains only for the first season, 3 for the first and second, and 4 for all three seasons.

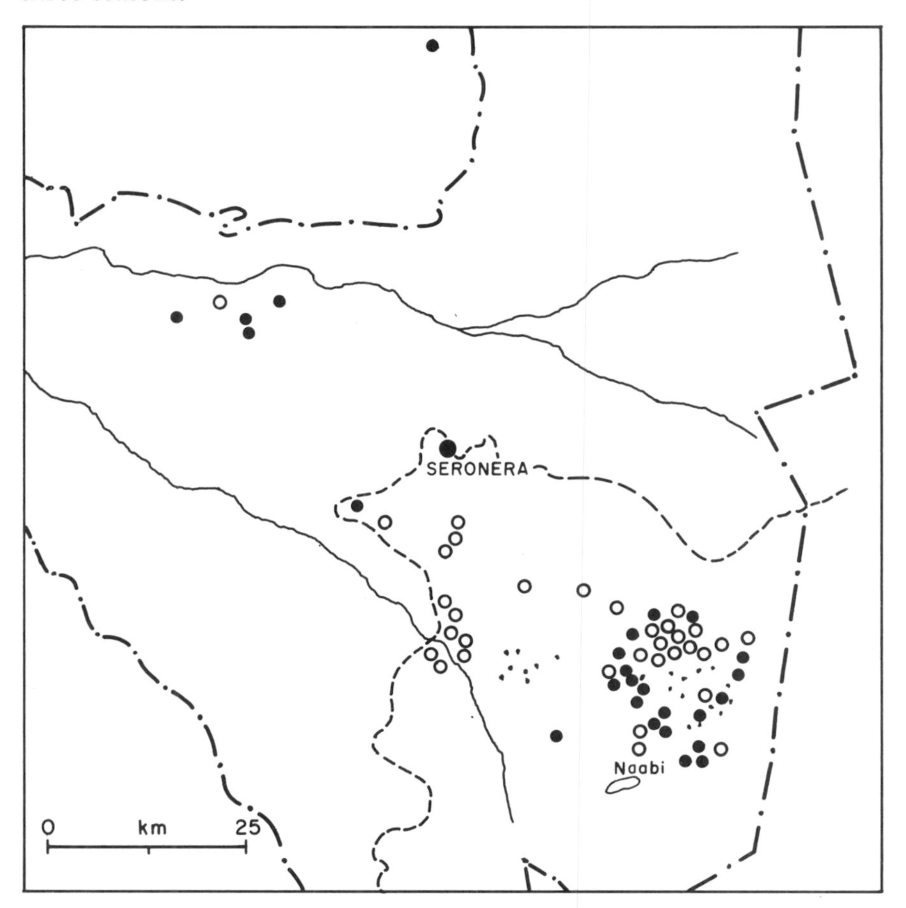

Fig. 14. The extent of movement of two nomadic males. Sightings of male No. 54 (open circle) were made between October, 1966, and January, 1969; of male No. 57 (closed circle) between November, 1966, and November, 1968. The dashed line represents the woodlands–plains boundary.

Figure 14 gives an idea of the distances moved by two nomadic males. Male No. 57 was tagged in the woodlands in November, 1966. Between February and March, 1967, he travelled east some 75 km to the Simba and Gol kopjes where he remained until June. He then went back to the woodlands briefly and by December was on the plains again. There he roamed widely until May, 1968. A hunter shot him the following November outside of the park. Similarly, male No. 54 was along the edge of the woodlands between October, 1966, and February, 1967, before he moved to the plains. After staying there until at least July, he traveled west into the woodlands only to return to the plains for the 1967–68 rainy season. I last saw him and his companion (male No. 110) in January, 1969, again at the edge of the woodlands.

Figure 15 shows the movements of a subadult male after he left his

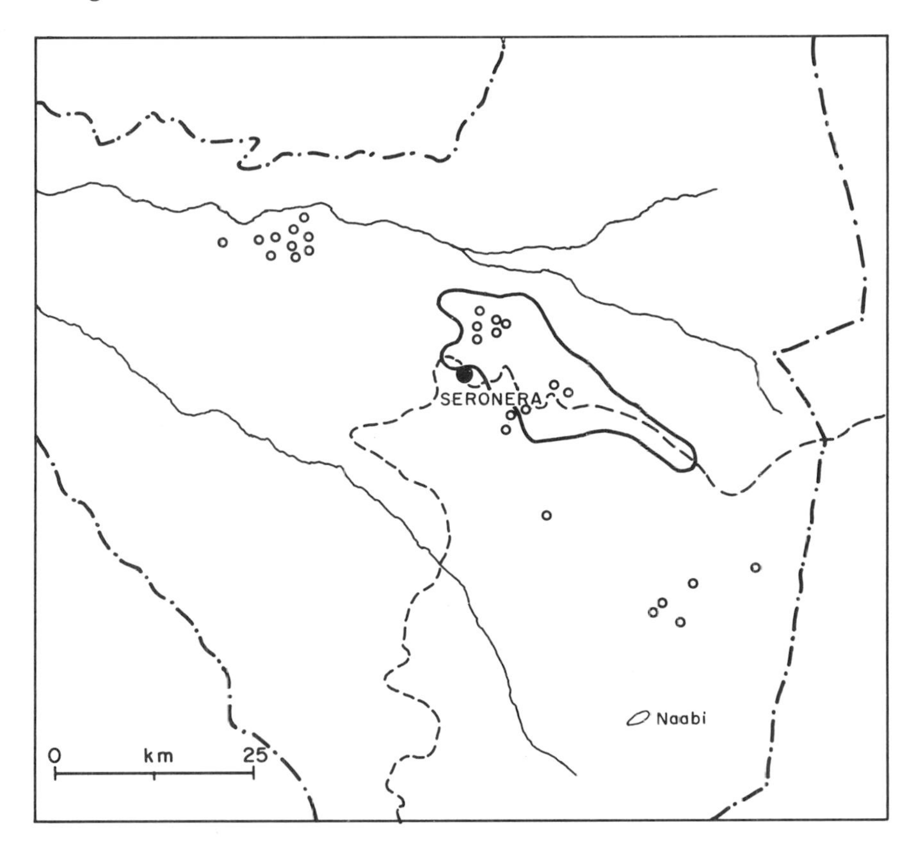

Fig. 15. The extent of movement of subadult male No. 47 before and after he left the Loliondo pride area (heavy black line). Sightings were made between October, 1966, and April, 1968.

pride area. On several occasions until December, 1966, I saw this male, as well as two other males of similar age and a male cub almost two years old, with the Loliondo pride. All four males then moved west to the Musabi plains area, returned to the vicinity of Seronera by March, 1967, and by August had once more gone west where they stayed until November. All traveled to the Gol kopjes in the plains in December. I last encountered them there in April, 1968.

Nomads traveled considerable distances between their dry and rainy season areas. Some 88 km (55 miles) separate the Musabi plains from the Gol kopjes, a trip made by, for example, males Nos. 47, 57, and 54. One subadult male seen on the Ndoha plains in February, 1967, was near Naabi Hill, 70 km away, the following May. The distance from the Gol kopjes to where male No. 57 was shot is 100 km. The total nomadic range of some animals is therefore large. Male No. 57 was known to roam over at least 4,700 sq km (1,800 sq miles), male No. 134 and female No. 135 over 2,100 sq km, and numerous others over 1,000 sq km and more. These minimum estimates indicate that some nomads use areas at least ten times as large as those of prides.

The movement patterns of nomads on the plains are of two types. Most animals roam widely and remain in one locality only a few days to, at most, a few weeks. Male No. 57 and female No. 78 were examples of this type. Other nomads settle down in one locality for months. They go to the woodlands for the dry season, but return to the same area during the following rains. Female No. 147 centered her activity around Naabi Hill for two consecutive seasons, and female No. 154 for three seasons. Female No. 157 occupied the vicinity of a waterhole at the eastern park boundary also for three seasons. The use of the same valley during the rains by females Nos. 69 and 102 in 1967, 1968, and 1969 was described earlier. Another female, No. 135, was tagged in January, 1967, in the woodlands near Mukoma plain. By March she had moved to the Barafu kopjes, and by November she was back in the woodlands, this time near Banagi. In March, 1968, I saw her near Naabi, but by May she was back at the Barafu kopjes where she had joined Male No. 107, his companion Halfear, and a female with cubs. The same group occupied the kopjes during the rains in 1969.

Males showed a similar tendency to occupy one locality for several seasons. Four pairs and a group of five males occupied about 900 sq km of plains within the study area in 1968 (fig. 5). Male No. 107 and Halfear were at the Barafu kopjes for three seasons, and male No. 58 and his elderly companion were in Cub Valley for a similar period.

Male No. 159 and his companion used about 200 sq km of terrain around Naabi Hill between April and July, 1967. When I saw him in the area again in January, 1968, he was without his companion. He now wandered widely, to the Gol kopjes, the eastern park boundary, and elsewhere. During the rains in 1969 he was back on the plains, still a solitary roamer. The Naabi Hill area was occupied by five males, including Nos. 149 and 150, in 1968. They took over the Magadi pride late that year (see fig. 6), and, in 1969, no males seemed to center their activity around Naabi Hill. Two males occupied the plains about 10 km east of Naabi Hill from December, 1967, to May, 1968, but I could not find them there in 1969.

Between December 10 and 18, 1967, the behavior of a male was observed in detail as he established himself at the Gol kopjes. I describe the process at length not only to show how this was done but also to convey a general impression about the life of a nomad on the plains. Male No. 134 was tagged in the woodlands about 10 km west of Lake Magadi in January, 1967, when he was about 3½ years old—a solitary, handsome animal with unscarred face and a short yellow-brown mane. By February he had moved to the plains where he associated with a variety of females and males. In July and September he was in the woodlands near Banagi. Female No. 60, a prime adult of stocky build, was tagged as she wandered alone south of the Moru kopjes in November, 1966. The following February she ranged around Naabi Hill and the Gol kopjes with a young adult male, her constant companion—an unusual association between a male and female. This was the extent of my knowledge about the history of male No. 134 and female No. 60. I saw these two together for the first time on December 9, and they remained companions until at least the end of 1968.

On December 9, male No. 134 and female No. 60 rested on a kopje beside a young adult male, and half a kilometer away was a lioness with a male cub about 7–8 months old. At 1915 the two males slapped each other, then roared. At 2005 male No. 134 and his female companion left the kopje and 15 minutes later they scavenged a wildebeest from several hyenas. We tranquilized the male and placed a radio around his neck. The following morning the pair was in a nearby kopje and we began to observe it continuously (see also fig. 28 and table 14).

December 10–11. Both left the kopje roaring repeatedly at 1935. They alternately rested and walked until at 2110 some hyenas called in the distance. The lions trotted 2 km to where hyenas had killed a wildebeest and there fed intermittently until 0105. They puttered the rest of the night visiting several kopjes. At 0610 several hyenas attacked and killed an adult wildebeest. The pair walked 1.5 km to the carcass and appropriated it. Coming from the opposite direction was the lioness and cub seen in the area two days before. When the male spotted her at 100 m he trotted 30 m toward her, then returned to the kill. Female No. 60, however, chased her vigorously. She fled but suddenly turned, charged 20 m, swiping the air with a forepaw, and retreated once more. When female No. 60 came close, she attacked again, clouted her on the head, and followed this by a rush during which she bit her in the neck, drawing blood. Female No. 60 rolled over, but as she tried to rise repeated swats twice knocked her down again. Both females then walked parallel and 10 m apart. Suddenly the mother of the cub stopped, sniffed the ground and roared softly, apparently looking for her offspring which had run off alone. She returned to the carcass and there lay down beside the male.

Meanwhile female No. 60 followed the scent trail of the cub to a kopje where the cub had hidden itself in a dense bush on top of a boulder. Although the female tried to reach the snarling cub, she could not penetrate the branches and finally reclined at the base of the boulder, a position she still occupied at 2200 hrs.

December 11–12. Male No. 134 and the mother of the cub stayed together all day. After 1840 they moved short distances, but she made no search for her cub. At 2135 they met a second female but for 2 minutes only. After a largely inactive night they returned to their old kill and ate for $\frac{1}{2}$ hour. The male roared frequently and was answered by another male at 0505. This male—the same one with which male No. 134 associated on December 9—approached at 0605. Male No. 134 growled and rolled on his back; the other male walked off into a kopje, now followed closely by male No. 134. Suddenly they attacked each other: male No. 134 was knocked down with a blow but rose quickly and charged the other, which fled. Both then walked 15 m apart, roaring. The new male chased the other once, then both lay down, but at 0640 they tussled again. At 0648, male No. 134 moved toward a kopje, trailed by the other, but No. 134 stopped, faced his antagonist, and both reared up on their hindlegs and slapped. The new male crouched while male No. 134, his ear bloody and a cut in his scrotum, stood over him. Male No. 134 rejoined the mother of the cub while the vanquished one left, his ear also bloody and minus a tuft of mane. However, he remained in the area at least until December 19.

At 0700 I found female No. 60 still by the kopje at which I had left her the previous evening. She now rested beside the partially eaten cub which she apparently had caught and killed only an hour or two earlier. She was joined by male No. 134 at 0750. At the same time the mother of the cub circled the kopje, looking up at female No. 60, until the latter approached slowly. The mother fled. The male discovered the carcass, grabbed it, but instantly dropped it, having discovered what it was. Female No. 60 returned and by 0845 had devoured the cub.

December 12–13. At 1805 many roars emanated from the kopje about 1 km from the one in which male No. 134 and female No. 60 had been lying all day. They answered many of the calls, and at 1905 they met males Nos. 103 and 104. These were companions and probably brothers, about 4 years old. During 1967 they roamed widely over the plains south of the Moru kopjes, at Lake Magadi, around the Simba kopjes, and in other places. The male and female charged the pair, at first bounding but then slowing to a trot for 3.2 km, while roaring repeatedly. Thereafter they either rested or wandered through several kopjes, giving the impression of searching for something. Suddenly the male walked rapidly .8 km where he met a young adult lioness chewing on a wildebeest head. She retreated 30 m; the male strutted and she fled. Female No. 60 arrived and gnawed on the head several minutes. They visited several more kopjes, finally staying in one of them for the remainder of the day.

December 13–14. In early evening both lions repeatedly answered distant roars, and they continued to roar as they made a rough circuit which by 0500 brought the male back near the kopje in which he had rested the previous day; the female had left him an hour earlier. At 0515 he joined two subadult lionesses and female No. 60 on a wildebeest they had scavenged from hyenas. The young lionesses left. Female No. 60 walked alone 1.5 km from the kill and was joined there at 0810 by the male. They greeted, sniffed each other's anal area, and together moved to a kopje where they rested.

December 14–15. The pair moved little during the night, only 5.2 km. At 0535 they appropriated some wildebeest remains from hyenas, but only the lioness ate, and at 0630 they began their daytime rest period.

December 15–16. One of the young lionesses the pair had met two nights previously arrived at 1655 and reclined 170 m away. The male rose, scraped the ground with his hindpaws, and reclined again. At

1820 female No. 60 trotted toward the other female, followed by the male. The newcomer turned and headed at a fast walk or trot across the plains for 5.3 km to a kopje where her companion was beside the partially eaten carcass of a zebra. Male No. 134 and female No. 60 followed her the whole way. The young lionesses left the kill at 0130 but the pair stayed with it all night and the following day.

December 16–17. The pair abandoned the zebra remains at 0245 and then wandered around until at daylight they entered a kopje and rested once more. Most of the neighboring kopjes were also occupied: the two young lionesses were in one, an old lioness in a second, the male with which male No. 134 once fought in a third, and 2 km away males No. 103 and 104 were in a fourth.

December 17–18. The pair left the kopje at 1830, walking steadily and roaring frequently. When they halted at 2015, I darted the male and removed the radio, which had ceased to function. At 2045 both left, still heading away from the kopjes. They rested from 2345 to 0025 and from 0100 to 0240. Suddenly they walked rapidly, sniffing the ground, looking around alertly, and occasionally breaking into a trot. After about 1.5 km they met male No. 88, a solitary adult, which traveled widely between Moru and Gol kopjes during 1967 (and also during 1968). Both chased him for 3.5 km, sometimes the female in the lead, sometimes the male. With a burst of speed the female caught up with the stranger but was slapped by him. When male No. 134 arrived, male No. 88 crouched with chin resting on the ground and growled. The pair then reclined 10 m away, and a few minutes later all three roared. When the lioness approached and lay down beside male No. 88 he growled; when male No. 134 came up and strutted he miaowed. The pair then left the male, rested 30 m from him, and at 0535 departed for a kopje.

On December 19, male No. 134 and his antagonist of December 12th were .8 km apart and had fought again—at least the latter had a fresh cut over his eye. Male No. 134 courted with the lioness whose cub was killed on December 12. Female No. 60 was not in sight. On January 9, 1968, the pair, together again, looked alertly at three young lionesses in the distance. The male walked to them and sat beside them without eliciting a reaction, but they snarled at the approach of female No. 60. She stayed nearby and after 10 minutes suddenly charged. The young lionesses fled and the male followed them. On January 10 and 11 he courted with one of the two lionesses he had met on December 14.

During the night of January 11–12 male No. 134 must have had an encounter with males No. 103 and 104, which he had chased away on December 12, for I found him walking restlessly and alone east of the Gol kopjes until 1020. That night he roared often—49 times between 1800 and 2400—and ambled aimlessly without approaching the kopjes. At 2110, when males No. 103 and 104 roared ahead of him, he turned and rapidly walked away. Male No. 134 and female No. 60 remained companions at least another year, but they never reoccupied the Gol kopjes or established themselves in a particular locality. They obviously had been evicted from the Gol kopjes by males Nos. 103 and 104, for these two remained around in the area at least until April, and during the rainy season in 1969 they were there too.

The behavior of male No. 134 and female No. 60 indicated that they had attempted to claim the Gol kopje area for themselves. The male roared repeatedly—an average of 30 times per day as compared to 6.5 times per day by nomadic male No. 159, which was established for a season around Naabi Hill. They were aggressive toward other nomads—male No. 134 particularly toward other males, female No. 60 toward other females—a pattern similar to that shown by pride members toward strangers within their territory. This, together with the facts that some males occupied the same locality season after season and other males seldom tarried there, suggests that the areas were territories. Male No. 134 was clearly unable to establish himself alone, even though he claimed only about 90 sq km, and was replaced by a pair of males, again reminiscent of behavior observed in prides. Territories in the plains were only temporary, however, and all males left them during the dry season and became fully nomadic until the return of prey enabled them to reoccupy their former sites in some cases. I could not find out if males established temporary territories in the woodlands where space was already occupied by prides.

The presence of territorial nomads in the plains appeared to influence the movements of other males; at least, most males used the study area and plains immediately surrounding it so sporadically that some were seen only once or twice during the rainy season.

Territorial behavior on the part of nomadic lionesses was not clear, mainly because few interactions were seen, and those that occurred were difficult to interpret without knowing the history of the animals. Female No. 60 behaved in territorial fashion, but the lionesses around Naabi accepted an occasional female into the group. It was my impression that many of the localized lionesses were not territorial but that their mere presence in an area caused other nomads to move away. Nomads sometimes concentrated near the migratory herds, but after a few days the lions dispersed in spite of the fact that food remained abundant.

Interactions between Nomads

The responses of nomads to each other depended, among other factors, on whether or not the animals were territorial. The aggressive behavior of territorial male No. 134 and female No. 60 toward other nomads, particularly toward individuals of their own sex, was described above. I have no comparable observations for other lions, except that male No. 159, which had a territory around Naabi Hill in 1967, on three occasions attacked a subadult male. One morning, for instance, an adult and two subadult lionesses and this subadult male approached male No. 159. He ignored a subadult female that attempted to rub heads with him and walked toward the subadult male which retreated, circled, and a few minutes later reclined 15 m from the group. Suddenly the adult male rushed, grabbed the youngster by the rump and bit him in the back. The young male twisted around, and, flailing with his hindlegs, clutched the adult by the shoulders with his front paws. They rolled over twice, separated, and the subadult slapped but his claws became tangled in the other's mane; he then crouched while the adult charged the lionesses, two of which quickly rolled on their side. The animals settled down after that, the young male still 20 m from the group.

Interactions between other nomads were highly variable. On several occasions lions saw each other at a distance, stopped and looked, and sometimes ventured to within 100 m of each other before parting. Often nomads joined casually, but usually I knew that they were either companions that had been temporarily separated or former acquaintances. At other times, I was uncertain whether or not the animals knew each other. On eleven occasions, however, the animals appeared to be strangers, and five of these occasions are described below:

On January 17, 1968, after male No. 134 and female No. 60 had been evicted from their territory, they reclined 100 m from two adult males. The lioness approached to within 10 m of one male, then suddenly trotted to him and rubbed her head against his. The male merely pulled his lips back, exposing his teeth. Male No. 134 advanced to within 50 m, where he was joined by the lioness, and later both walked away.

Three subadult males and a subadult female, which I previously saw 40 km to the west, were near female No. 52 and her two grown cubs. The subadults advanced in single file, the female leading, and crowded around female No. 52, who crouched snarling. The subadults almost touched her, then lay down 3 m from her, but 10 minutes later walked off.

On February 7, 1967, male No. 134 rested 50 m from female No. 147 and another lioness near Naabi Hill. He rose three times between 1335 and 1820 and approached, but each time the lionesses faced him and snarled when he came to within 20 m.

Two lionesses took a wildebeest kill from wild dogs at 1825. Twenty-five minutes later two subadult males and a subadult female spotted the kill and trotted to it. One male rushed to the carcass with a growl. One of the lionesses retreated 10 m and the other ceased to eat; then both left.

Tailless, an adult male, was courting when I found him at 1450. He stayed with the female until 1930, at which time he walked off alone, roaring repeatedly until he received a distant answer. At 2045 he joined his usual companion, Cutlip. Fifteen minutes later he suddenly rushed and caught a sick, solitary zebra. The zebra screamed and struggled for 6 minutes before it died. Attracted by the sound, another adult male approached to within 100 m. Tailless and Cutlip attacked and chased him 2.1 km. When they returned to their kill, female No. 78, her cub, and male No. 55 were eating. Also present was an old, emaciated male. Tailless first approached the old male in a strut, then ate too. When the old male slapped at him, he merely growled. Cutlip, having tarried near a kopje, arrived back at the kill a few minutes later. The old male lunged at him with a snarl but when Tailless grabbed at his exposed rump he backed off. Cutlip, instead of attacking the old male, swatted the cub, leaving claw marks on its rump. All fed after that, but at 2250 the male that was pursued earlier again came near the kill and was driven away once more. The old male left at 2335. Tailless chased male No. 55 away from the carcass and appropriated it. At 0830 Tailless and Cutlip went to a nearby kopje and rested.

The most striking aspect of interactions between nomads was their tolerance of each other. Although an animal was often antagonistic when approached too closely by a stranger, especially if a kill was involved, and entry into a group was at first discouraged, a nomad was often accepted or at least permitted to partake of a meal. Nomads were sometimes chased vigorously by others, but it was my impression that if the stranger had persevered in its approach instead of fleeing it would have been allowed to join. On one occasion four adult males and two adult females, all usually solitary, shared an elephant carcass for several days. Instances of males permitting other males to feed with them have already been described. The intermittent peaceful meetings between casual acquaintances provided further proof that not only did nomads tolerate others in the same area, but also that they were not averse to having social contacts with them. Nomads differed strikingly from prides and from territorial nomads in their behavior toward lions which did not belong to the group. This behavior seemed to be directly related to the kind of land tenure system: lions which defend no land generally accept strangers.

PLATES

Plate 1. At least 15 members of the Masai pride rest in the shade of an acacia at the edge of the plains.

Plate 2. Several members of the Masai pride drink at a puddle, a typical instance of social facilitation.

Plate 3. Two males rub a third one in greeting.

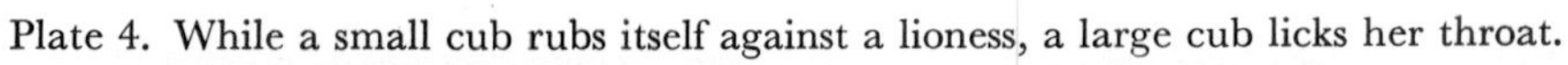

Plate 4. While a small cub rubs itself against a lioness, a large cub licks her throat.

Plate 5. One lioness sniffs the anal area of another lioness after meeting her. Note the tail positions.

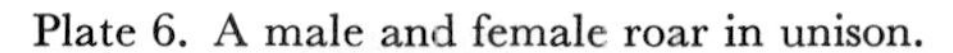

Plate 6. A male and female roar in unison.

Plate 7. A male lightly rebuffs a cub by exposing his canines.

Plate 8. With bared teeth a lioness faces a playful cub which rolls on its back in a defensive posture.

Plate 9. A gorged nomadic male sits panting beside a kill, while a small cub takes advantage of the shade cast by his body.

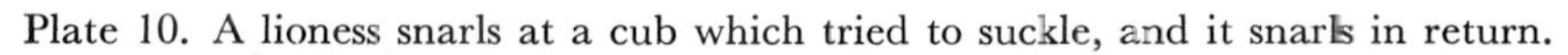

Plate 10. A lioness snarls at a cub which tried to suckle, and it snarls in return.

Plate 11. A lioness threatens a male in a dispute at a kill.

Plate 12. A male copulates with a female.

Plate 13. A copulating male dismounts abruptly as the lioness whirls around to face him.

Plate 14. A lioness sits beside the grassy depression in which she has hidden her three newborn young.

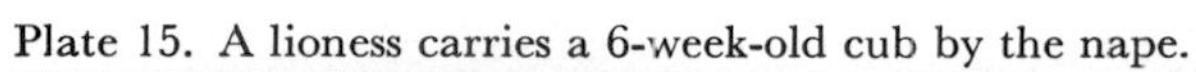

Plate 15. A lioness carries a 6-week-old cub by the nape.

Plate 16. A nomadic male carries a cub he has killed.

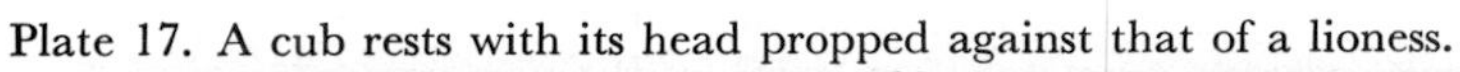

Plate 17. A cub rests with its head propped against that of a lioness.

Plate 18. Two cubs suckle.

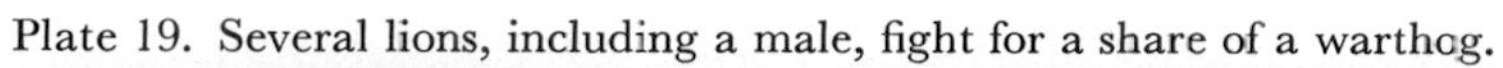

Plate 19. Several lions, including a male, fight for a share of a warthog.

Plate 20. Males and females crowd flank to flank around a zebra kill.

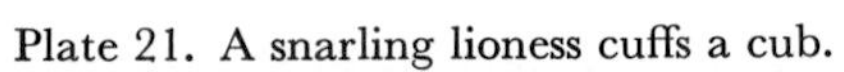

Plate 21. A snarling lioness cuffs a cub.

Plate 22. After having driven several lionesses from a topi kill, two males share the remains with some cubs.

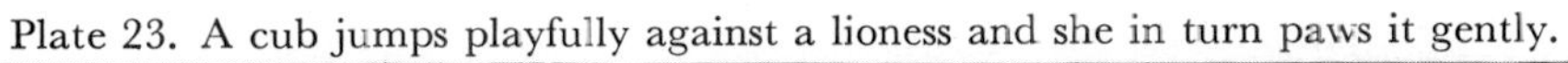

Plate 23. A cub jumps playfully against a lioness and she in turn paws it gently.

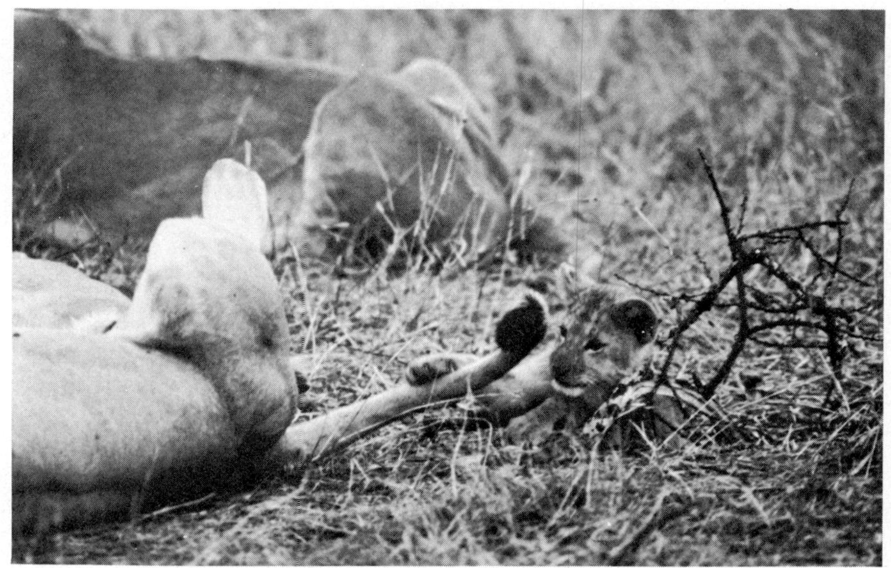

Plate 24. A cub plays with the tassle at the end of a lioness' tail.

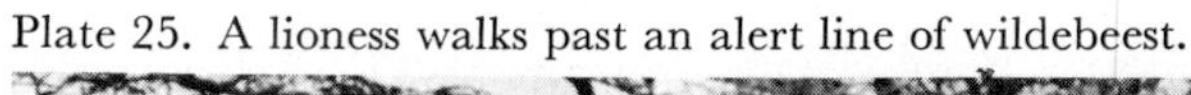

Plate 25. A lioness walks past an alert line of wildebeest.

Plate 26. Wildebeest crowd around a waterhole, a position which makes them highly vulnerable to predation.

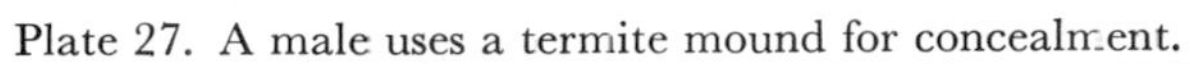

Plate 27. A male uses a termite mound for concealment.

Plate 28. A lioness stalks through high grass toward two Thomson's gazelle.

Plate 29. A lioness grabs a fleeing zebra by the thighs with her forepaws.

Plate 30. A lioness pulls a zebra down by clutching its neck and biting its nape.

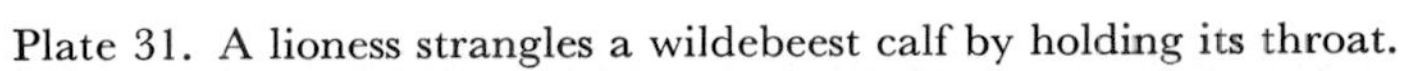

Plate 31. A lioness strangles a wildebeest calf by holding its throat.

Plate 32. A lioness suffocates a wildebeest by covering its muzzle with her mouth.

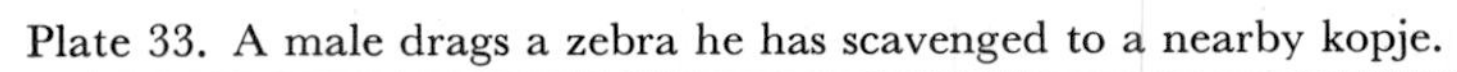

Plate 33. A male drags a zebra he has scavenged to a nearby kopje.

Plate 34. A lioness eats the intestines of a freshly killed zebra.

Plate 35. A lioness holds a piece of wildebeest with her paws while she cuts the skin with her carnassial teeth.

Plate 36. Three subadult males feed on a scavenged wildebeest while three spotted hyenas and a blackbacked jackal wait for the remains. Note the ear tags in one lion.

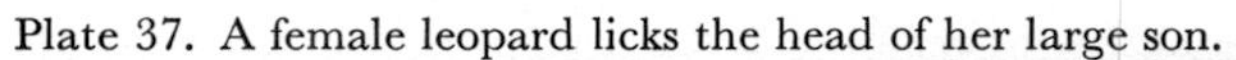

Plate 37. A female leopard licks the head of her large son.

Plate 38. A female cheetah strangles a Thomson's gazelle and at the same time drags it toward the shade of a tree.

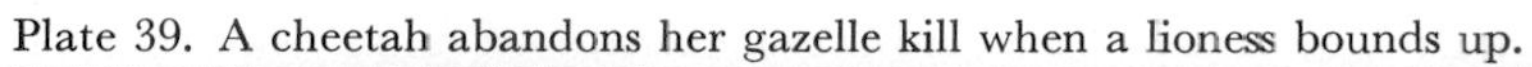

Plate 39. A cheetah abandons her gazelle kill when a lioness bounds up.

Plate 40. Wild dogs greet by nuzzling each other's mouth before setting out on a hunt.

Plate 41. Large pups eat a gazelle while the adults form a circle around them, behavior which prevents other predators, such as the distant hyena, from snatching the kill.

Plate 42. While one dog holds a zebra foal by the nose, others tear at the belly and anus.

Plate 43. Several hyenas have attacked and hamstrung a wildebeest bull.

5 Behavior within the Group

The successful functioning of a society is entirely dependent on its communicatory system: in the case of lions, on the vocalizations, facial expressions, scents, postures, and so forth by which these animals signal. A society is an adaptive unit, its structure formed both by environmental and social pressures. To understand how the lion's society operates, it is necessary to analyze its various components, to describe aggressive and mating patterns, the relationship of the young to the group and others on which selection pressures operate.

Communication

In lions and other cats the expressive movements are highly developed, as the detailed studies of Leyhausen (1960, 1965b) have shown. "There are few animals in whose faces a knowledgeable observer can so clearly read a prevailing mood and predict what actions—friendly or hostile—are likely to follow" (Lorenz, 1964). The vividness of the lion's expressions are enhanced by its color pattern: tawny except for the black tassel at the end of its tail, a black area on the back of each ear, and black lips and nose, all parts of the body which are used to convey emotion. This expressiveness of cats is in some ways surprising, for most are solitary animals which, it could be assumed, do not need such an array of complex signals. In their communicatory system they resemble the social primates and canids more than they do such relatively asocial animals as bears.

Signals are of two main types. Long-range signals are fairly stereotyped and consist largely of olfactory and visual markings in the environment and the roar. Signals used at close range, between members of a group, are exceedingly complex, representing without doubt the most intricate patterns in the lion's behavioral repertoire. Using its teeth, lips, ears, tail, and other parts of the body, a lion conveys a constantly changing signal pattern, augmented and intensified by vocal, olfactory, and tactile messages, all part of one functional unit. Some sounds and expressions constitute graded systems, which change constantly with the intensity of an interaction, and others represent conflicting emotions.

A signal is not determined by its properties alone, and the same signal may attract or repulse depending on who emits it and who perceives it. Lions use and presumably need such complex signals to regulate their society, indicating that a full analysis of the lions' methods of communication is a requisite for an understanding of their social system. A study wholly devoted to this topic would be as rewarding in the lion as it has been in primates (see Altmann, 1967).

While conspicuous patterns of communication, those of high intensity, are most apparent to the human observer, subtle signals or even the absence of specific ones are also important in regulating the lions' daily routine. The mere fact that animals are resting or moving casually indicates that all is well, that no immediate response is required, although some such innocuous actions as drinking, urinating, and clawing trees may stimulate others to behave similarly.

Senses

Of relevance to communication is the ability of an animal to perceive signals—its acuity of sight, hearing, and smell. Lions have excellent eyesight, judging by their response to such distant small moving objects as descending vultures, but in daytime at least their vision does not seem superior to mine. While their response to moving prey is rapid, they seem to have difficulty in discerning stationary animals in unusual positions. For example, on several occasions lions looked directly at gazelles that leopards had stored in trees, apparently without seeing them.

I found it difficult to judge the hearing ability of lions because mine was usually restricted by being in a car. At night lions sometimes jerked to attention, and in most instances I too was aware of the cause, usually a distant roar. Once a male suddenly ran 1.3 km through high grass to where two jackals had just disemboweled a gazelle. He could only have been attracted by hearing the gazelle bleat, yet I had perceived nothing. This and similar incidents based on subjective inference make me suspect that the hearing of lions is superior to that of man, although part of their seemingly greater acuity is probably the result of their being more conscious of the nuances of sounds in their surroundings.

Many writers have commented on the lion's sense of smell, with some, such as Stevenson-Hamilton (1954) and G. Adamson (1968), noting that it has a good nose, and others that it does not. "I have, however, never seen any indication that a lion has the keenness of scent such as the dog possesses," wrote Akeley and Akeley (1932). My observations indicate that lions have an excellent sense of smell. For example, one night two males left a kill and rested 1 km away. A third male fed another 20 minutes, then put his nose to the ground and followed the erratic trail of his companions until he reached them. On another occasion, a lioness walked across the plains, sniffing the ground, zig-zagging, veering off some invisible trail then angling back to it until, after 2.4 km, she reached her group, which had been resting out of sight behind a ridge. On a third occasion, when a lioness returned to a thicket to fetch her cub to a kill, she met another lioness. This lioness sensed that a kill had been made and with nose to the ground trotted back over the trail made by the other female until she reached the carcass. Lions are surrounded by many scents which convey specific information. Their whole daily routine is undoubtedly influenced by scents—a lioness in heat, the track of a stranger, an old kill—yet it is an aspect of their life which unfortunately remains almost closed to the human observer.

Tactile Communication

Lions touch each other briefly and often inadvertently in a variety of situations, such as play, fights, and when crowded around a kill. But they also seek physical contact. Resting lions often drape a paw over the shoulder of a neighbor, lie back to back, or touch in other ways. Cubs may suckle as a means of initiating contact. Head rubbing and social licking are the two most common peaceful tactile gestures. These two patterns, more than any other, help to promote group cohesion; they serve the same purpose as grooming does in many primate societies (Sparks, 1967).

Head rubbing. One lion rubs another by approaching from the front, side, or back and touching cheeks gently in passing, or by first languorously rubbing its forehead, then face and neck, against the head of the other, eyes often closed (plate 3). Occasionally it slides its body sinuously along the side of the other with tail raised almost vertically but tipped limply toward the animal being greeted; in extreme cases, the tail is draped over the back of the other. As it passes by, it often swivels its rump slightly toward the head of the other lion. At times a lion circles another one closely once or twice without touching, licking its lips before rubbing heads. Although the movements are usually gentle, an adult sometimes pushes so vigorously with its forehead into the neck or shoulder of the other animal that it stumbles sideways. A soft humming or moaning sound often accompanies rubbing. Small cubs characteristically rub first the top of their head against the chin of a sitting or standing adult, then do so with their whole back and tail, leaning into the chest of the other lion, as if to establish as much physical contact as possible. After rubbing, an animal sometimes rolls on its back or even on top of the other animal. The behavior is directed almost exclusively at the head and neck except that cubs sometimes rub the leg or rump of a lioness. Rubbing is a relatively common activity, much more so than licking (table 7).

Although an animal is sometimes discouraged from rubbing by bared teeth or a growl, in most instances the gesture is reciprocated. When one lion sees another approach casually, with head slightly turned and often lowered as if to rub, it shifts its head so that the cheeks can meet and, once the cheeks touch, returns the pressure. After that both may indulge in a bout of mutual rubbing lasting as long as a minute. Once two lionesses approached a subadult male which raised his head and presented his cheek, as if anticipating a rub, but they merely walked by. On many occasions an animal intimated rubbing only with a slight gesture such as flicking its raised tail toward the other lion or barely shifting its rump in the other's direction.

All group members readily rub heads with each other but their age and sex affects the frequency with which they do so. Adult males tend to rub primarily with other males and seldom initiate such contact with a female or cub. All lionesses and cubs readily rub each other without evidence of a hierarchical system. Seven lionesses and 13 cubs, 7 to 9 months old, were seen to interact 469 times in March and April, 1969, rubbing heads with the following frequencies:

Female rubs cub	11%
Female rubs female	19%
Cub rubs female	62%
Cub rubs cub	8%

These data show that cubs use rubbing mainly in contacts with females, and that females primarily rub each other.

To obtain some idea as to how often members of a pride rub each other, I noted such interactions for three months in the Masai pride. As table 8 shows, males seldom rub others, but members of the group sometimes initiate contact with them. Males often rest at the periphery of the group, and attempts by others to lick or rub them are sometimes discouraged with a growl. Out of 15 rubbing attempts by cubs I recorded, the male reacted aggressively 3 times. Lionesses readily interacted with each other, except female R, which was a cripple, and N, which, for reasons unknown to me, seldom rubbed.

Cubs readily rubbed adults, particularly females L, M, O, and Q, but few of these reciprocated. During 47 hours of observation, females L, M, and O were rubbed a total of 191 times by the four cubs (table 9). The small cubs rubbed less frequently ($p = <.001$) than the large cubs. Although litter mates tended to rub other females less frequently than their mother, the differences were not statistically significant except in the case of the large cubs interacting with female *M* ($p = <.001$).

Head rubbing occurs in several situations, but particularly after animals have been separated. Cubs run to their mother and rub after her return from a hunt. After joining a group, a lioness often rubs several members in succession, and when several return there may be a veritable orgy of rubbing and licking lasting for as long as 5 minutes. A joining pride member sometimes trots to the nearest lion and rubs, as if trying to establish social contact without delay, as if there were some uncertainty about the nature of its reception. Once one lioness approached while another stood with her head aggressively lowered. The former trotted up and rubbed. Each tried to sniff the anus of the other and they then rubbed once more. Another lioness joined the group and rubbed the first two animals it reached, a cub and a female. When a male glanced up, she fled 15 m, quickly rubbed another lioness, and then rolled on her back. If a resting lion changes its position, such as from the sun to the shade, it often rubs the animal beside which it reclines. Two lions passing each other may detour and rub briefly. At dusk, before setting out on a hunt, group members rub frequently and intensively.

Lions may also rub as a means of establishing social contact before proceeding with some other behavior. Cursory cheek rubbing occurs before the lion presents a part of its body to be licked, and cubs rub a lioness before proceeding to suckle. On several occasions a lion rubbed before taking meat from another. Once a cub rubbed a feeding male; although the male swatted it once, he then left the remains to the cub. Another time, a lioness rubbed and took a wildebeest tail from a male, but as she turned away he slapped her on the rump. She growled and rolled on her back, but with the tail still in her possession.

Rubbing is also prevalent after a fight or other tense situation. When a lioness coughed at three other lionesses that had been hitting her playfully, five cubs ran over and rubbed against her. On another occasion three lionesses attacked a male. After he retreated, three cubs rubbed him. Our pet cub often rubbed us conspicuously after he had been irascible during a feeding session or after he had been punished for some misdemeanor.

Judging by the situations which elicit rubbing, the behavior functions mainly as a greeting (Leyhausen, 1960). A greeting symbolizes that the intentions of the animal are peaceful, that it belongs to the group and has been accepted by it. Such acceptance is particularly needed from the powerful males. This helps to explain why females and cubs rub the males quite often whereas males seldom return the gesture. In this context the greeting inhibits aggressive patterns. After a fight the greeting helps to reestablish amicable relations and before a communal endeavor it integrates the activity of the group and tightens social bonds. It is also possible that each lion pride has a distinct group odor, like that described by Schultze-Westrum (1965) in the flying opossum, which helps to unite the group. Head rubbing would tend to spread the odor among the group members.

The greeting pattern, including circling, rubbing, raising the tail, swiveling the rump toward the other animal, and rolling on the back have a striking similarity to the sexual behavior of the female. Although head rubbing usually contains fewer components than the sexual pattern, there seems to be little doubt that rubbing was derived from it, that the greeting is a ritualized form of sexual behavior (Leyhausen, 1960). On several occasions greeting between males led to intense caressing and finally to homosexual mounting and thrusting, intimating a close relationship between the two patterns.

Social licking. Lions lick one another in several situations. When, after a meal, the muzzle and throat of an animal is bloody, a group member often licks the fur—behavior of obvious utilitarian function, for the head and neck are the only parts of its body a lion cannot easily clean by itself. Social licking is particularly prevalent after a heavy rain. At such times lions lick each other mutually for several minutes, not just the head and neck but over much of the body. Such licking may help to dry an animal, particularly the woolly coat of a cub, and it may also provide fresh water to drink. The wounds of one animal are sometimes licked by another, again a useful gesture. When, for example, the nomad Tailless was seriously bitten around the eye, his companion sometimes licked the exudate away. Males occasionally lick estrous lionesses, and cubs lick the nipples of lactating females.

The final circumstance, and the one of concern here, is social licking based on no obvious stimulus other than the presence of another animal. During rest periods, particularly at dawn and dusk, two lionesses frequently approach each other, rub heads, and lick head, neck, throat, shoulder, or other part with steady strokes of the rough tongue. Sometimes the licking is reciprocal, at other times mutual, with both animals active at the same time as they sit or lie. The recipient obviously finds pleasure in being licked for it remains quiet, often with eyes closed, as it slowly turns its head and neck to expose different parts for attention (plate 4). Occasionally mutual licking involves a cluster of three to four adults. A cub usually lies on its back or side while a lioness licks much of its body. The tongue is so powerful that standing cubs are sometimes pushed off their feet. While the lioness licks, the cub often bats playfully at her face, and she in turn may then paw it gently in return. Many licking interactions are brief, a stroke or two of the tongue, but others may continue for as long as 5 minutes.

Occasionally a lion invites licking of a certain part of its body by placing it in front of the muzzle of another lion which either responds to or ignores the gesture. Once a lioness presented her neck to another lioness and held that position for half a minute, glancing up several times as if trying to ascertain the lack of response. Any member of the group may approach any other member and initiate a licking bout without evidence of any hierarchical pattern among females (see table 18). Males seldom lick females except when courting, and I saw them lick cubs only four times. Sometimes an individual discourages contact with a snap or growl.

The head and neck receive by far the most attention (62%) in social licking interactions (table 10). The back and abdomen are licked infrequently except by lionesses licking cubs. The orientation of social licking toward the head of the other animal is particularly striking when compared to the parts of the body which an animal licks for itself (table 11). A lion typically licked several parts of its body in succession. The forepaws were licked most often (44%), whereas the hindpaws received little attention (6%). Yet the forepaws were seldom licked as a social gesture (.4%). As is evident from table 7, social and self-licking occur about equally often, but scratching is an infrequent form of body care (fig. 16).

To obtain some idea if animals of a certain age and sex licked socially

R. KEANE

more often than others, I recorded all interactions in the Masai pride during a specific period (table 12). Since all pride members were seldom together, the results are adjusted according to the number of hours an individual was present; this varied from 10 to 34. Males licked or were licked much less often than any other group member except female R which was a cripple. Females that interacted with cubs had the highest licking scores: M and Q were the mothers and L was a close companion of these two. The large cubs were licked 10 out of 33 times by their mother and 20 times by female L. The small cubs were licked 15 out of 25 times by their mother. Thus mothers tend to lick their cubs more frequently than do other females except in the case of a companion such as female L. The large cubs licked others (50% of the interactions) as often as they were licked, but small ones seldom licked others (7%), indicating that at some period between the ages of 5 to 10 months a cub becomes more socially oriented in its licking behavior. In this group, the cubs were the focus of much licking behavior by three lionesses; 36% of all such interactions were directed at the cubs even though they comprised only 19% of the animals. In contrast, during June, 1969, I observed 101 instances of social licking among the 7 lionesses and 13 cubs, 7 to 10 months old, of the Seronera pride. Lionesses licked other lionesses proportionally twice as often as they did cubs, suggesting that among these lions the cubs provided less of a social stimulus than among the others. Possibly the persistent attempts of these hungry cubs to suckle whenever they came near a lioness inhibited licking behavior. Cubs licked females proportionally twice as often as they did other cubs, usually, it seemed, as a means of establishing social contact with a female before trying to suckle.

Social licking is often used in conjunction with head rubbing, suggesting a close functional relationship. One example illustrates this: a lioness joins a group and rubs heads with another lioness, and they then lick each other's neck; the newcomer then rubs a second lioness, briefly licks the top of her head, and joins a third lioness which she both rubs and licks. Although of obvious use in body care, licking undoubtedly helps to reinforce the social bonds in the group, supplementing the rather brief contacts made during head rubbing.

Fig. 16. Body care activities of lions: upper left—a lioness rakes her claws down a tree trunk; upper right—a lioness stretches, tearing the sod with her claws; middle—a male cub yawns and stretches by arching his back; lower left—a lioness licks her groin; lower right—a lioness scratches her head.

Anal sniffing. When two lions meet, one often sniffs or actually touches with its nose the anal area of the other, or both do so mutually while standing side by side, their tails raised and bent toward the other (plate 5). Such behavior is conspicuous when one lion seems uncertain about the identity of the other and needs to verify it. When five lionesses of the Seronera pride met a Masai pride male one night, the male and one female first sniffed mutually and then a second and third female sniffed him. On another occasion a lioness approached a male from another pride, greeted him and circled him closely three times before they sniffed mutually. The animal that is being sniffed sometimes lowers its rump and turns it away, or reacts more forcibly by snarling or even slapping. Anal sniffing lends support to the suggestion made earlier that lions can recognize each other by odor, but whether they do so individually, by a group smell, or by both is not clear.

It is probable that males frequently sniff the anal area of a female to determine whether or not she is in estrus. Since a greeting by a female is so similar to sexual behavior, his sniffing in this context may often be a means of evaluating her intentions.

Visual Communication

Since a signaling animal is usually visible to group members, all or part of its body may be used in communication (Marler, 1965). The whole body of the lion is highly expressive, but the most intricate repertoire of signals is concentrated in the head, where the eyes, lips, teeth, and ears provide a graded system of expression. According to Fox (1970) the more solitary species among canids have uncomplicated stereotyped visual signals when compared to the social species, but this generalization does not seem true for the felids. The solitary house cat is as expressive as the lion, judging from the work of Leyhausen (1960). Differences that exist seem more quantitative than qualitative.

Facial expressions. Leyhausen (1960) distinguished between aggressive threat and defensive threat in the facial expressions of the house cat. In aggressive threat the mouth is almost closed, the ears are erect and twisted so that their backs face forward, and the pupils of the eyes are small. In defensive threat the canines are exposed and the lips withdrawn, the ears are flattened, and the pupils are large. Various gradations in expression exist between these two extremes. Other cats, including the lion, use similar facial expressions, and Leyhausen's work provides a useful framework on which to base more detailed descriptions.

The facial expressions of monkeys and apes and of carnivores are in many respects similar even though the face musculature differs to some extent (Bolwig, 1964). Fox (1970) compared the facial expressions of several canids with the expressions of primates as described by Van Hooff (1967) and found that all except two of the expressions and the situations evoking them were similar in the two groups. Cats, too, resemble both primates and carnivores. I use here Van Hooff's descriptive terminology whenever possible to draw attention to similarities and to facilitate comparisons.

Relaxed face[1] (fig. 17). The features are in a "neutral position" (Van Hooff) with ears erect and pointing laterally, the eyes closed or partially open, and the mouth either closed or hanging slightly open with the lips drooping flaccidly. An unwavering look with a bland expression conveys mild aggression, as it does in primates. "Lions are sensitive creatures and dislike being stared at. To look directly into the eyes of a lion for any length of time causes the animal visible discomfort" (G. Adamson, 1968). Consequently, lions seldom focus for long on another individual. When they do, their eyes have that peculiar detached gaze which give the human observer the feeling that the lion is looking around and through him rather than at him.

Relaxed open-mouth face (fig. 17). The ears and eyes are either relaxed or fairly alert, and the mouth is open with the lips drawn back and often slightly raised at the corners. The teeth are not exposed and no vocalizations are emitted. Lions use this expression in play. It possibly indicates an intention to bite gently (Van Hooff, 1967).

Alert face (fig. 17). An alert lion has its ears cocked, and its eyes are opened more widely than when its face is relaxed; the lips are either closed or slightly parted. There is a noticeable tension in the face while the animal looks fixedly in a certain direction. Many situations elicit this expression, particularly when the animal watches other group members, when it sees or hears other lions or objects of interest in the distance, and when it spots, stalks, and chases prey. Other lions usually respond to the gesture by becoming alert too.

Roaring face (plate 6, fig. 17). Roaring lions raise their muzzles slightly and with eyes and ears relaxed emit the sound through a partially opened mouth. The lips are brought forward, almost as far as in the tense open-mouth face, but they are held less rigidly. The

1. This expression, the relaxed open-mouth face, the alert face, the tense open-mouth face, and the bared-teeth face, are the same or similar to expressions described for primates by Van Hooff (1967).

RKcane

expression used by lionesses when calling their cubs is similar except that the mouth is barely opened. Any response by other group members is to the vocalizations rather than to the expression.

Yawning face (fig. 17). A yawning animal raises its head and opens its mouth so widely that all teeth are exposed. Its eyes are closed and the tongue protrudes a little past the lips. Although the gesture is prominent, it is neither directed at a specific individual nor does it elicit a response, as it does in baboons, which use their exposed canines during yawns as a visual display in agonistic encounters (Hall and DeVore, 1965).

Grimace face (fig. 17). After sniffing certain things a lion often grimaces, a gesture termed *flehmen* in the German literature (see Verberne, 1970). The animal opens it mouth, but not as widely as when yawning, thereby exposing its teeth; it also raises its muzzle and wrinkles its nose. The eyes are almost completely closed and the ears remain in a relaxed position. The tongue usually does not protrude past the lips, in contrast to grimacing tigers in which the tongue hangs out of the mouth. The expression resembles superficially the bared-teeth face except that the eyes are closed, the ears relaxed, and no vocalizations are emitted. Lions grimace after having sniffed urine and scent marks (their own as well as those of other lions), wounds, estrous lionesses, as well as a dead turtle, mongoose, and lion cub. Any powerful odor seems to elicit the response, especially if it is unusual. Other lions ignore the gesture or sniff at the same site.

Tense open-mouth face (figs. 17, 18). In this expression the ears are twisted so that the black markings on the back of them face forward, the eyes are large and round, and the mouth is partially open with the corners brought so far forward that the lips are almost in a straight line. The teeth are not visible or barely so. The lion stares at the opponent in the head-low posture and emits growls or coughs. It may lash its tail. Such an expression is prevalent when one lion for some reason does not tolerate the proximity of another, particularly at a kill. Females also use this facial expression and vocalization during copulation,

---

Fig. 17. Facial expressions of lions: top left—relaxed face of a dozing lioness; top right—relaxed face of a lioness sitting by her kill; upper middle—relaxed open-mouth face of a cub as it pounces playfully on another cub; lower middle (left)—alert face of a lioness watching prey; lower middle (right)—roaring face of a male; bottom left—yawning face of a lioness; bottom middle—grimace face of a male after sniffing scent; bottom right—tense open-mouth face of a lioness while guarding her kill.

R. KEANE

except that their ears are often erect as during the alert face. This pattern represents the highest intensity of aggressive threat, to use Leyhausen's term (1960), and attack is almost certain unless the other individual retreats.

Bared-teeth face (fig. 18). This expression, termed defensive threat by Leyhausen (1960), consists of a graded system which is often influenced by components from the tense open-mouth face. The ears are partially to wholly retracted, the eyes are usually visible as slits, and the teeth are displayed in varying degrees with the lips pulled back; the nose is wrinkled. Growls, snarls, miaows, or a combination of these accompany the expression. The animal faces its opponent if features of aggressive threat are prominent, or stands with head turned partly to one side if features of defensive threat predominate. When an animal faces away from an opponent, it not only avoids a direct stare but also exposes its teeth conspicuously in profile, a display of weapons which no doubt functions as intimidation.

The bared-teeth face is common in all agonistic encounters. At a low intensity, as when a male discourages a cub from rubbing heads, the lion may merely growl and lift its upper lip, exposing a canine, on the side facing the other animal (plate 7). At a somewhat higher intensity, again a common response by both males and females toward cubs, the corners of the lips are retracted, showing the row of teeth, but the mouth itself remains closed or is only slightly opened; the ears are laid back, although this part of the expression is obscured by the mane of males. The animal growls and may jerk its head at the other one. In many agonistic interactions the mouth is open but not fully so as the animal snarls, coughs, or growls. Sometimes the corners are pulled back and the lips somewhat everted, as if to expose as much of the teeth as possible; at other times components of aggressive threat predominate, and the teeth are brought forward so that only the canines and incisors are exposed (plate 8). At a high intensity the mouth is open, exposing all teeth, and the expression is reinforced with lunges, bites, and slaps if

---

Fig. 18. Facial expressions of lions. The bared-teeth face: top left—a lioness bares her teeth in response to being greeted by a cub; top right—a male bares his teeth slightly while being attacked by a lioness after copulation; upper middle (left)—a male opens his mouth in response to a greeting by a cub; upper middle (right)—a bared-teeth face with strong components of the tense open-mouth face by a male guarding a kill; lower middle—a copulating pair, the male with a bared-teeth face, the female with a tense open-mouth face; bottom left—a lioness deters a cub from suckling with her open mouth; bottom right—a lioness hisses at a male.

the other animal persists with the contact. Essentially the same expression, but with mouth even wider agape, is displayed when lions hiss.

These expressions are common at kills, during squabbles when females discourage cubs from suckling, and in other agonistic situations. They are also typical of courting lions, displayed by the female before and after copulation and by the male during it. The accompanying behavior of the animal depends on the situation and may vary from attack to such defensive gestures as rolling on the back. Exposed teeth and other features function as a warning essentially defensive in nature but with the implication that the weapons may be used.

It is noteworthy, when considering the possible communicatory value of these expressions, that several of them elicit no response even though they are conspicuous. These include the grimace and the yawn. Neither is reinforced by a vocalization, and the social context in which they are given is innocuous. The alert face has signal valence primarily if the whole body, not just the head, becomes rigid. The most varied expressions are displayed in agonistic situations. The bared teeth alone elicit responses in particular social situations; usually, however, vocalizations are so much a part of the functional unit that the effect of the expression in itself is difficult to measure.

In passing, I would like to point to the expression of lions when in pursuit of prey. As mentioned earlier, the cat uses the alert face in such a situation, not a bared-teeth face as is usually indicated in museum exhibits. On the other hand, lions attack man with bared teeth. The exposed teeth represent a defensive reaction, whether in response to another lion, man, or an attacking prey animal; in other words, they contain an element of fear which probably is not the expression an intrepid hunter wants his trophy to convey.

The photographs and descriptions of primate facial expressions published by Van Hooff (1967), Marler (1965) and others are almost identical to those of lions except that cats have more mobile ears than monkeys. Similar expressions include the alert face, relaxed open-mouth face, tense open-mouth face, and bared-teeth face. I may have overlooked some subtle but distinct expressions, and the cats may therefore share with primates as large a repertoire as the latter do with canids. The fact that many expressions in primates, cats, and dogs not only are elicited by the same type of social situation but also have the same function suggests that similar selective pressures operated on all their societies to help produce mobile lips and eyes and prominent teeth as mediators of certain emotions.

Body postures. Whereas some signals rely primarily on facial expressions and vocalizations to convey the intensity and nature of emotion, others use posture to a marked degree either as the main element or in conjunction with other patterns. Most of the postures used by lions in their daily routine, such as those of walking, resting, drinking, and so forth are important in the integration of group activity. They convey that all is well and provide the baseline against which situations requiring a higher intensity of response can be compared. The evolution of postures has been conservative and many of them serve more than one function. The response of other members to a particular posture depends on the general circumstances in which it is displayed and on other signals that may be transmitted at the same time. Postures used in courtship and play are described in a later section. Here I mention several prominent ones used in hunting and in agonistic encounters.

The stalking walk, crouching walk, and crouch. When a lion spots some prey or, at times, another lion it assumes the alert face and then watches with body held rigid, either lying, sitting, or standing. Other lions respond to this posture by looking in the same direction. Sometimes the lion first walks toward the object in this alert stance, then attempts to approach undetected by using the stalking walk. It advances at a walk or trot, depending on the situation, with legs slightly bent, nose straining forward, ears cocked, and neck somewhat lowered. After a certain point it advances in the crouching walk, abdomen almost touching the ground, neck and head held low, as it creeps silently forward only to halt at intervals and cautiously raise itself to glance ahead, each movement conveying intense concentration. At times it halts and crouches with legs flexed beneath its body, even making slow treading movements with the hind paws as if to rush, and chin so low that it may touch the ground. In the case of a hunt, other lions usually become aware of prey as soon as one lion orients toward it, and they may then stalk in similar fashion (fig. 19).

The same three postures are also used to avoid contact with other lions or with man. At first the animal may trot away with its body slightly lowered, looking back over its shoulder occasionally; if pursued, it may attempt to escape into high grass or a thicket in a crouching walk. Finally, it may crouch facing the danger and emit growls and other agonistic expressions if approached closely. Once, for example, a nomadic male approached a resident lioness. She fled .3 km, but he followed. She finally crouched 50 m from him and with bared teeth growled sharply, each time jerking up her head, her tail lashing up and

R Keane

30219000133645

down. Lions also crouch when confronted by a superior opponent. They then assume the bared-teeth face in a head-twist posture and miaow or snarl. On the whole, stalking and crouching postures make the animal inconspicuous by presenting as little body surface to the environment as possible.

599 Sch 15

The strut. This is in many ways the behavioral opposite of the crouch. The animal stands with its legs stretched and neck raised almost perpendicularly, and the chin is tucked in slightly (fig. 23). The tail may be held high, looped over the back. No sounds or striking facial expressions accompany the display. The whole posture makes the animal conspicuous, with the mane prominent, especially since a male often stands broadside or walks stiffly around another lion. In most instances, the display lasts only a few seconds. Strutting is almost wholly confined to adult males, although I saw females intimate the display by a desultory stretching of the legs on several occasions, and cubs sometimes use the posture in play. Males direct the action mainly at females. When a nomadic lioness rejoined her male companion, he strutted in front of her, then scraped the ground with his hind paws. When a pride male found a lioness on a kill, he strutted broadside 5 m from her while she trotted up and rubbed heads with him. On yet another occasion, when three lionesses approached a male aggressively, he displayed broadside but with his head turned away from them. Once after they had inadvertently approached each other, two males from different prides strutted. The males of many vertebrates display conspicuous parts of their body to an opponent as a means of asserting rank. The strut of the lion probably functions similarly.

The head-low posture. This posture is assumed in conjunction with the tense open-mouth face. Standing with its forelegs spread apart slightly more than usual, the lion holds its head and neck obliquely downward, growling or coughing while doing so. Its gaze is directed steadily at the other animal. The thorax seems to slump, as Leyhausen pointed out to me, causing the shoulder blades to protrude. The posture is used by lions when attempting to discourage others from approaching, as when they are guarding their kill or their cubs. It is an aggressive threat of high intensity, and, as noted earlier, attack may follow.

73.11-1001

Fig. 19. Walking and stalking postures of lions used during a hunt; from top to bottom—a lioness walks in her usual relaxed posture; she walks in an alert posture; she assumes the stalking walk; she crouches; and she rushes at her quarry.

HIGHLAND COMMUNITY
JUNIOR COLLEGE LIBRARY
HIGHLAND, KANSAS

The head-twist posture. During agonistic interactions one lion often faces the other with teeth bared and head twisted slightly sideways while emitting a harsh miaow. It may then swat lightly with a paw. At times it crouches, placing its face sideways on its forepaws. In some situations, particularly when a lioness and cub display the bared-teeth face to each other, both animals may assume the head-twist posture. At times, one animal turns its head and miaows at the mere approach of another. The head-twist is an intention movement to roll on the back, for the animal occasionally keeps turning until its shoulder is on the ground and it falls over. When one lion swats another the latter may literally flip on its back so fast that the blow becomes ineffective. A lion sometimes rolls over merely in response to a growl or lunge, occasionally urinating as it does so much like a submissive dog, behavior such as this being prevalent at kills (plate 20). Lorenz (1964) described a similar pattern in the house cat when he wrote: "The more frightened the animal becomes, the more sideways becomes its position, until finally one paw is raised from the ground. . . . Should the fear of the cat mount still higher, this reaction leads to the last desperate means of defence which the animal has at its disposal: it rolls right over on to its back and turns all its weapons towards its aggressor."

I have never seen a lion attack another that has rolled on its back, and even the head-twist while crouching tends to terminate a fight. When, for instance, a male rushed at a lioness, she assumed the crouch and head-twist and he desisted. Similarly, when a subadult male placed his face on the ground after having been chased by an adult, he was not harassed further. As Lorenz (1964) and Leyhausen (1960) noted, the head-twist leading to a backroll is a defensive posture, an interpretation suggested by the facial expression alone. "A true submissive attitude does not occur in adult cats," as Leyhausen (1960) pointed out, but, as he rightly noted, the head-twist functions as one in that it tends to terminate the interaction, both because the posture conveys a lack of aggressive intent and because lions are generally disinclined to fight an opponent which faces them with weapons exposed.

Tail positions. The tail of a lion normally hangs quite limply with the end of it turned up somewhat, but various social situations influence that position. During head-rubbing and anal-sniffing contacts the animals raise the tail so that it either arches over their back or tips toward the other animal. When a lioness has cubs or a courting male at heel she often has her tail looped up to the level of her rump or higher and she flicks it occasionally, an action which may help cubs to keep her in sight in dense vegetation. Fighting lions and those chasing prey tend to hold their tail fairly straight somewhere between the horizontal and vertical. No special position is used by stalking lions except that the only movement of crouched ones may be an occasional flick of the tailtip or ears. In agonistic situations, particularly in those containing a large element of aggressive threat, the tail is often lashed up and down, the only tail movement pattern I found to be unambiguous. Tail positions alone convey various levels of excitement which require in most instances further signals to elucidate the precise emotion.[2]

Vocalizations

Lions are vocal animals which seem to emit a bewildering variety of snarls, moans, growls, grunts, and roars. However, the repertoire is not as large as it first appears because discrete signals, such as those typically found in birds, are uncommon. Instead most sounds belong to three graded systems linked by intermediates not only within the same system but also between them, thereby enabling the animals to communicate subtle changes in the intensity and nature of their emotions. Most sounds are harsh, without purity of tone, and resemble in many respects those of primates as described by Marler (1965): "In the higher forms there seems to be a predominance of sounds that have a relatively wide frequency spectrum, with most of the energy in relatively low frequencies (below four or five kilocycles per second), without much intricate changing of frequency; the changes in pitch that do occur are slow and are often blurred by the 'noise' that is also created."

Given the graded nature of most sounds, it is difficult to evaluate the size of the lion's repertoire. Most vocalizations fall into the basic categories of miaowing, roaring, and growling-snarling, categories which include at least nine more or less distinct expressions. Purring, puffing, bleating, and humming complete the repertoire. By the age of one month, lion cubs emit in rudimentary form all sounds that adults do, although the full roar is not displayed until subadulthood.

Not included in this discussion are nonvocal sounds such as sneezing

2. Numerous authors have commented on the horny nail which is attached to the skin hidden at the tip of the tail in the tassle (see Guggisberg, 1961). Some lions lack this appendage and in others it is a mere scale. The longest nail I saw was 7 mm long. It has no known function.

and panting and some inadvertent ones such as the grunt forced out of an animal when it flops on its side in order to rest.

Puffing. Our pet cub occasionally emitted one or two gentle "pfff-pfffs" through his lips just before he rubbed cheeks with me. George Adamson told me that his tame lionesses grunted very softly prior to greeting males. I have not heard this sound in the wild, probably because I was too far from the animals. Lions puff in the same kind of situation that elicits "prusten" in tigers. Those cats puff repeatedly through lips and nostrils as they approach each other, to signify friendly intentions, much as house cats do in purring (Leyhausen, 1950).

Purring. When I cuddled our cub it sometimes purred softly but very briefly. One male emitted several purrs as he rolled on his back in the grass. According to George Adamson his lionesses purr while courting. As Leyhausen pointed out to me, the purring of lions differs from that of the house cat in that the sound is produced during exhalation only rather than continuously while breathing in and out. Purring appears to signify contentment, but compared to *Felis* the vocalization is unimportant in the social life of lions.

Bleating. A small cub which was dying of starvation bleated—a nasal *äää-äää*—when I picked it up. Other cubs bleated only in agonistic situations, such as when squabbling at a female's breast, but the sound was then harsh, containing elements of snarling and miaowing, and the lionesses responded by moving away or snapping at the young. I was not able to identify the bleat in adults although some of their vocalizations resembled it, and, in fact, it may be a variant of the miaow rather than a category by itself.

Humming. Two lions often hum softly when they rub cheeks, lick each other, or play gently—*mmm* with mouth closed or *uuuu* with it slightly open. Suckling cubs and resting adults may also hum. The sound signifies contentment, and other lions never responded visibly to it.

Roaring. Numerous authors have written about roaring, often inaccurately. Kearton (1929), for example, described three kinds of roar: "One is a kind of grunt repeated again and again when he is hunting. The second is an angry and blood-curdling sound used when he is killing. The third is really a roar . . . that indicates triumph and self-satisfaction and is mostly used when he has slaked his thirst after the kill." Roaring represents a graded system, ranging from barely perceptible grunts to full roars, a fact recognized by Schenkel (1966a). Although the sounds differ in loudness, duration, and frequency of emission, all are of a low frequency with fundamentals at less than .3 kHz but with overtones up to 1.0 kHz.[3]

Woofing. An abrupt, loud *wuu* is occasionally emitted by a startled animal. For example, when a cub leaped on a resting lioness, she reared up with a woof, and, on another occasion, a cub woofed after a lioness slapped it suddenly at a kill. Once a monitor lizard lashed its tail at the face of a lioness and she woofed as she jumped back.

Grunting and soft roaring. Roaring at a low intensity consists of one or more hollow-sounding grunts which, if the animal receives no response, often grade into a suppressed roar, a moaning *uuoo*, *aauu*, or *aaouuu*. Figure 20(A) shows a grunt of moderate intensity followed by one of low intensity, and figure 20(B) illustrates a soft roar. All are brief, .2 to .5 seconds in duration, and of low frequency (.75 kHz or less). The grunts are predominantly used by lionesses when calling their cubs. Sometimes a lioness begins to grunt at intervals half a kilometer from the place at which the cubs are. When the young hear her, usually when she is no more than 75 m away, they scamper out of hiding and run to her. Lionesses also emit the sound if cubs stray from their side. Once several cubs crawled beneath my car. A lioness grunted but they ignored her. She walked 120 m away, returned and called once more and this time the cubs followed her. On another occasion, two small cubs fled at the approach of the car, but when their mother grunted they rejoined her. In most instances the cubs merely run toward the lioness without answering, but occasionally one miaows or grunts in response. Adults sometimes roar softly while walking alone seemingly in search of other group members. On one occasion, a lioness called, and two others nearby looked up, their heads raised above the grass. The lioness joined them and all rested together. Cubs grunt and moan when they lag behind the moving group. One cub, for instance, miaowed then grunted harshly at the others 50 m ahead but was ignored. Yet on some other similar occasions a lioness waited. In general the various intensities of soft roaring function as contact calls with the animal drawing attention to itself in a variety of situations.

3. 1 kHz equals 1,000 cycles per second.

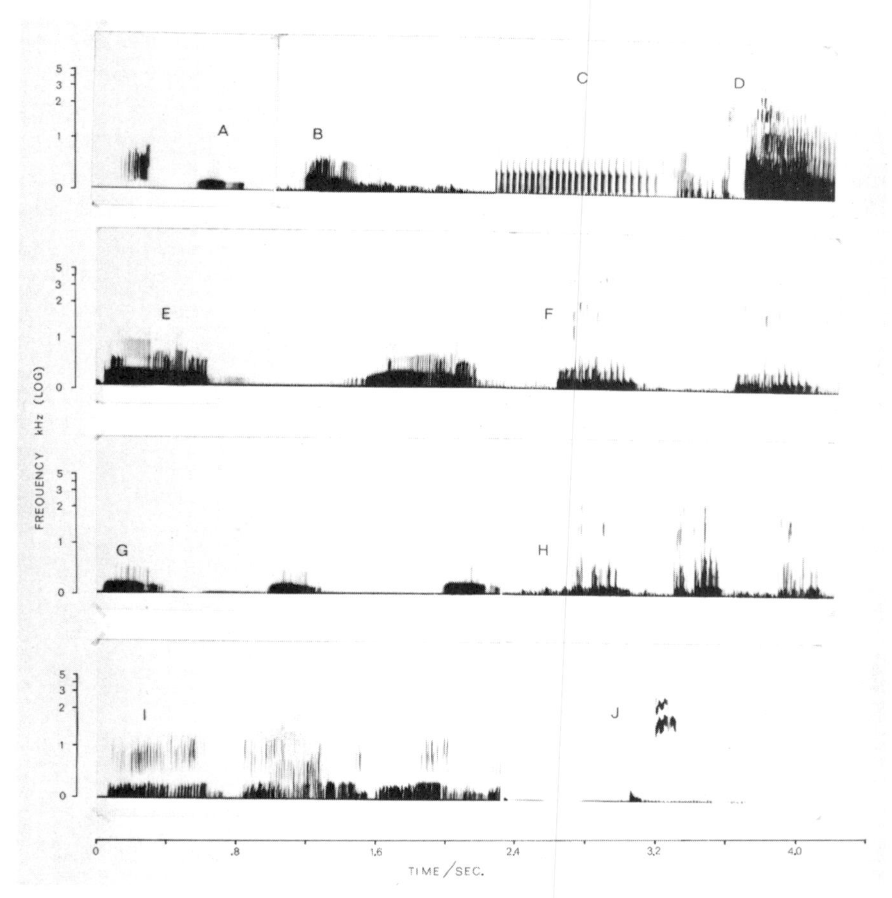

Fig. 20. Sound spectrograms of lion vocalizations (all by adults, except J). A—grunt of lioness calling cub; B—soft roar; C—rolling growl; D—snarl; E—loud roar; F—panting growl; G—grunts following a loud roar; H—harsh growl; I—snarl; J—miaow by cub.

Loud roaring. The loud roar is the lion's most conspicuous gesture, one that can be heard on any night, although the animals call little in districts where they are much persecuted (Percival, 1924). Among my most pleasant memories of the Serengeti are the nights I spent on the plains with everything silent until suddenly a roar filled the void—first one or two moans, then full-throated, thundering, until it died slowly in a series of hoarse grunts. Of 31 roaring episodes I counted, the average number of roars was 9 (4–18) followed by an average of 15 grunts (0–57). Each episode lasted about 40 seconds (25–60 sec). Ulmer (1966) found in his captives that "a typical chorus of our five lions lasts an average of 36.6 seconds."

Figure 20(E) illustrates two individual roars by a male. Each of these roars was about .6 to .7 seconds in duration and about .8 seconds elapsed between them; each of three grunts following the roar of another male (fig. 20[G]) was about .3 seconds in duration with an interval of .6 to .7 seconds. Spectrographically, these grunts resemble those used by lionesses to call cubs.

Lions usually roar while standing, but they also do so readily while lying on their belly or side, as well as while sitting, walking, or running (plate 6). The abdomen sucks in visibly as the animal inhales and then expands as the lion exhales to produce each roar. Males and females roar similarly except that the calls of the male are somewhat deeper in tone and louder than those of the female. When hearing an animal roar in the distance at night, it is not only difficult to judge its sex but also to estimate how far it is because of the varying loudness and almost ventriloquial quality of the sound. I have on several occasions heard roars from a distance of 3 to 4 km, and it is possible that they can be detected from as far as 8 km (Stevenson-Hamilton, 1954). Roaring by one group member often stimulates others to join, and the ensuing concert is surely one of the powerful and impressive animal sounds in nature. When two lions roar, the individual sounds are often synchronized in such a way that they overlap only a little or not at all. This may enhance the signal value of the roars by adding to the time during which noise is actually produced. Cubs generally do not participate in communal roaring. Once a cub, 5 months old, accompanied the calling adults with a soft *ooo*, as did several cubs 9 to 18 months old. The full roar is not evident until the animal is at least 2½ years old. Similar communal vocalizing has been described in several other social mammals including the wolf (Murie, 1944) and howler monkey (Carpenter, 1965).

Lions call mostly at night as shown in figure 21, which is based on the number of single or communal roaring episodes in 56 whole days of observation. Generally roars are rare before 1700 and after 0800. A tape recording of a roar played at midday may not stimulate a lion to raise its head, whereas the same call at night may elicit an immediate reply. It was my impression that lions roar particularly often just before dawn. Such activity is evident in figure 22 for the animals in the Seronera area and for nomadic male No. 134 but not male No. 159. Other peaks show little agreement. The Seronera animals roared often between 0100 and 0200, male No. 159 between 2100 and 2300, and male No. 134 between 1900 and 2000.

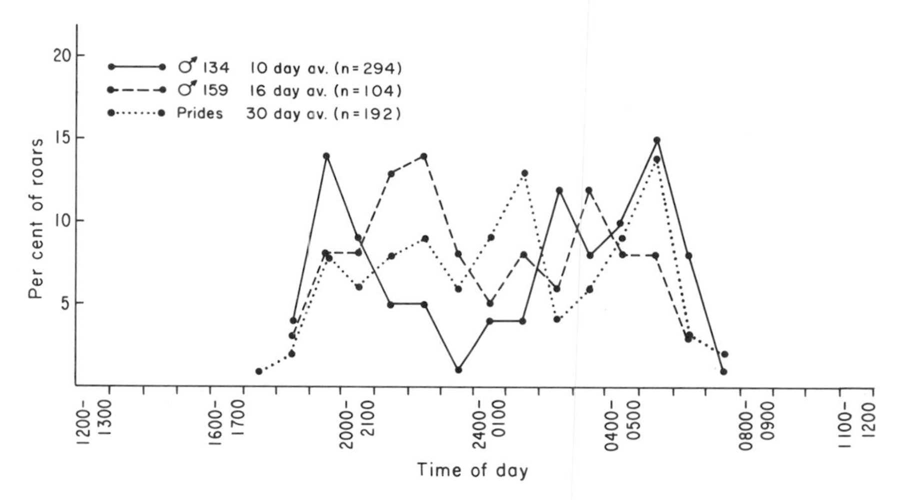

Fig. 21. Time and frequency of roaring by lions.

The number of times a lion roars in the course of a day varies. Male No. 159, a nomad on a temporary territory, had 2 to 16 roaring episodes per day with an average of 6.5; male No. 134, a nomad in the process of establishing a territory, averaged 29 episodes. Other nomads, none of them territorial, that were followed for a total of eleven nights averaged only 3 episodes. Several females and cubs of the Seronera pride which I observed for three consecutive days roared only once on two of the days and 12 times on the third. The average number of single or communal roaring episodes by this pride on fourteen nights observed was 6.2. Similarly, several lionesses and cubs of the Masai pride were followed for four consecutive days and these roared twice the first night, 11 times the second, not at all on the third, and once on the fourth. The average for the pride in sixteen days of observation was 6.5. Since members of a pride are often scattered, with each one calling during the night, the total number of roars heard in an area may be considerable. I was unable to detect a seasonal difference in the amount of roaring. It was my impression that lions roared more during rainy nights, especially if these had been preceded by a dry period. In October, 1966, the Seronera lions seemed to roar more than usual, possibly because many nomads were in the area.

Lions roar in various situations. Most of them roar spontaneously, it seems, or in reply to distant calls. I was unable to determine the stimulus that elicited roaring in 75% of the 300 episodes that involved pride members, although the animals possibly heard sounds at times which I did not. The other 25% of the episodes were in response to distant roars. When lions hear such calls they may ignore them, or jerk to attention, or answer. Of 190 distant roaring episodes heard while I was observing resident lions, 40% were answered by the lions, sometimes by only one or two animals, at other times by the whole group. Occasionally a lion replies with a soft roar audible for only 100 m or so. One dying nomadic male gave several grunts in response to distant roars even though he was too weak to raise his head. During four consecutive days of observation, several lionesses and cubs of the Seronera pride heard distant roars (more than .5 km away) on 42 occasions and answered 12 (29%) of these; they also heard 23 roars close to them and answered 17 (74%) of these, indicating that proximity affects the response. Resident lions seldom move toward roars they hear even when these are given by nomads or other lions intruding into a pride area. However, when a lion is searching for its group members it responds immediately. One lioness, for example, roared and then looked around alertly. Ten seconds later lions called in two places. She roared once more, was answered once, and headed in that direction.

Fighting sometimes elicits roaring. When a lioness slapped a male, several others as well as the combatants roared. On another occasion, one lioness accidentally hit another with her tail. The second one snarled, which in turn excited the first lioness to clout a nearby male on the nose. He merely roared and the whole group did too. Harsh, short roars are given by lions as they chase others, particularly when nomads pursue residents or vice versa. However, roars in agonistic situations were infrequent. Of 192 roaring episodes that I tallied around Seronera (see fig. 21), only 2.6% were of this type. Occasionally the whooping of hyenas stimulated lions to roar. Ulmer (1966) noted that the rumble of freight trains caused his captives to call. Once a lioness grunted softly to her cubs and a nearby male answered with a full roar. Roosevelt and Heller (1922) stated that hunting lions ". . . deliberately try to stampede the animals by roaring . . ." but I found no evidence of this.

The full roar, like the soft roar, advertises the animal's presence. It denotes "Here I am," and in this capacity has several functions as a long-distance signal. First, it helps lions find each other, although their propensity to ignore calls makes this somewhat difficult. Once, for example, a male lost sight of a lioness he was following. He roared but received no response and finally tracked her by scent. Second, roaring enables lions to avoid contact, by, for instance, delineating the pride area, as Schenkel (1966a) pointed out. Although trespassers into a territory may not flee from a roar, their subsequent movements may be influenced. Third, roaring enhances the physical presence of an animal by making it more conspicuous. During agonistic encounters the roaring may help to intimidate the opponent. Fourth, communal roaring, like most social endeavors, may help to strengthen the bonds of the group.

Communication between lions would be greatly enhanced if they could recognize each other by their roars. Wolf-howls, for example, have "harmonic characteristics . . . that would distinguish individuals on the basis of any one howl" (Theberge and Falls, 1967). I could distinguish the roars of a few males, and lions possibly can do so too. Tapes of roars of both nomads and other prides were played to the Seronera residents without eliciting responses different from those of their own pride.

Miaowing. Small cubs emit a yippy *ia* or *iau* which reaches a frequency of 4 kHz (fig. 20[J]). One cub, about two weeks old, miaowed as it crawled near its mother (fig. 22[L]). The calls were about .3 seconds long with the fundamental frequency between 2 and 3 kHz. When the cub lost its balance on a rock, its calls grew harsher and more prolonged (.5 sec) and intense noise was produced over a wider range of frequencies (0–4.0 kHz) as illustrated by figure 22(M). Marler (1957) pointed out that pulsed sounds with a sharp onset and a wide frequency range are good location signals. The constant miaows of small cubs as they move around and behind the group probably help lionesses keep in contact with them. Cubs also miaow as they approach an adult before greeting it or when being licked roughly by a lioness.

Prolonged, harsh, snarly miaows, lasting as long as 2 to 3 seconds are often given by cubs when they want to suckle but are not permitted to do so by the lioness and when they squabble with each other in competition for a nipple. These miaows are often emitted in the bared-teeth face. Figure 22(A, B, E) illustrates three such harsh miaows by cubs. All have a wide frequency range (up to 5 kHz) and last half a second or longer. They are exceedingly irregular sounds with respect to their energy levels. B has most of its energy at about 2 kHz, and A has its energy widely spread between .3 and 2.0 kHz. Figure 22(C, D) shows a harsh miaow by a cub which ends in a soft roar. The staccato syllables of snarls are not evident in these harsh miaows.

The miaows of adults almost always sound snarly and are emitted

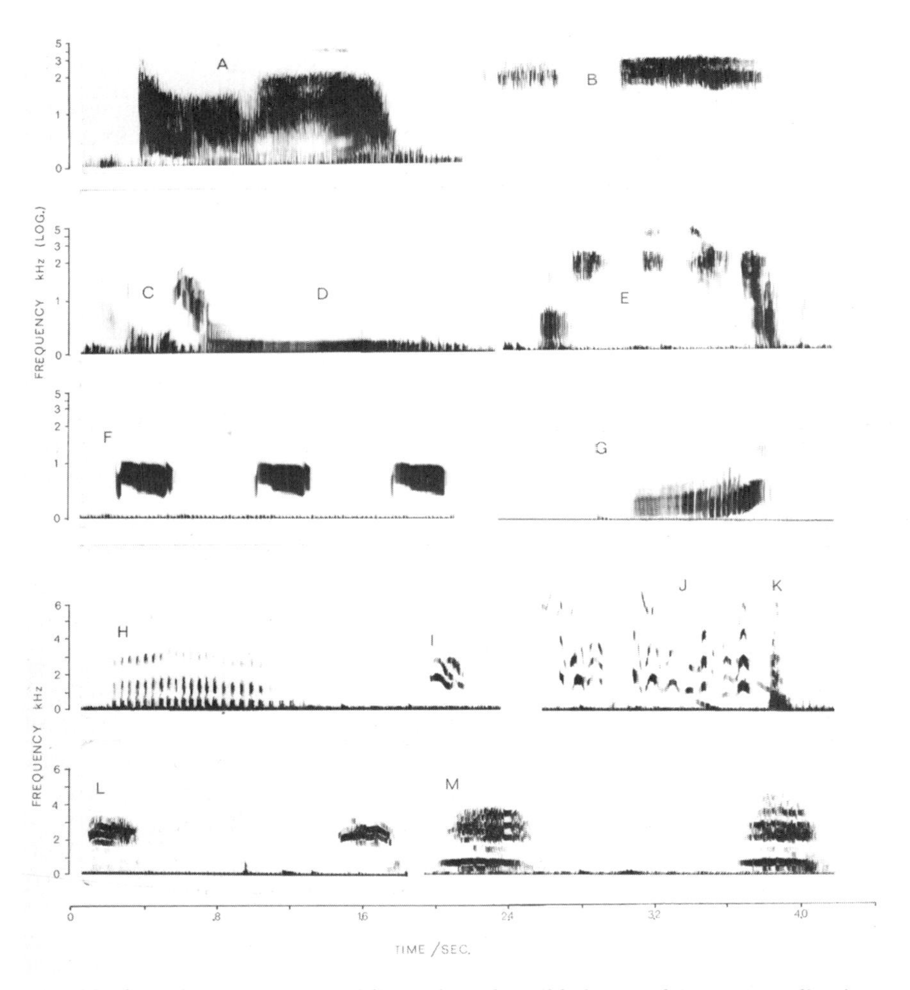

Fig. 22. Sound spectrograms of lion, cheetah, wild dog, and hyena vocalizations. Lion: A, B, and E—harsh miaows by cub; C and D—harsh miaow by cub ending in soft roar; L and M—miaows by cub. Cheetah: H—chirr; I—chirp. Wild dog: F—hoot; J—whine; K—bark. Hyena: G—whoop. The cheetah, wild dog, and hyena vocalizations were made by adults.

with bared teeth, sometimes in the head-twist posture. Females use this vocalization often as they approach or are being approached by a male. When a nomadic male walked toward a resident lioness, she first growled, then miaowed harshly. When a male accidentally rolled on a lioness, she reared up with her teeth exposed and miaowed. Males also miaow during copulation.

Lionesses usually ignore the miaows of a cub or merely look in their direction. Once, however, a cub wandered alone into dense grass and called from there. Its mother walked toward it roaring softly. The harsh miaow with the bared-teeth face is so much part of the agonistic situation that the response of others to the vocalization alone is not clear, other than that it draws attention to the animal.

Lions typically miaow when they are lost, unsure of themselves, or prevented from reaching a desired goal, all conditions of light distress. Often the miaow contains components of the snarl, a compound sound both of distress and defensive threat.

Growling, coughing, snarling, and hissing. Judging by the tense open-mouth face that usually accompanies growling and coughing and the bared-teeth face that is seen in conjunction with snarling and hissing, the former two are essentially aggressive sounds and the latter two defensive. During social encounters, however, the emotions of the lion vacillate so much that growling and snarling often become indistinguishable; indeed, spectrographic analysis shows that the sounds are similar in structure (see fig. 20[C, D, F, H, I]). The harshness of a snarl when compared to a growl seems partly due to the fact that the animal has its mouth open.

Growl. The intensity and duration of growls vary considerably. Low, almost continuous, rolling growls are emitted by lions trying to avoid close contact, as when guarding a kill. Such growls are staccato in nature, of low frequency, with most of their energy below .5 kHz but with overtones almost reaching 2.0 kHz (fig. 20[C]). Occasionally lions produce a series of short growls together with a snuffling or panting noise. Growls of this type in figure 20(F) are about .5 seconds long with about a .5 second interval; most sound is at a frequency of less than .5 kHz but overtones reach 4.0 kHz. Such a growl is common when two lions bite into the same piece of meat and dispute it. Short, harsh growls are prevalent at kills, often accompanied by a slap, and when lionesses attempt to deter cubs from suckling. Harsh growls are spectrographically similar to the panting growl except that they are briefer—.2 to .3 seconds—and the interval between them is only about .3 seconds (fig. 20[H]). Most growls are produced during expiration, but on a few occasions an animal at a kill vocalizes while inhaling as well, to produce a sound which resembles the "sawing" of leopards.

Lions often responded promptly to the growl of another animal even if that animal was obscured by vegetation. One lioness on a kill growled at another one, which promptly reclined 20 m away. On another occasion, a cub approached a reclining lioness. She growled once and the cub veered aside and sought a drink elsewhere. Certainly to a man on foot, the growl of a hidden lion provides a strong incentive for reversing the direction of travel. The growl means "watch out, do not come closer."

In addition to these growls, there is a barely perceptible *grrr*, a single note, which I heard only a few times and only when a member of a resting group spotted danger, such as once when a park visitor got out of his car 80 m from some lions. The others respond immediately by leaping up and looking around.

Coughing. The cough is a violently expelled growl emitted once or several times in succession. Attacking animals cough during the actual charge, adults occasionally cough at a cub to deter it from some action, and all of them use the vocalization occasionally while squabbling at a kill. Coughs are also directed at hyenas, leopards, people, and cars. During an attack the cough may at times grade into a full roar. The response of animals to a cough is similar to the response to the growl, and the sound functions as a threat of high intensity.

Snarling. Harsh snarls—*aarrr-aarrr*—resemble the growl spectrographically except that the staccato syllables are slurred (fig. 20[D, I]) and that at a high intensity much noise over a fairly wide frequency range is produced. The same general situations elicit both vocalizations; the main difference between them is not so much in the sound as in the accompanying facial expressions and body postures. These indicate that the animal is threatening while on the defensive. The responses of the other group members resemble those given to the growl. One sound, seemingly a snarl of an exceedingly high intensity, was heard only three times, after males attacked lionesses on a kill. The females emitted loud, high-pitched, long-drawn *aaaa*s as they scattered.

Hissing. Occasionally a lion opens its mouth as if to snarl, but only a hissing sound emerges. If expelled abruptly, the hiss becomes a spit. Quite specific circumstances elicit the hiss, most often the sight of a strange lion. When, for example, a nomadic male approached two resident lionesses, they hissed at him. On another occasion a lioness saw the male of her pride 65 m away. She hissed as if at a stranger then approached in a head-low posture, but at 30 m she suddenly seemed to recognize him and casually lay down. On several occasions a lion hissed at water before drinking, possibly after having seen its reflection. Kruuk (pers. comm.) saw a lioness hiss at a spitting cobra. Occasionally lions hissed at a vehicle after it approached them closely. Hissing

R.KEANE

may also occur in circumstances which usually elicit a growl or miaow, but, on the whole, the sound seems to be a defensive threat in response to a somewhat unusual situation. Other animals respond to the hiss as they do to the growl or snarl.

Olfactory and Visual Marking

Any lion leaves tracks, urine, and feces in its environment, and these may serve as signals to other lions passing later. Schenkel (1966b) felt that an "open unconcerned manner of moving in their own territory and the 'proud' posture of looking around" was a visual territorial pattern of marking among male lions, but I was unable to note a consistent difference in posture between residents and nomads. In addition, lions use several distinct methods of long-distance signaling that leave evidence of their presence and in effect become physical extensions of the animals themselves.

Spraying and scraping. As a lion walks, he stops at intervals by a tuft of tall grass, a shrub, or a sapling and with raised muzzle rubs his face in the vegetation. His eyes are closed and his motions languorous, as if the experience is pleasurable. Then he swivels around, sometimes wriggles his rump closer to the bush, and with tail raised ejects from one to as many as twenty squirts of fluid from the caudally directed penis (fig. 23). The fluid consists of urine mixed with scent from two anal glands located near the base of the tail. It is squirted upward at an angle of 30 to 40° in a powerful jet that may propel it 3 to 4 m but usually splatters it among the leaves. The musky odor can be detected by a person on foot from a distance of at least 5 m. Spraying is predominantly male behavior among lions; among tigers females also spray commonly. I saw it only on seven occasions in lionesses, usually in perfunctory manner with rubbing brief or absent and urine ejected in weak spurts.

Males rub their heads in the vegetation before rather than after squirting, except on rare occasions. They may also fail to rub or only show their intention to do so by a slight raising of the nose. Rubbing often occurs at sites where others have squirted previously. Males thus pick up odors which later may be passed to other pride members, as

---

Fig. 23. Strutting and scent-marking postures of lions. From top to bottom: a male struts before a lioness; a male scrapes the ground with his hind paws; a male squirts from his caudally directed penis; and a lioness urinates.

Leyhausen pointed out to me. In this fashion, a pride possibly acquires a distinctive odor which is based primarily on that of the males. The general motions, including the raised tail, resemble those used in greeting, which, as was mentioned earlier, are derived from courtship behavior.

While spraying is solely an olfactory signal, scraping is a multiple one. A scraping lion hunches its back slightly, lowers its rump, and alternately rakes the ground some 2 to 30 times or more with the claws of its hind paws, tearing the sod and leaving the soil scarred (fig. 23). The scrape is obvious when the soil is soft, and the paws undoubtedly leave a concentration of scent. In addition, the lion often urinates while scraping or immediately afterward. The flow of urine is sometimes steady, as in normal micturition, but usually it appears in weak to moderately strong spurts. The urine often wets the animal's hind legs. Judging by the odor, the fluid consists of urine containing little or no scent from the glands. Occasionally a lion squats and urinates (fig. 23) and then scrapes the soil a few times as if in an afterthought. Scraping is common in both males and females. The puma also marks by scraping and urinating, though in a somewhat different manner. It "scrapes together leaves, twigs, or pine needles into mounds four to six inches high, then urinates on them" (Hornocker, 1969). The tiger scrapes and uses urine in the same manner as the lion. However, tigers often deposit their feces on the scrape, and these markings are often placed in conspicuous locations such as along roads and passes between hills. Lions never defecated on a scrape, and the casual way they drop feces around the terrain suggests that these are not used as an integral part of the marking system. Resting lions move to the periphery of the group to defecate, but usually not to urinate, the only circumstance in which I have seen them refrain from defecating where they are.

The frequency with which lions spray or scrape varies with the habitat. Males prefer to spray against vegetation or a tree trunk at least one meter high. Consequently they spray seldom while on the eastern plains except around kopjes. One male there scraped five times in a 24-hour period. Pride males around Seronera squirted or scraped in a ratio of about 1:3. For example, one male squirted once and scraped five times in 24 hours, another squirted once and scraped three times. Animals marked equally often at all times of the year although it was my impression that the Seronera pride did so more frequently than usual after one pride male was killed.

Spraying and scraping by one lion frequently stimulate others to behave similarly. Once four lionesses each scraped at the same spot, and on another occasion two males squirted where a lioness had just done so. In yet another instance a male sprayed a sapling. A lioness went to the same tree, first scraped beside it, then sprayed once. Two more lionesses came and each rubbed her face in the branches and afterwards scraped.

Because of this tendency to mark a previously used site, some bushes along roads, kopjes, and other landmarks past which lions frequently travel become scent posts similar to those established by, for example, wolves (Pimlott et al., 1969). One night in the plains two males visited seven widely scattered waterholes and marked the edge of each one, just as they had done on previous visits to some of them. G. Adamson (1968) noted that his tame lions "will never pass a chosen bush without stopping and going through the marking ritual."

Even though one 8-month-old male cub had a pungent orange-colored substance in its anal glands, marking with scent was not observed in lions until they were almost 2 years old. One lioness, aged 22 months, scraped and urinated, the youngest animal that was seen to do so. J. Adamson (1960) noted that a tame lioness sprayed trees at the age of 18 months. I saw no spraying by males until they were $3\frac{1}{4}$ to $3\frac{1}{2}$ years old.

Nomadic and resident lions spray or scrape in several circumstances. (1) Animals often mark in no particular context, but it is probable that, in most instances, they respond to scent already present. For example, one male sprayed seven times in 1.9 km of travel while walking along a road. (2) When two lions approach each other, one or both may mark the site. Once, after two pride males joined after a night's separation, one sprayed and scraped and the other just scraped. A nomadic male walked toward and then scraped 5 m from a resident lioness; after she moved, he sniffed the place where she had been lying and scraped once more. On two occasions, when a cub playfully approached a male, he rose and scraped. A male may also mark after he has been greeted by another animal. (3) Agonistic interactions frequently elicit scent marking. After a male bit a lioness in the back and she in turn clouted him in the face, he retreated 10 m and sprayed. On another occasion, several lionesses threatened a nomadic male, and he sprayed and scraped before lying down at the periphery of the group. Once a pride male first chased a nomad from the vicinity of his kill and then sprayed bushes repeatedly during his return to it. (4) Male lions commonly scrape and spray beside estrous lionesses. (5) Both males and females mark the vicinity of kills, particularly after they have finished eating and are ready to abandon the remains. (6) Lions sometimes deposit scent after agonistic encounters with other predators, such as hyenas. On two occasions, males scraped after I approached them closely with the car.

Spraying and scraping are used interchangeably. The two patterns complement rather than supplement each other. As pointed out earlier, lions are able to follow each other's tracks. Additional marking would therefore seem to be redundant except that it conveys specific information (whether the lioness is in estrus) and it provides a signal which because of its high concentration lasts longer than mere tracks. Judging by the situations that elicit scent marking, the behavior denotes a claim to ownership, a right to be there whether the issue is a piece of land, a kill, an estrous lioness, or a dispute within the group.

The function of scent depends on the status of the individual perceiving it. In discussing the scent marking of tigers, Schaller (1967) wrote: "The scent serves a number of possible functions: (1) it could enable tigers to follow each other in the forest, using the odor as trail markers; (2) it could delineate the extent of the range, indicating to others that the terrain is occupied, which may attract or repel the visitor according to the circumstances; (3) and it could communicate specific information to others using the range, such as the identity of the individual, the amount of time that has elapsed since it passed, and, in the case of a tigress, whether or not she is in heat." The same possibilities apply to the lion. The actual response of lions to scent is in most instances difficult to evaluate because it may be delayed. A nomad, for example, may ignore the scent of a pride, but the knowledge that others are in the area will enable it to prevent an encounter. Pumas establish scent stations. "On a number of occasions an animal tracked to one of these sites abruptly changed course, sometimes retracing its route for a considerable distance. Invariably it was found that another lion or family of lions was in the area" (Hornocker, 1969). In this way, scent would function importantly as a spacing mechanism, a conclusion also reached by Leyhausen and Wolff (1959) and Hornocker (1969).

Dung rolling. Lions roll on their back in several circumstances, notably in play, while resting, and in self-defense. Sometimes an animal squirms indolently on its back in dust or sand, possibly a form of body care; at other times it rolls over quickly, twists back and forth, and walks away. Lions also roll in dung. During the dry season many buffalo bulls concentrate along the Seronera River, and their fresh, soft droppings are common on the roads. Lions rub the side of their face in the dung until it is soiled and then slowly roll over, writhing with eyes closed, as if enjoying the experience. One lion rolled on a dry buffalo chip, another on an elephant dropping, and a lioness and cub rolled on the feces of a zebra. The behavior was common in 1967 but, for some reason, infrequent in 1968 and 1969; I did not see it in 1966. A male and a female rolled once for several minutes on a low brushy plant, *Lippia javanica* (*Verbenaceae*), which has a strong mint-like odor. Some of the behavior patterns used by dung-rolling lions resemble those of courting females, a comparison which was also made by Palen and Goddard (1966) after they had watched house cats rolling in catnip (*Nepeta cataria*). Thom (1934) noted that tigers may roll on elephant droppings.

I am not certain that dung rolling belongs in this section on marking. Roaring, scraping, or spraying sometimes precede or follow that behavior, and, of course, the body odor of the lion is rubbed over the ground, but just what advantage, if any, a lion derives from a face and mane smeared with feces remains unknown.

Tree clawing. Lions of all ages sharpen the claws of their forepaws by raking them down the trunks of trees. Squatting (78% of the instances) or standing on the hindfeet (22%), the animal hooks its claws into the bark and pulls them down slowly. Its muzzle is raised; its eyes are closed. Other group members may join a clawing animal until as many as five or six crowd around the tree. In the plains, where trees are rare, lions claw the ground while standing with forepaws stretched forward and rump raised (fig. 16). Some trees are scarred repeatedly just as some bushes are frequently sprayed, and there seems little doubt that they may serve as sign posts. However, I have not seen trees lacerated by lions to the same extent as has been reported for jaguars: "In front the bark was worn smooth, as if by the breast of the animal, and on each side there were scratches, or rather grooves, extending in an oblique line nearly a yard in length" (Darwin, 1845, quoted in Wynne-Edwards, 1962).

THE DAILY ACTIVITY PATTERN

A visitor to the Serengeti usually encounters a group of lions as it rests utterly relaxed, "poured out like honey in the sun," as Lindbergh (1966) phrased it. Visitors assume that the lions are resting after a night of prowling in search of prey. This supposition is not quite accurate; lions rest most of the night too. To obtain data on the lion's activity pattern, a total of 62 days were spent with a particular animal on a 24-hour basis, sometimes continuously for several days, at other times for only one. Either an assistant or I remained with an animal throughout the night, but occasionally we left it for periods of an hour or two in daytime as it rested. We followed and recorded the behavior of only one particular animal even if several were present, a necessary technique because groups often split and animals performed such activities as feeding for different durations. Most lions were followed on moonlit nights, but, judging from data obtained when tracking two nomads by radio on dark nights, the moon had little effect on their activity pattern. I recorded three main activities, those of walking, feeding, and resting, and the last included not only sleeping animals but also those that groomed themselves or were otherwise active but stationary. Hunting as a distinct activity was difficult to measure because lions were, in effect, always alert to a possible meal but only during an actual stalk was their behavior distinctive.

Activity Times

According to Rainsford (1909) and others, lions are active only at night in areas where they are heavily hunted, and, according to Kühme (1966), they are "strictly nocturnal" even in the Serengeti. Both the season and the habitat influence the lions' activity. In the plains and in other areas where there is little cover behind which the cats can stalk their prey, they are largely nocturnal (fig. 24), but around Seronera and in the woodlands they often hunt in daytime, particularly during the dry season when prey gathers along river courses. Figure 25 shows such a seasonal shift in activity time on the part of Masai pride lionesses.

Although lions readily hunt, feed, and mate at all times of day, their activity peaks are after 1700 and before 0800. They usually have a brief flurry of activity around dusk at 1700 to 1900, as they stretch, defecate, groom themselves and each other, play, and rub cheeks. More often than not they then recline again for an hour or so before another bout of greeting precedes their actual departure from the rest site. Spurts of walking seldom last more than an hour, as they sleep, sit, or otherwise remain stationary at frequent intervals. As figures 24

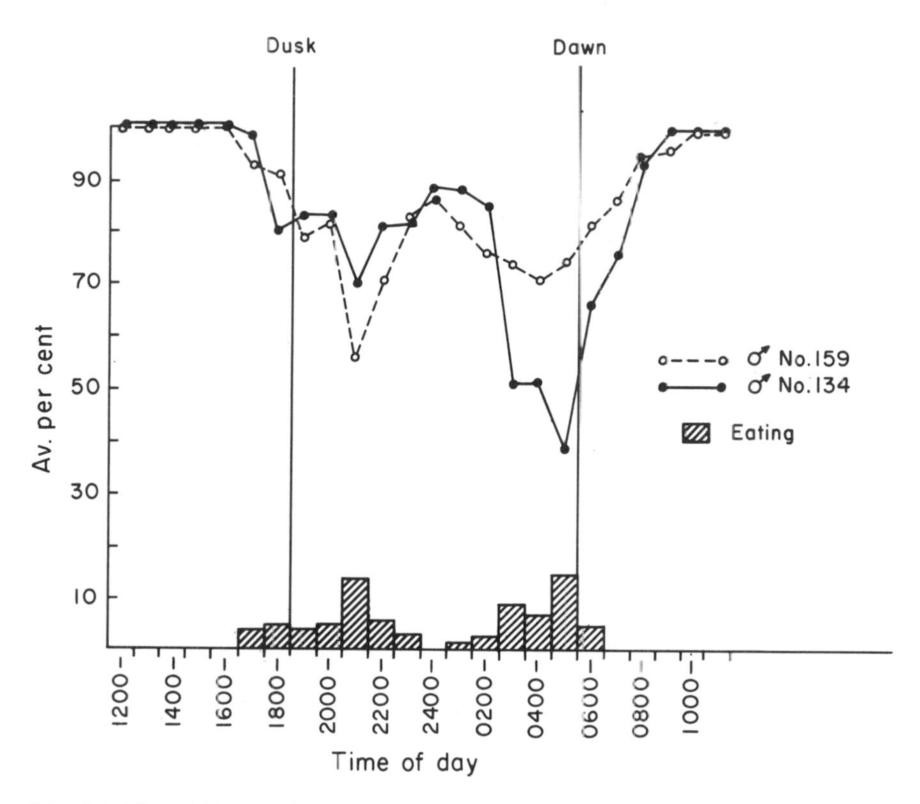

Fig. 24. The daily activity pattern of two nomadic males showing average percentage of each hour that the animals were resting. Data from both males were combined to show the feeding schedule. Male No. 159 was observed for 16 whole days and male No. 134 for 10 whole days.

and 25 illustrate, lions tend to roam more in the early evening before 2200 and toward dawn than at other times of the night. Shortly after dawn the animals often groom and play prior to settling down for the day. When the sun grows warm around 0900 to 1000 they often shift their rest site to a shady spot nearby.

Lions feed whenever they have the opportunity to do so. Most eating is at night, with peaks shortly after dark and before dawn in the case of two nomads (fig. 24), but prey obtained in daytime is also consumed immediately (fig. 25).

On the average, lions spend about two hours a day walking, ranging from a few minutes to 6 hours 15 minutes, which was the longest period noted. About 40 to 50 minutes of the day are spent eating, but, of course,several days may elapse without feeding only to be followed by a night during which 4 to 5 hours may be spent at it. The rest of the day,

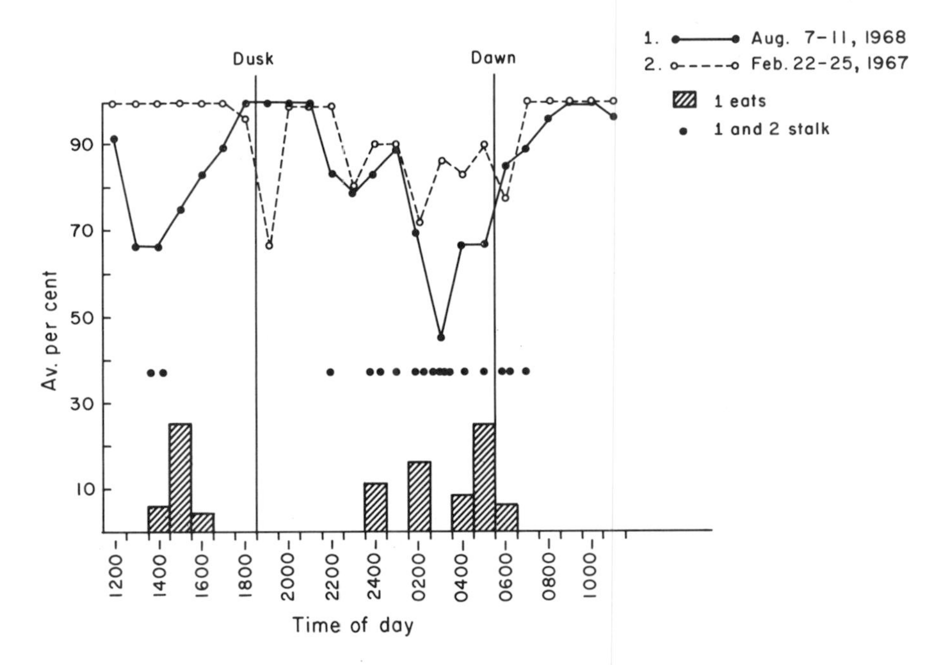

Fig. 25. The daily activity pattern of two pride lionesses observed for 3 and 4 continuous days, respectively. The figure shows the average percentage of each hour the females were resting, the time of day they attempted to catch prey, and the average percentage of each hour one spent feeding.

about 20 to 21 hours on the average, the cats are largely inactive. The activity patterns of nomads and residents are remarkably similar (table 13). During the dry season, when cubs were dying of starvation, lions were no more active than at other times of the year: they were obviously sparing in their expenditure of energy.

Forms of Activity

I discuss feeding, mating, play, and other forms of activity in detail later, but several general forms require further comment here.

Resting. A lion sometimes chooses a prominent resting place such as a rock, a hill side, and even the branch of a tree. At the age of three months our pet cub began to climb into chairs and other high spots and surveyed the activity around him from there. A lion rests on its belly or side, or on its back with hindlegs spread when the sun is hot, with frequent changes of position in the course of a day. One nomadic male, for example, walked from 0830 to 1020, an unusually long period of activity so late in the morning, then reclined on his side. At 1235 he drank at a nearby puddle and continued his rest, this time on his belly. After that he shifted his position frequently—15 minutes on the side, 15 on the belly, 15 on the side, 30 on the belly, 10 on the side, 10 on the belly, and finally 220 on the side. At 1845 he raised his head, roared, then licked his forepaw. During heavy rains, the animals sit hunched or they may crouch.

Lions in a group rest either widely scattered or close together, depending, among other factors, on their need for shade. All may crowd into a patch of shade when it is hot and shift their position throughout the day with the position of the sun (plate 1). A group sometimes splits to use shade provided by two or more trees, and the animals then may rest 50 m or more apart. Adult males often rest alone as far as 100 m from the group although they too recline with the others on occasion. I estimated the distance between bodies of reclining subadult and adult lions, principally those of the Masai pride, during December, 1968. The average distance between lionesses was 4 m, but their bodies were actually touching in 24% of the 144 observations made. The average distance between males and females was 19 m and bodies were touching in only 10% of 41 observations made. Males were 15 m apart on the average but my sample consisted of only 14 observations.

Adult lions are extremely tolerant of the sun. They frequently spend the whole day in the open even though shade is readily available one kilometer or less away; yet on another but similar day they may move into a kopje or beneath a tree. Of 791 lions classified on the plains between January and April, 1968, only 12% rested in a kopje; the others were in the open, 15% of them by a waterhole, 14% by a kill, and the rest merely on the plains. Lions in the sun often rest with open mouth, and their breathing rate tends to be some 4 to 6 times faster than their normal 10 to 15 times per minute. They pant primarily after unusual exertion or a large meal. When gorged, they are obviously uncomfortable in the sun: they may sit or stand rather than recline, saliva drips from their lips, and occasionally one crawls beneath another to take advantage of the spot of shade there. Cubs are less tolerant of sun than adults and seek shade cast by an adult or by a car when older lions seem oblivious to the heat (plate 9). Lions were never seen to rest in water, something tigers often do on hot days.

In cooperation with H. Baldwin the temperature of three lions in the plains was monitored continuously by telemetry. A temperature probe was placed in the fascia beneath the skin on the back of the neck of a lioness. She rested throughout the day in the open except for three brief periods of walking. Her subdermal temperatures varied at least 3°C, being highest when she was in the sun and lowest at night, except after she had walked for 25 minutes (fig. 26). A male lion was monitored for 2½ days, and his temperatures during the second day are also presented in figure 26. He showed similar but less violent fluctuations than the lioness probably because he copulated every few minutes and because the probe may have been placed inadvertently in the muscle rather than the fascia. The many minor fluctuations were not obviously related to environmental temperatures or the activity of the animal.

It is difficult to determine how long a lion actually sleeps each day.

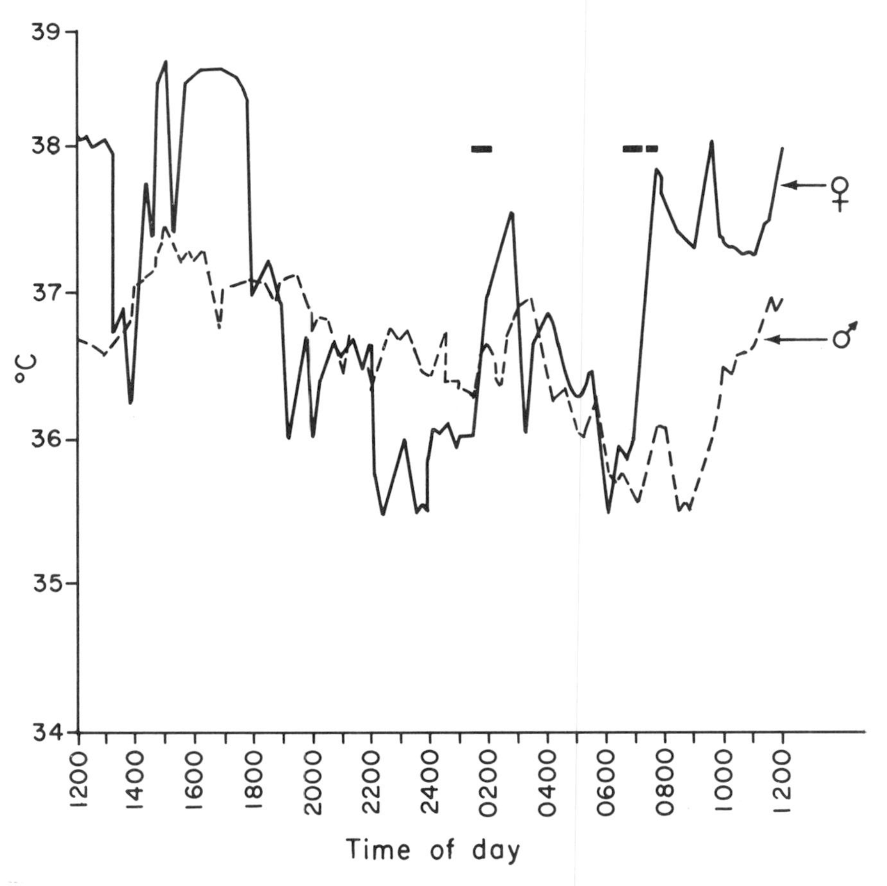

Fig. 26. Continuous subdermal temperature recordings of a male and female during a 24-hour period. The male copulated several times per hour; the female rested the whole day except for three brief walks (indicated by black bars). (The data were summarized by H. Baldwin.)

The animal often lies with its eyes closed, yet the movements of the ears suggest that it is awake. According to Haas (1958), lions in zoos sleep 10 to 15 hours per day. Measurements of brain waves of house cats in soundproof cages indicated that they spend, on the average, 8.4 hours per day awake, 12 hours in light sleep and 3.6 hours in deep sleep (Jouvet, 1967).

Moving. When leaving a rest site or a kill, lionesses often do so in an irregular line or loosely scattered, the cubs trailing behind or occasionally running beside and around the adults, and with the males usually bringing up the rear. If their route follows a trail or road, they proceed in single file sometimes for 1 km or more, an observation at variance with that made by Kühme (1966), who stated that lions never move in this formation. Their normal walking speed is about 3 to 4 km per hour.

Wright (1960) attempted to find out how far lions travel during the night by measuring the distance between consecutive day rest-sites. He obtained an average of slightly over 2 km. This method of deriving data is inadequate because lions often move extensively during the night, and in the morning return to the same rest site as on the previous day (see fig. 28). The actual distances that various animals traveled during 24-hour periods was measured on the car odometer and varied from 0 km to 21.5 km. One nomadic male, No. 134, walked 108.4 km (5.0–21.0 km) in 9 continuous days, an average of 12 km per day (table 14), but he was an exceptionally active animal. A lioness accompanied him during most of his travels. Another male, No. 159, a nomad on a temporary territory, walked about 199.5 km (0–21.5 km) in 21 days, an average of 9.5 km per day, usually in company with another male (table 15). We lost contact with the animal for a total of about 15 hours during this period because the radio did not function well, and the distance he traveled was therefore only known approximately on several days. Resident lions walked similar distances. In a sample of 11 days, the Seronera pride averaged 5 km (1.8–10.0 km); in a sample of 14 days, the Masai pride averaged 4.5 km (1.3–9.0 km). The Masai pride males traveled between 2 and 14.5 km with an average of 6.5 km in the 5 days I observed them. In the semidesert Kalahari Gemsbok Park one group of lions traveled 34 km in a day (F. Eloff, pers. comm.). Sometimes lions moved several kilometers without stopping. One lioness walked steadily for at least 5.6 km when returning to her cubs, and another did so for at least 8 km in a similar situation. Usually, however, the progress of the animals is interrupted by other activities, as illustrated by one group of 7 lionesses and 13 cubs which I followed on June 28 to 29.

1200–1715	All rest.
1715–1745	Much mutual licking after a shower.
1745–1815	Most rest.
1815–1835	Much licking and greeting; cubs try to suckle, and adults urinate and defecate.
1835–1900	Most rest.
1900–2010	The group crosses the Seronera river and follows a thicket that extends into the plains. They stop for 5 minutes at 1950 then continue on, the lionesses scattered, the cubs tagging behind.
2010–2310	All lie without sleeping, listening to distant roars, to reedbuck whistling in alarm.
2310–2400	Five lionesses spread out and slowly move toward a reedbuck, which bolts. They then sit alertly and occasionally move a few meters.
2400–0210	All lie.
0210–0235	They travel slowly.
0235–0300	All lie.
0300–0305	The animals walk 100 m, then look in the direction from which a reedbuck calls.
0305–0610	All lie except for a 5 minute walk at 0515.
0610–0625	Ignoring a giraffe, the group crosses the thicket and enters high grass in the plains.
0625–0650	All rest.
0650–0700	They walk back toward the Seronera River.
0700–0720	When the lionesses spot about 100 Thomson's gazelle, they spread out and encircle the herd but fail to catch any.
0720–1110	All lie.
1110	Two gazelle wander toward the lions in the grass but escape the one that rushes them.
1110–1200	The group rests, having traveled 5.5 km that day.

The daily movements of lions were so erratic that I was unable to predict them. On the whole, the animals were conservative, remaining in an area for days and even weeks, as long as prey was available, and sometimes even longer. Individuals made no journeys that could be interpreted as a search for prey concentrations. Males sometimes roamed widely into terrain not currently used by the lionesses, but such movements seemed to be patrols related to the maintenance of

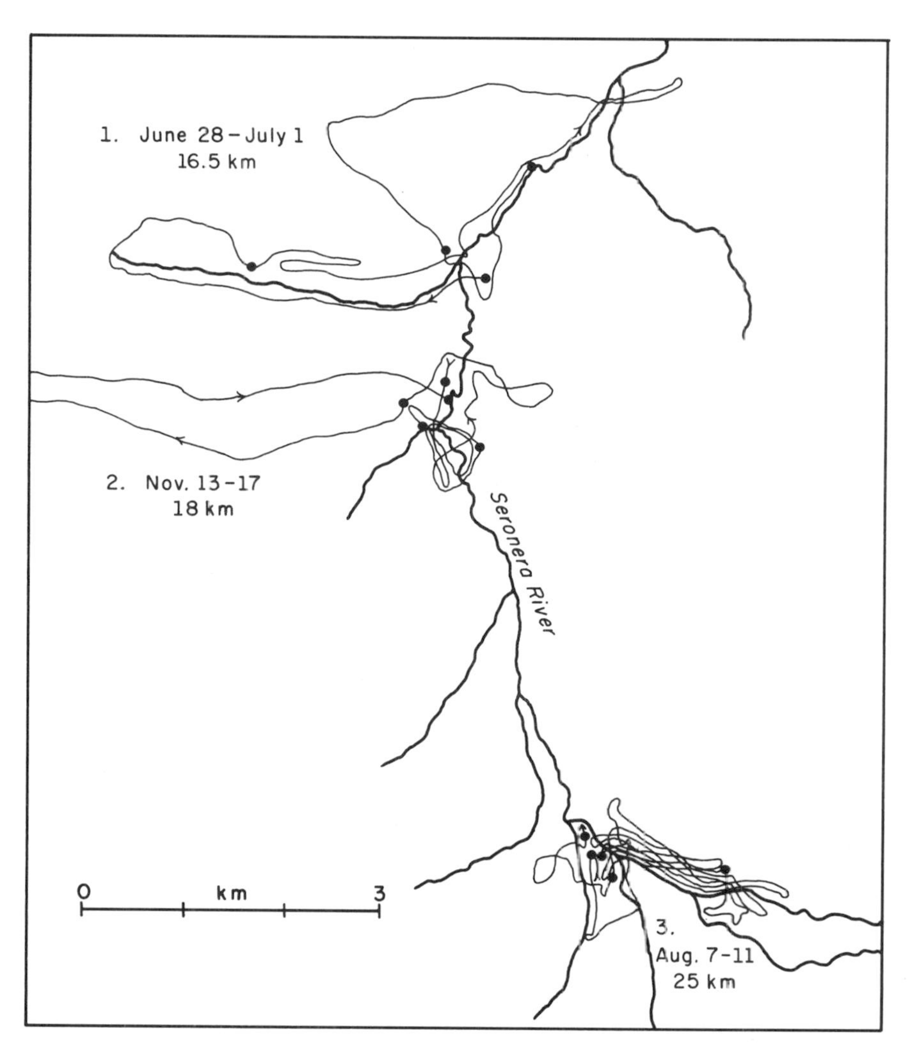

Fig. 27. The daily routes of the Seronera (1 and 2) and Masai (3) prides for 3, 4, and 4 consecutive days, respectively. The black circles represent the main daytime rest sites.

the pride area. During the dry season, lions usually centered their activity near rivers, as is evident from figure 27, which shows the routes taken by two prides during this season. Nomads were less sedentary than residents but their movement patterns were similar. The territorial nomads centered their activity around a particular landmark, as, for instance, male No. 159 at Naabi Hill and male No. 134 at the Gol Kopjes. The latter animal crisscrossed the area around the western kopjes, spending most of his daytime rests in two of them (fig. 28).

Many factors, other than the tendency to remain within the confines of a certain area, determine the direction taken by a lion in its travels. The observations on male No. 134 illustrate some of these. Hyenas on a kill stimulated him to make one eastward loop (see fig. 28); he followed two lionesses to a kill at the most southernly kopje; his large circle to the southwest was apparently in response to hearing another male there; and his thrust to the north was in pursuit of two intruders. Possibly he also avoided areas where he heard or smelled other lions. Movements of residents are strongly influenced by habitat conditions, in that animals prefer to hunt near cover, such as along a river course.

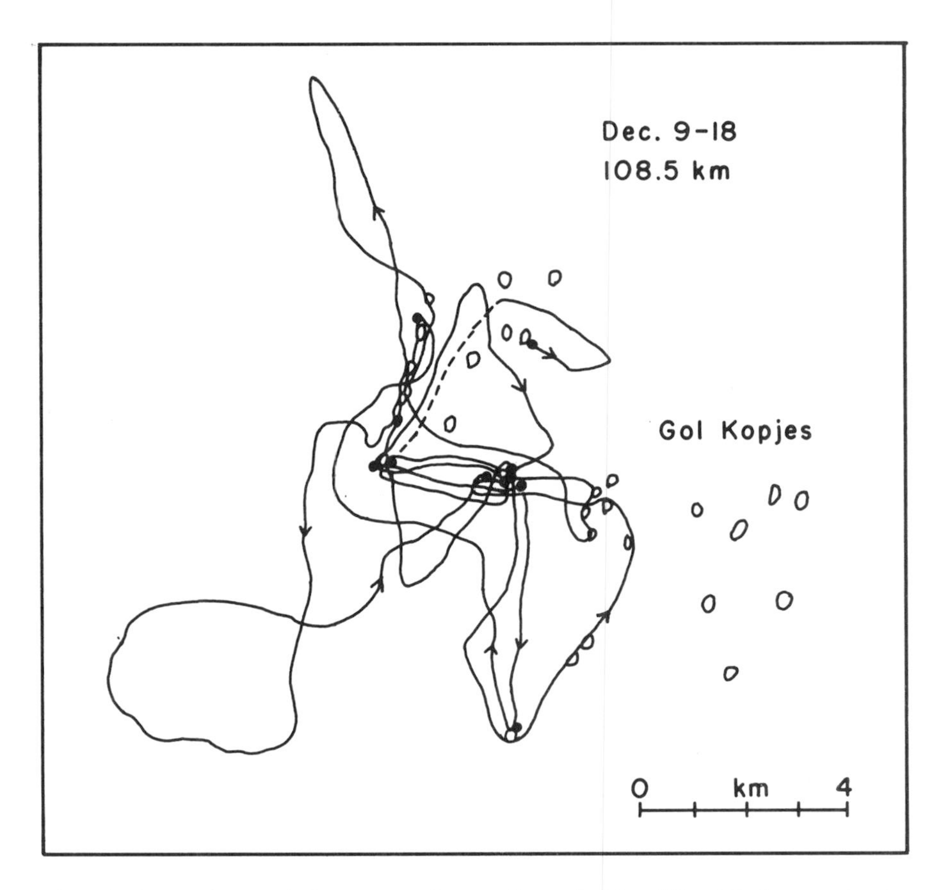

Fig. 28. A continuous 9-day route of nomadic male No. 134 while he was in the process of establishing a territory. The black circles represent his daytime rest sites, the uneven white ones are kopjes.

Tree climbing. The environment of the lion in the Serengeti is essentially two-dimensional with trees far less important to the animal than, for example, to a leopard, which may spend half of its life lying in them. Lions readily jump or clamber into acacia and *Kigelia* trees and there lie on the broad horizontal branches. They are not adept climbers, and their descent appears clumsy as they slide head-first in a shower of bark down the trunk and finally leap to the ground. I observed lions climb one meter or higher into trees on 127 occasions, a total which indicates the climbing only of vertical trunks, not fallen trees. Eighty percent of the animals ascended to a height of 2 to 5 m and none went higher than 7 m. Lions climbed for several reasons, most often in play or to rest (table 16). Females in particular sometimes ascended trees briefly and there looked around as if scanning for prey. Several lions escaped into trees when chased by a buffalo and when I attempted to tag them; Kruuk told me that two lionesses climbed trees when harassed by hyenas. Males ascended trees proportionally less often than did females and cubs.

The lions in the Lake Manyara National Park climbed trees far more often than those in the Serengeti. They were resting in trees on two-thirds of the occasions on which we encountered them during the day, usually in one of their preferred *Acacia tortilis* and *Balanites aegyptiaca* trees (Makacha and Schaller, 1969). The Chemchem pride was seen in trees on 202 occasions, 80% of the instances in 17 different trees even though many others just as suitable were available. One male chose 15 different trees for 93% of his climbing. The reason why Manyara lions rest in trees so often is unknown. Fosbrooke (1963) noted that lions in the Ngorongoro Crater ascended trees during an epidemic of biting flies, but this was an unusual situation not reported for other areas. The vegetation in the various parks is in many respects so similar that no correlation between it and tree-climbing is evident. The Manyara lions sometimes escaped from buffalo and elephant by climbing trees, but there would seem to be no reason for lions to remain in them all day because of the remote chance that they might have to climb one. I think that the behavior represents a habit, one that may have been initiated by, for example, a prolonged fly epidemic, and has since then been transmitted culturally.

Drinking. Various opinions have been expressed about the lion's need for water. "After satisfying their hunger, lions invariably made straight-way for water" (Wells, 1933). Brocklehurst (1931), on the other hand, found that in the Sudan "as there is no water, except in isolated wells of a depth of 30 to 40 feet, they must obtain the moisture, necessary for their existence, from the stomachs of the antelopes which they kill."

The Serengeti lions make no special effort to drink immediately after having eaten, even if their prey is near a river. In 30 whole days of observation, the Masai and Seronera prides ate 25 animals and only once or possibly twice did the lions drink within an hour afterwards. Much of their drinking was casual, a few laps in the course of crossing a stream at night, even though they had not been near water in the daytime. It was difficult to find out how often lions drank because they could easily go to water briefly at night without my being aware of it. During the 30 days of watching the Seronera lions, I was uncertain if the animals drank on 9 of these and rain fell on 5, which possibly affected their behavior because moisture can be obtained by licking the pelage. On 7 days the lions did not drink and on 9 they did so once. One nomad drank twice in a day, once for 6 minutes 10 seconds at 2250, after eating, and again at 0830 hours. Generally lions probably drink at least once every day or two if water is readily available. Male No. 159 was on the plains when water was confined to a few scattered pools. Twice he went without drinking for five days and on the sixth moved several kilometers to the nearest pool (table 15).

Group Organization

Social mammals tend to have some form of group organization which determines the outcome of competitive situations and which helps to maintain cohesion. In this section I discuss social order and leadership.

The Social Order

The structure of societies has usually been analyzed within the conceptual framework of dominance (DeVore, 1965; Mech, 1966; Schaller, 1967). Evidence for the existence of a hierarchical system depends on the outcome of competitive interactions, notably with respect to food, mates, and any other resource in short supply, but the results are often ambiguous, as noted by Gartland (1968). In the lion and other dimorphic species a potential dominance system is built in: in a test of strength any male lion would be superior to a lioness. However, such combat is infrequent and interactions depend less on the behavior of the male in asserting himself than on the response of the female to him. Lions interact directly when rubbing cheeks and licking each other, during agonistic encounters, particularly at kills, and, in the case of males, when a lioness is in estrus. I examine here each of these patterns from the standpoint of the response of individuals to each other.

Rubbing and licking interactions. Among certain primates, such as in *Papio* (Hall and DeVore, 1965), adult males are often approached and groomed by females seemingly as an indication of the males' high status. As mentioned earlier, male lions tend to remain aloof from interactions in the group, except insofar as food is concerned. They seldom lick or rub others, and when females or cubs interact with them the gestures are often hurried as if the animals are uncertain of the response. Females readily rub and lick cubs and vice versa (see tables 8 and 12) without indication that the activities are related to dominance per se. It may be hypothesized that the animal which initiates an interaction with another of the same age and sex is the subordinate one. Tables 17 and 18 show that head rubbing and social licking interactions between the various lionesses of the Masai pride follow no particular pattern: any female will rub and lick any other female. The actual number of interactions have little significance because individuals associated for different lengths of time. The same general pattern was true for males and for cubs. There is thus no evidence that friendly contacts between lions are directly related to a hierarchical system.

Competition for mates. Since several males tend to share a pride, and nomads often travel in pairs, a hierarchy, if present, should become evident in competition for estrous lionesses, especially since other males are usually around such a female (see table 20). The three Masai males acquired the following number of estrous lionesses: Black Mane, 15; Brown Mane, 12; Limp, 11. The differences are not statistically significant. Two observations are of interest in this context:

1. A male and female copulate repeatedly between 0955 and 1200 while a second male, No. 112, lies 100 m away. I tranquilize first the female and a few minutes later the courting male, and tag both. When the female raises her head, male No. 112 approaches, sniffs her anal area, licks her rump, and then both walk 200 m side by side and lie down. The other male revives, sees the female appropriated by the No. 112 male, but merely reclines 100 m from the pair. At 0700 the following morning he is still there.

2. A male lies beside an estrous lioness and 80 m away is a second

male. After I dart her and the drug has taken effect, he nudges her inert form, then remains by her side. I shoot him too and he collapses 10 m away. The second male walks up and sits by the female. A few minutes later the revived male approaches, but the usurper growls and swipes the air once with a paw. When the female walks off, she is followed closely by her new escort while the former one watches.

Brief fights over lionesses in heat may occur, but they are uncommon. In most instances males do not dispute the possession of a female.

Agonistic interactions not at a kill. For the most part lions live amicably together. For example, seven lionesses which we observed continuously for four days interacted aggressively only once when not at a kill. Fights as a rule are short—a slap or two accompanied by much vocalizing and baring of teeth—and biting is infrequent. In spite of their smaller size, lionesses not only retaliate when males harass them but actually initiate most aggression—frequently with impunity. When a male approached a lioness with small cubs, she rushed him and he trotted off; when a male sniffed the anal area of a female, she clouted him on the head and he moved back 5 m; when a male came near a lioness, she hit his face without obvious reason and he merely shook his head and scraped. During an actual chase, the male is usually in pursuit of the female, but as he draws close she whirls around and hits him once or twice, thereby discouraging him from further action. Several lionesses may attack a male and put him to flight, a cooperative action also observed in baboons (Hall and DeVore, 1965). Reciprocal aggression is common between lionesses and cubs, particularly when the latter want to suckle (plate 10). A bared-teeth face and a swat by a lioness may elicit the same reaction in a cub. Fighting within an age or sex class follows the same general pattern, although some individuals seldom retaliate. In the Seronera pride, the two peripheral lionesses, J and K, usually fled when approached by others, behavior which was related to their tenuous status as pride members rather than to a specific hierarchical system.

Agonistic interactions at kills. A kill has a most disruptive influence on lion society. The lions seem to become asocial, as each animal bolts its meat while snarling and slapping at any group member that seems to threaten its share. When the kill is small, such as a Thomson's gazelle, the males often appropriate the carcass from a female even though she may object vigorously. Several examples illustrate such behavior.

1. A lioness pounces on a gazelle and a male appears almost immediately. He stops 10 m from her, then rushes. She flees 10 m, dragging the gazelle, but he grabs it and they lie side by side, each holding the carcass and growling. After 5 minutes the female suddenly jerks the gazelle from him and runs 50 m with it, the male close behind her. She drops it and he eats.

2. A female catches a gazelle and runs with it across a shallow creek and along a thicket for 200 m, closely pursued by three lionesses. She retreats under a bush and growls while facing the others. They hesitate. Suddenly a male bounds up, crashes into the bush, and attempts to take the carcass. The female retains her hold on it, and for 15 minutes—growling, pushing, and pulling—the two crouch side by side without eating. Suddenly the body rips in half and each obtains a share.

3. Three lionesses and five cubs feed on a zebra foal at 0835 when a male appears. One lioness charges him and the others, too, assume a threatening attitude. At that, the male lies 10 m from the kill; when he rises two minutes later, another lioness rushes at him. He roars in response but reclines once more. At 0955 he moves to within 6 m of the kill, and a lioness there hits him over the head. He roars again. Suddenly at 1005 he rushes to the remains and takes them over, permitting only one cub to eat with him.

The extent to which males parasitize lionesses during the dry season, when gazelle are often the only available food, is shown by the Seronera and Masai pride males. These males ate gazelle 60 times when I knew the source of their meat. In 12% of the instances the males had killed their own prey, in 12% they had taken it from another species of predator, and in 76% they had appropriated it from a lioness. On the few occasions when a male killed for himself, he shared with no one except once when another male forcefully took a portion. I saw only one instance in which a lioness took a whole kill from a male, and she was in estrus. The male watched her eat for 1¼ hours and mated with her 4 minutes after she finished. Males and females share large kills but the former may take over the remains.

The relationship between males and cubs with respect to food is complex. A male sometimes takes a piece of meat from a cub or hits it while they are crowded around a kill. On the other hand he frequently permits cubs, but not lionesses, to share the remains.

Lionesses commonly take meat from cubs. Once two small cubs each snatched a gazelle leg and fled, but two lionesses rushed up, bowled them over, and grabbed the food. On another occasion, a starving cub trotted to a lioness which had taken the remnants of a gazelle from a cheetah. She hit the cub several times, but it crawled back after each blow and finally took a bite. The lioness then sat with her full weight on the cub, tore the piece away, and retreated into a thicket with it. Cubs, in turn, snarl and slap at any female which tries to take meat from them, and, if the portion is small, they may defend it successfully. I quantified the aggressive interactions among females and small cubs at several kills for a total of 14 female-hours and 30 cub-hours of observation. Cubs outnumbered females by a ratio of about 2:1. Of the 259 aggressive acts noted, 25% consisted of jerking the head in the bared-teeth face at another lion, and the rest involved slaps at the head and neck. In addition to growling and snarling almost continuously, each animal behaved aggressively on the average of 6 times an hour:

Cub aggressive at female	4%
Cub aggressive at cub	55%
Female aggressive at female	6%
Female aggressive at cub	35%

These percentages show that cubs were aggressive toward females far less often than toward other cubs and that females attacked cubs proportionally about three times more often than other females. It appears that lions tend to avoid aggravating those lions which can retaliate most effectively.

If a male has a large piece of meat, such as half a wildebeest, he usually permits his male companions to share it with him, though reluctantly. For example, while one male fed, another approached casually and reclined 2 m away. Seven minutes later he suddenly lunged for the carcass and both ate. Twenty-five minutes later a lioness tried to join but was slapped. A second lioness then crept to the carcass and surreptitiously tried to pull a bone away, but a male immediately took it from her. Similarly, a male ate while four lionesses and several cubs waited. Another male arrived and joined the male on the kill. Although he was swatted three times, he remained and ate. The three Masai pride males were present at the same large kill on 36 occasions. On 33 of these all fed together, and on 3 of these Limp failed to join, at least once because he was already full. When a male obtains a small portion of meat, others in the vicinity usually do not dispute it with him. A male of the Masai pride took most or all of a gazelle from a lioness 14 times when a second male was nearby, yet only once did the other attempt to take a piece. Black Mane had the meat to himself 3 times, Brown Mane 4 times, Limp 6 times, and once the first two males tore a carcass in half.

In one typical situation Limp took a gazelle from a cheetah. Five minutes later Black Mane arrived but lay down 22 m away when Limp growled. Ten minutes after that Brown Mane showed up and first reclined 5 m from the kill, then retreated to 10 m when Limp growled and lunged. He waited there one hour, until Limp left the site, before picking over the scraps. Black Mane obtained nothing.

Once a nomadic male had a portion of a wildebeest which he dropped and relinquished immediately when his larger companion walked toward him, but such a response was rare. If, however, a male does not have complete possession of a piece of meat, his companion may try to take it. On one occasion, for example, two males chased a female which had just killed a gazelle. One male took it from her, and the other male then grabbed the carcass too and they tore it into two pieces. At such times they may lie side by side, growling and pulling as they hold the kill without eating for as long as 15 minutes, until finally it tears apart.

Lionesses share meat from a large kill, but, like the males, they do so unwillingly. The one that arrives late is usually greeted with a flurry of slaps as she tries to crowd in. Small kills are often disputed even if a lioness has full possession of it; in general, females fight each other more readily than males. In one typical instance, a lioness had killed a gazelle when three others and four cubs arrived. All attacked immediately. Two lionesses, including the one that caught the prey, each emerged from the melee with over a third of a gazelle, a cub obtained a leg, and the rest got scraps only. After that, those with small portions were not molested by others.

Cubs compete vigorously for milk and meat without evidence of priority; they vacate positions at a nipple or kill only when forcefully removed by a stronger rival.

The various observations indicate that lions do not have a hierarchical system in the sense that some animals are clearly dominant and others accept their subordinate position in competitive situations without dispute. Similarly, Leyhausen (1960) noted: "Being solitary animals, cats do not establish rigid social orders if kept by numbers in a fairly large room." There is a tendency for the larger of two animals to win an encounter, such as a fight over a piece of meat, primarily because it is stronger. This fact is recognized but not necessarily accepted by the others. Consequently, the weaker animal may retaliate; it fights for its rights, so to speak, and this too affects the interactions, for it often makes the stronger animal reluctant to attack (plate 11). Each lion in a group knows and responds to the fighting potential of every other member. It is a system based on the amount of damage each animal can inflict on an aggressor; it is a system based on a balance of power, something not unknown in human affairs. A tense peace is thereby maintained and is broken only by sporadic clashes which though noisy and intimidating tend to do little damage. Other factors also decrease strife. When one lioness attacks a male, other lionesses may also enter the fray against him, and even powerful males are unable to withstand such a concerted effort. Lions will avoid conflict, especially when an animal has assumed a defensive posture. As noted by Schenkel (1966b), lions have few inhibitions with respect to slashing and biting; the best way for a lion to avoid serious injury is to stay away from an opponent's teeth. The claws may do damage but the loosely attached skin and, in the case of males, the mane help to decrease the chance of injury. Although cubs may be slapped and bitten hard, adults tend to treat them more gently than they do other adults.

Leadership

There is no consistent leader in a lion pride. When most or all members are together an adult female usually leads the procession, and in the course of a kilometer several different ones may do so. Subadults infrequently are in the lead in such a situation and cubs characteristically follow. Males may move with the group at times, but more often they tag behind the rest. Lionesses and cubs respond little to the movements of males, which usually come and go without being followed. One or more lionesses frequently split from the group without concern from the others; in fact, constant joining and parting of members is typical of prides, and following behavior often depends on such special social relationships as two or three animals being companions and cubs remaining with their mothers. The size of the available prey also influences the readiness with which lions follow each other. When large prey is being hunted the animals tend to form larger groups than when they have only gazelle to eat. In general, cubs, young subadults, and males orient much of their activity with respect to the movements of adult lionesses, any one of which may initiate activity.

Some small nomadic groups have more or less definite leaders. Male No. 159 characteristically followed his companion, and female No. 78 led her usual entourage of two or three males. On the other hand, when male No. 134 and female No. 60 traveled together, neither led consistently except that the male always went ahead when there was food in the offing.

A rigid leadership system would be of little advantage in a society such as the lion's, and, in fact, would prohibit the society from adjusting to varying prey conditions and would create a lack of initiative during the hunt.

MATING BEHAVIOR

Although sexual behavior of house cats has been well described (Leyhausen, 1960) few detailed data are available for lions (see Cooper, 1942; Kühme, 1966). Sexual activity in these cats is conspicuous but it is almost wholly confined to adults; cubs and young subadults rarely display it, not even in play, and I saw only three instances of it in lions less than 3 years old. One 17-month-old male mounted a 12-month-old female while biting her neck; another male, 22 months old, bit a lioness of the same age in the nape while mounting. When one subadult lioness walked in a wobbling, semicrouched position after I darted her, and just before she collapsed from the effects of the drug, a 26-month-old male mounted her, probably because her posture resembled sexual presentation.

Homosexual Behavior

I saw homosexual behavior on three occasions, all among the males of the Masai pride. Once Brown Mane circled Limp in a semicrouch, as if presenting sexually, and the latter then rolled on his back, his penis unsheathed. On another occasion, the three males rubbed cheeks, rolled on each other, and Brown Mane mounted one of his companions briefly. Rubbing and rolling also preceded the third occasion; after that Brown Mane and then Limp mounted Black Mane, who in turn briefly mounted one of the others.

Heterosexual behavior

A male sometimes shows sexual interest toward a female by sniffing her vulva and attempting to mount, but if she is not receptive his advances are either ignored or rebuffed with a growl or slap. One lioness seemed oblivious to a male which mounted her three times as she lay on her side, but when he tried to mount a fourth time she snarled at him.

An estrous lioness is restless, lying down for a few minutes only to jump up and walk rapidly some 50 m and there roll over and twist on her back, behavior also observed in the house cat (Leyhausen, 1960). The accompanying male remains near her, moves, when she does, sniffs her vulva and grimaces, occasionally licks her nape, back, or rump, something males rarely do at other times, and attempts to mount. At first the lioness may not accept him, even if she has presented herself to him, and his advances are received with a swat which may or may not be returned. At other times she playfully avoids him, as the behavior of an elderly lioness during a period of 25 minutes illustrates.

> She trots to the male and turns abruptly in front of him, walking away with her tail raised. She does this three times in succession, rolls over twice, and finally crouches by his head with her rump slightly raised. But when he steps toward her, she rolls on her back, then walks away. He follows. She bounds back to him, circles him closely twice, runs a little ahead and lies. At his approach she races off, but he merely reclines. She returns, lies facing him, noses almost touching, then rushes away. He rises and she comes back, touches his nose with hers, then presents sexually. But before he can react, she bounds 20 m, rolls on her back, and then abruptly rakes her claws on a tree trunk. Running back to him again, she circles him twice and presents. When he noses her vulva, she leaps ahead, growling. He pursues and mounts but she snarls at him while twisting aside. Finally both lie side by side.

A male usually stays within a meter of a lioness in full estrus. He trails her when she walks, his head almost touching her rump, or he marches parallel and less than 3 m from her, holding himself erect. She walks rapidly with tail looped high and lashes it occasionally. He sometimes tries to stop her by stepping in front of her, but she merely walks around him. Either the male or the female initiates copulation. Initiations by females outnumbered those by males by a ratio of about 3:2 (table 19).

A male characteristically approaches the female and stands by her rump, or he licks her or nudges her gently with his nose. In either case the lioness crouches, her forelegs flat on the ground from elbow to paw, and her rump slightly raised with tail turned to one side. Occasionally she first walks 10 to 20 m or circles him, greets him, or slaps him lightly. Her rump is typically oriented toward his head, but at times she crouches at right angles to him. He either shifts his position or paws at her hindquarters until she moves. When the lioness initiates the mating, she often rubs cheeks and sinuously moves her body against his as she circles him with tail raised almost vertically; at other times she simply crouches in front of him or trots in a semicrouch around him.

The male squats over the female, his forelegs by her side and his hindlegs by her rump, and thrusts at the rate of about 3 times in 2 seconds for a total of 10 to 30 and more (plate 12). House cats typically bite into and hold the nape of the female throughout copulation (Leyhausen, 1960). As Ewer (1968) pointed out, the young of many species become immobile when held by the scruff—useful behavior when being carried by the mother—and it is probable that the neck bite in mating cats induces temporary passiveness in the female. However, the neck bite in lions has become more ritualized than in the house cat. The male may ignore the nape or lick it; sometimes he only gapes widely over it, and at other times his quivering jaws touch the neck of the lioness. One or more actual bites, strong enough to fold the skin, were delivered in 57% of the copulations observed (table 19). Such bites are light and brief although on rare occasions the neck may be bloodied. Some males mouth the neck within a few seconds after mounting but most bite it only when close to or during what appears to be the physiological climax, a time when the rump and thigh muscles flex. There may be individual variations between males in their tendency to grip the neck. One male ignored the nape in 14 out of 17 consecutive copulations observed, another bit or touched it during all 14 copulations seen. In 6% of the copulations the male licked the neck before or after biting it.

While a male is mounted, the lioness emits a deep rolling growl almost continuously, which together with her facial expression represents a gesture of aggressive threat. The male usually makes no sound at the beginning but then gives one or more yippy miaows with teeth bared which merge into loud, harsh, prolonged *iaauuu* as he reaches the climax (fig. 18). The miaow is a call of light distress, but in this context it possibly serves to decrease the aggressive tendencies of the females by evoking other responses, although these are not evident. Occasionally the male growls and the female miaows once or twice.

Contact after the climax terminates abruptly. As the male miaows, the growls of the female turn into a snarl; in half of the copulations the male dismounts at this point (table 19). In the other half he jumps aside, sometimes with a noise that seems intermediate between a roar and a snarl, as the lioness twists her head around with an explosive snarl, her paws sometimes raised as if to strike (plate 12). In 4% of the copulations she wheeled around and slapped the male. The duration of copulations varied from 8 to 68 seconds (av. 21 sec.) in 93 that I timed. After the male dismounted, the lioness rolled on her side or back in about 75% of the instances observed. At other times she either rested on her abdomen or walked to another site. The male, too, sometimes rolled on his back.

Lionesses in full estrus copulated frequently throughout the day with the time between matings varying from one to 108 minutes in 198 intervals timed (excluding the male mentioned below). The average number of minutes between matings was 12, but individual couples varied considerably from this figure with the result depending on the stage of the female's estrus and the amount of initiative by the male. The intervals of one couple averaged 7 minutes in 14 matings observed, those of another couple 35 minutes in 8 matings observed.

H. and D. Baldwin and I observed one courting nomadic male continuously for 2½ days. When we found him at 1025 on December 21, 1967, he was with four lionesses, one of which presented to him but did not permit him to mount. We placed a radio on the male at 1140. After that he walked 20 minutes and lay down within 40 m of another male, his usual companion, who was courting with another female. He remained by the courting pair until 2005, then returned to the other lionesses and mated repeatedly with the estrous one. At 0730, December 22, a second female in estrus presented to him, and he mated alternately with both of them until 1510. After that he ignored the second female even though she presented. Earlier, at 1145, the other courting couple had joined the group, and this male too spurned that female when she presented to him. The male with the radio and the one lioness continued to mate at intervals throughout the day. At the end of the first 24 hours the male had done so 86 times, 74 times with one female, 12 with the other. During the second 24-hour period he mated 62 times with one lioness. He then copulated 9 more times. At 0312, December 24, he left the female abruptly, joined her again at 0325 but left on a run at 0345. We lost sight of him at 0400, and when I found him again at 0740 he was not beside the female. Thus in 55 hours he copulated 157 times. The average interval between copulations was 21 minutes (1–110 min). He copulated at all hours, except between 1100 and 1200, with a slight peak in activity between 1200 and 1700. He did not eat during the three days, even though the other lionesses had a wildebeest within 100 m of him.

A male accompanied by two estrous lionesses was also noted on three other occasions. In two hours of observation one nomad, for example, mated 14 times with one female and once with another. Yet he seemed to prefer the latter. He followed her and stayed by her side and only when importuned by the other did he copulate with her, sometimes reluctantly. Once she circled him and crouched in front of him 3 times in succession and elicited only a growl.

While courting lions seemed mostly concerned with the business at hand, they also indulged in other activities. One male caught a gazelle and permitted the lioness to eat it. On another occasion, a lioness spotted some gazelle in the high grass nearby. She crouched and began to stalk, but the male misinterpreted the posture and mounted her. Suddenly she rushed out from under him and chased the gazelle unsuccessfully. Courting couples joined other pride members on kills on several occasions.

The tendencies to retreat, attack, and mate are evident in the sexual behavior of the lion, just as they are in the courtship of most vertebrates. At first almost wholly aggressive, judging by her facial expressions and vocalizations, the female becomes defensive toward the end and also shows an inclination to move away from the male. The responses of the male are ambivalent, but the components of defensive threat generally predominate over those of aggressive threat.

The Social Milieu of Courting Lions

Because courting couples travel little and seldom hunt, other pride members usually drift away from them with the result that they remain alone about half of the time (table 20). In this table the lionesses which were definitely in estrus are analyzed separately from those which did not copulate while I was watching them. False heat and other factors could have skewed the data but, as is evident, they are similar for the two categories. A male often waits some 30 to 200 m away while his companion copulates. About 20 to 25% of the courting males had one other male near them and 25% had several pride members, including a male, within 200 m.

Males are said to fight vigorously over estrous lionesses (see Guggisberg, 1961), but in my experience they seldom did so. Courting males are often irrascible, lunging at any male that approaches to within about 5 m and sometimes even at a female, but such intrusions are generally avoided. Once two males walked side by side then reared up and clawed each other before one of them returned to an estrous lioness. On another occasion, I saw a lioness with one nomadic male on one day and with another the following day. Both males had recent cuts on their bodies possibly as a result of having fought over the female. In most instances, however, the extra males waited near the courting couple, on several occasions for at least two days, without making an attempt to appropriate the lioness. If a male leaves the side of a lioness on his own volition his companion may then fill the void. It is also possible that in some instances the extra male is waiting to rejoin his companion rather than to mate.

An estrous lioness may accept several males in succession if she is not attended closely. One female in the Magadi pride mated with three different males in 12 days. One female of the Seronera pride mated with a male of the Kamarishe pride in the evening but by morning she had switched to a Masai pride male; another female of that pride mated with three Masai pride males and a Kamarishe pride male in the course of eight days. Four nomadic companions courted with four estrous lionesses—females nos. a, b, c, d—over a period of three days as follows:

	Male			
	I	II	III	IV
April 2	a	c	—	—
April 3	b	d	a	c
April 4	b	d	a	—

The choice of mate is often determined by the lioness, as illustrated by one example. Male No. 18 courted with the same lioness for four days. On the fifth day she was with his companion, male No. 21, while No. 18 was in a kopje some 50 m away. Suddenly she left this male and entered the kopje, shortly thereafter reappearing with male No. 18. She grunted softly, circled him, and presented. He fled when male No. 21 approached at a trot, but the female followed her first partner and rested by his side, disdaining male No. 21.

Other pride members ignore courting couples even if these mate within the group, in strong contrast to captive wolves where the dominant female "will violently attack any less dominant female that allows or entices a male to mount her" (Woolpy, 1968). In one instance, two courting couples were within 10 m of each other. Estrous lionesses show no possessiveness toward their male. Once a lioness approached a courting couple and presented, and the male copulated with her while the other female rested beside him.

Relations of Young to Pride

The response of a lioness to her cubs is so finely balanced between care and neglect, between her own desires and the needs of her offspring, that the survival of the young ones is threatened whenever conditions are not at the optimum. As an intensely social creature, a lioness prefers to be with other pride members rather than separated from them with her small cubs, and as a voracious eater she bolts any meat herself rather than share it. To describe the way in which lions adjust to these conflicts is the main purpose of this section. Most comments refer to lionesses and cubs; males have little direct influence on the raising of young although their presence provides indirect benefits such as protection. Cubs sometimes seek out a male and sit by him, rub him, or attempt to play with him, but such advances tend to be rebuffed or treated with indifference.

The First 2 to 3 Months

Lionesses usually give birth and keep their cubs in thickets or kopjes where it is not possible to observe them. The Masai and Seronera prides commonly used the kopjes and dense stands of young acacias that bordered the Seronera River as places in which to hide their cubs. One lioness of the Cub Valley group had cubs for two successive years in a grassy hole about 1½ m deep and 2 m wide (plate 13); another nomadic one had her cubs in a patch of reeds one meter high.

Cubs are born almost helpless, with eyes closed and weighing about 1.2 to 2.1 kg. Their eyes generally open between the third and fifteenth day after birth and the first incisors erupt on about the twentieth day (Crandall, 1964; Carvalho, 1968). Although able to crawl within a day after birth, cubs cannot walk well until the age of about 3 weeks. Females sometimes shift small cubs to another site, one lioness doing so five times in a month (G. Adamson, 1968). Lionesses usually carry cubs by picking them up with their teeth by the neck, or in the case of very small cubs by the back and shoulders. The cubs become passive when held in this fashion, hanging loosely with hindlegs drawn up. One lioness in the plains moved her litter of three cubs at least three times during the first two weeks of their life. Root watched her once as she carried a cub to a new place, fetched the second, but seemingly forgot the third. Five minutes later she moved it too, then made a fourth trip and checked the site as if making certain that none had been forgotten. Cubs older than 6–7 weeks were not carried; on three occasions when I saw a lioness attempt to do so they rolled over and squirmed until she desisted.

A lioness with newborn cubs is often separated from the rest of the pride not because she withdraws but because the others continue their usual routine while she has to tend her offspring. However, she maintains contact with others in several ways. Sometimes two or three lionesses have cubs at about the same time and may become companions which care for all cubs jointly. Lionesses also seek out other pride members and spend the day with them; up to twenty-four or more hours may then elapse between visits to the cubs. One morning I met two Masai pride females as they returned to their cubs hidden in a kopje. They stayed with them all day but after dusk returned to the pride. The next day, at 1125, they were still with the others 10.5 km away. I suspect that cubs are often abandoned in such situations by females unwilling to leave the conviviality of the group. When several lionesses have cubs in one area, that locality may become the focal point of the whole pride, a place to which members come to meet others.

A lioness makes little attempt to keep her newborn young isolated from contact with the rest of the pride. Lionesses without cubs readily enter thickets in which small young are lying, but I was unable to see what happened there. The tame lioness Elsa withdrew from her human foster parents and became highly secretive before and after the birth of her first litter (J. Adamson, 1960). Some Serengeti lionesses eluded me when I attempted to follow them to the cubs' hiding place, but they were not observed to behave like that toward other lionesses. The two males of the Cub Valley group usually rested within 50 m of several small cubs in a hollow; once, however, Root saw the mother lunge at a male that approached the young too closely. One lioness and her two cubs, only two weeks old, spent the day with the pride, the tiny young crawling among the massive adults, and that evening she carried them 30 m into a bush and left them there while she went hunting with the others. One night a lioness sat by a thicket in which her cub, about 12 days old, was hidden. When a male arrived, she snarled at him but made no effort to deter him from investigating the thicket. This he did and rejoined her two minutes later. While intruders into a pride area may kill cubs they find there, as described earlier and as was also noted by G. Adamson (1968), there was no evidence that resident males harm cubs. On one occasion an elderly lioness of the Seronera pride carried her three-week-old cub to the remains of a wildebeest kill and abandoned it there. It was so weak from starvation that it was unable to suckle, apparently because its mother had no milk. Two Masai pride males approached, attracted by the vultures in the tree above the kill.

One male found the cub when it bleated and gently picked it up by the nape but then dropped it. He lifted it once more, carried it 2 m, placed it on the ground, and after nuzzling it briefly walked off.

Although litters are highly vulnerable, their chances of being found by predators during the many hours that they are left unprotected are reduced not only because lionesses usually leave them where they can hide but also because they remain fairly immobile. When disturbed, they crawl into rock clefts, under grass tufts and similar places where they are difficult to find—for me at least. Very small cubs sometimes miaow, thereby making themselves conspicuous. Cubs hide themselves at the first intimation of danger, even when their mother is present and at no signal from her. When, for example, nine cubs, 2 months old, heard a plane fly over, they dashed into a kopje while three lionesses ignored the sound.

Occasionally a lioness is careless with her cubs, and in this context the fate of one litter is of interest. At some time during the night an elderly lioness of the Seronera pride killed a wildebeest. She left it to fetch her cubs even though they were only a week old, rather atypical behavior. While she was gone a female leopard scavenged on the meat. N. Myers arrived on the scene at 0930 and found the lioness with one cub and the leopard in the tree above them. Shortly before 1100, she started to leave, presumably to bring her other cub to the kill. But she was obviously nervous about the leopard: she headed away a few meters, then looked back and returned to the cub, a hesitation that continued for half an hour. Finally she left. When she was 100 m away, the leopard descended, grabbed the cub and fled with it. Coughing, the lioness ran up and the leopard dropped the cub, but it was dead, bitten deeply into the chest. She licked it for 15 minutes, pawed it, then carried it back to the kill. Around noon she left, and the leopard returned and ate, ignoring the dead cub.

Unaware of these events, I had been waiting 1.3 km away by the other cub since 0620 that morning. The cub had been lying on the road when I found it, but I had placed it in the grass 2 m away to prevent its being run over by cars. At 1300 the lioness arrived and grunted loudly when 100 m from the cub; she stopped, looked, obviously uncertain of the precise spot at which she had left her young. It miaowed at intervals, just as it had done all morning, and she finally located it. She carried it to the kill, stopping occasionally to readjust her grip. Spending the rest of the day and most of the night beneath the tree, she seemed oblivious to the dead cub beside her as well as to the leopard

5 m above her. The leopard fled at 0430. At about 0700 two Masai pride males joined her. One of them must have rolled on the living cub inadvertently for it was dead at 0800 and an autopsy revealed that it had been squashed.

When a lioness returns to the area in which she has left her young, she grunts or roars softly and the cubs run to her. While lions possess a vocal signal that means "come" they lack one that signifies "stay," except for the aggressive growl. Before leaving them, a lioness sometimes leads her cubs into a thicket then departs hurriedly. A rapid, silent movement away from cubs deters following, in contrast to a slow movement, which elicits it especially if reinforced with grunts.

"Three and a half months after a lioness has left the pride with one of the males on her nuptial excursion she will be ready to leave it again to have her cubs. She will take with her one of the older lionesses, too old to bear any more cubs of her own. . . . This lioness will help her to hunt and protect her cubs and act as nursemaid when they are older. These attendant lionesses are well known to all bush-whackers and are referred to as 'Aunties'" (Carr, 1962). J. Adamson (1960), Wright (1960), Denis (1964), Cowie (1966) and numerous other authors mentioned "aunties" as part of the lion's social life, but I found little evidence in support of their existence. On one occasion, an old female without cubs accompanied a lactating one over 3 km to where her cubs were hidden. Another time a childless adult remained all day with several litters in a kopje while the mothers were away. Cubs are usually left unguarded, but lionesses other than the mother may rest near them, possibly because it is a good place to meet other pride members. Such behavior resembled that ascribed to "aunties," but I knew that it was not repeated on subsequent days. In addition, some lionesses in the pride become companions. If one of these lionesses has a litter, the companionship may break up, as was the case with females J and K of the Seronera pride, but also it may persist and one lioness then fits the description of an "auntie." When females L and Q of the Masai pride had litters they became companions, remaining together even after the former lioness lost her cubs. Of course, any lioness near cubs, no matter for what reason, may function as a guard with obvious benefit to the young. Similarly, the activity of several lionesses in the vicinity of a litter increases the chances that one will capture prey. But I never saw such lionesses lead cubs to a kill nor did they carry food to them; any benefit childless lionesses provided was accidental.

During the first two months of life cubs subsist on milk which is gradually supplemented with meat. Quoting other sources, Wright (1960) wrote: "Their diet was entirely milk for the first two months. They were weaned on meat regurgitated by the female, and passed on to lumps of fresh meat brought to them from a kill." I never saw a lioness regurgitate meat for cubs nor was this observed by J. Adamson (1960). Small lumps of meat are not taken to cubs and only rarely is a gazelle or similar large piece carried to them. I examined several lairs without finding prey remains. Most cubs probably have no meat until the lioness leads them to the kill. The youngest cubs I observed feeding on meat were about 5 weeks old. Since incisors erupt at 3 weeks and canines at 4, it seems unlikely that cubs feed on meat before that age. Our pet cub ignored meat at 3½ and 4 weeks of age. Tested again at 5 weeks, he clutched the remains of a gazelle, licked it and ate from it, an abrupt and striking transition which suggested that a definite maturational change had occurred.

Full integration of cubs into the pride is a gradual process. Some litters may meet most members before they can walk and others may not do so until at the age of 5 to 6 weeks they begin to accompany their mother. Young are not seen regularly with the pride until they are mobile enough to keep up with the moving group at the age of about 8 weeks. Also at about that age, lionesses may combine their litters if they have not already done so. In December, 1968, five lionesses of the Seronera pride brought their litters together. The litters ranged in age from 6 weeks to 4 months and formed a group of 13 cubs. One lioness of this pride introduced her two cubs to the others when they were 6 to 7 weeks old; judging by their behavior it was the first time they had seen lions other than their mother:

> At 0945 the lioness and her offspring approach six cubs, all about 4 months old, lying together beneath a tree. As these run up, she growls and her two cubs flee, one of them 20 m, the other 50 m into a thicket where it remains alone most of the day. The lioness and her young then recline 4 m from the six cubs. These are obviously curious about the newcomer, craning their necks and attempting to inch closer. But it coughs and growls when they approach to within one meter. When the lioness moves 3 m away, two of the large cubs walk over to the small one. It rolls on its back, swatting and snarling; its mother jumps up and in the head-low posture coughs several times at the others. At 1105 the small cub leaves its mother's side and cautiously approaches another cub to within .3 m before dashing back to her and rubbing cheeks. At 1400 the cub still hisses and swats at any cub that attempts

contact, but 45 minutes later it hesitantly touches noses with one. At 1625 it lies flank to flank while suckling with another cub. The other small cub finally ventures from its hiding place at 1725 and suckles beside its sibling. One hour later two large cubs playfully bowl over a small one without eliciting an attack, and then a large cub suckles peacefully beside the two small ones, the task of integration well advanced. That night the lioness hid her cubs again and soon after that they disappeared.

Cubs with the Pride

Even after cubs are fully mobile they spend much of the day hidden or, as they grow older, just lying in the open until their mother returns. One lioness, for instance, merely left her 4-month-old cubs at the base of a tree at 1905 and returned to them at 0600 the next morning. A day in the life of five cubs, 4 months old, gives an impression of their routine:

> The Masai pride rests by a waterhole when I arrive at 1100 and continues to do so until evening, all members inactive except that the cubs suckle occasionally and once play for 5 minutes. At 1800 the group moves out, leaving the cubs behind. Stalking some zebra 2 km away, the animals capture one at 1915 and 35 minutes later the carcass has been pulled apart, each eating on its own piece, a male on the largest portion. At 2020 one mother heads back to the cubs and fetches all of them. They reach the kill site at 2130. The male is the only animal still with meat, and the cubs join him. At 2220 all eating stops; at 0120 the group moves on, again leaving the cubs behind. Between 0140 and 0330 the adults see some wildebeest and wait for the herd to drift closer; it does and one animal is caught. The carcass is torn apart in 15 minutes and at 0400 a mother—not the one that did so last time—departs and returns with the cubs 10 minutes later. They feed on scraps, suckle intermittently between 0600 and 0700, and then rest with the group, still doing so when I leave at noon.

Cubs interact peacefully with all pride members except that some adults, especially males, may repulse their advances (plate 16). Lionesses seem to pay little attention to their cubs as they roam throughout the group and sometimes play at the periphery of it, but they quickly react when young venture into a potentially dangerous situation. Once, when a 4-month-old cub approached a feeding male, a lioness roared softly 4 m away. The cub ignored her and she called again until it came, and she licked it. But the cub returned to the male and was slapped by him; at that the lioness called louder than before and the cub ran to her and suckled. On another occasion two cubs, 6 months old, played with a young lioness. She clouted a cub hard on the head and it rolled over with a snarl. Its mother rose immediately, first watched the young lioness in the head-low posture, then chased her 20 m.

The whole existence of cubs revolves around obtaining enough to eat. Their ways of doing so need to be discussed.

Suckling. Lactating lionesses permit small cubs of any litter to suckle on them, almost indiscriminately, and the cubs in turn respond to the calls and leadership of any such female so that I sometimes found it impossible to assign cubs to their actual mother. A cub may wander to three, four, five lionesses in succession in an attempt to drink, and cubs of several litters may occupy the four teats of one lioness. The ages of such cubs may vary considerably, as when once a 2-week-old young suckled beside another that was 7 months old.

Lionesses sometimes give their own cubs preference in suckling either by withdrawing with them from the group or by chasing other young away. On one occasion, when five litters of the Seronera pride were together, one lioness suckled her young 20 m from the group and another walked 25 m with her offspring before permitting them to feed. One lioness let her two cubs suckle but hissed at and mouthed a third cub, not her own, until it left. After her cubs had finished, she permitted two cubs from another litter to drink. When a lioness growled at three cubs that suckled on her, only the one that was not her own departed.

Such communal suckling may be of distinct advantage to a litter in that it can draw on the milk supply of several females in the event that its mother has little. In addition, when cubs of several ages are present, the large ones may obtain milk long after their mother has ceased to lactate. Several cubs of the Masai pride still suckled at the age of 12 months even though their mother had no milk after they were 7 months old; several Seronera pride cubs obtained milk for at least 3½ months after their mother's supply had dried up. If a lioness loses her litter, her milk contributes to the survival of other cubs in the pride. The cubs of both females B and H of the Seronera pride disappeared, but the lionesses continued to lactate for four more months, during which time thirteen other cubs suckled on them. If a lioness dies after her cubs are mobile, other lactating lionesses would presumably care for them. When female M of the Masai pride disappeared, her 7-month-old cub attached itself to females L and Q, both of which had cubs, and traveled with them for several months until it too vanished.

Communal suckling also has several disadvantages. Cubs are quarrelsome when suckling. A lioness may tolerate squabbling in 2 or 3 of her own cubs but when 6 to 10 cubs compete for her milk she often withdraws and none are permitted to drink. If a lioness has cubs hidden away, those already with the pride often suckle, thereby depriving the newborn ones. In competition over a nipple the smaller cubs are shouldered aside by larger ones unless the lioness interferes.

A female reclines on her side or back while the cubs suckle either by lying at right angles to her or by draping themselves over her side or abdomen (plate 17); occasionally one drinks from a standing lioness. Cubs seem to show no preference for a certain nipple. When milk is plentiful and competition for a place at the breast low, cubs suckle quietly, occasionally treading the abdomen with their forepaws. But more often they miaow and snarl, shove and slap and dig in their claws in an attempt to retain their place when others crowd in. The lioness at first growls or bares her teeth at them, mouths them, and, if they do not desist, she terminates all suckling by rolling on her other side or onto her belly, by pushing the young away with a hind foot, or finally by moving several meters. Cubs are persistent, however; they may try to burrow beneath her, or they may pursue her from rest site to rest site for as long as an hour until she sometimes permits them to suckle.

Cubs may suckle at any time of the day, whenever a lioness is available, but the behavior is particularly prevalent just after dawn and before dusk. When the cubs of females L and Q were 4 months old, I timed their suckling for 24 hours. They suckled on L for several minutes between 1200–1300, 2200–2300, 2400–0100, and especially between 0600–0700; they suckled on Q between 1200–1300, 1400–1500, 2300–2400, 2400–0100, and 0600–0700 hours. Cubs suckled on female L for a total of 137.5 cub-minutes and on female Q for 127, an average of 53 minutes for each cub. In addition they had meat twice that day. The same five cubs at the age of 7 months had 17 main suckling periods during four consecutive days in which we observed them. At least some suckling occurred each hour between 0900 and 1300, 1500 and 2400, 0400 and 0500, and 0600 and 0700. The amount of time that cubs actually suckle is difficult to discover because the young sometimes fall asleep or simply lie with their mouth over the nipple. They suckle only briefly at times, especially when gorged with meat, seemingly as a means of making social contact, a nonnutritive suckling also described in the house cat by Kovach and Kling (1967). Most suckling bouts lasted 1 to 10 minutes and a few as long as 20. Once, when a lioness returned to her 6-week-old cubs, two of them suckled for 7½ and one for 8½ minutes before stopping. Sometimes the nipples of a lioness were in use more or less continuously for an hour as several cubs drank in succession.

Some lionesses cease to lactate when their cubs are 5 to 6 months old, but most seem to retain at least some milk for 7 to 8 months, judging by the vigor with which cubs of that age suckle. However, milk for such large cubs is only a supplement, and meat is essential to their survival. Cubs 1 year old occasionally suckle briefly on their mother, presumably as a form of social contact. Once a 15-month-old cub attempted to suckle on a subadult lioness.

Obtaining meat at kill. With competition among lions at a kill severe, it is of interest to find out how much a mother helps her offspring to obtain their share. On 44 occasions I observed the behavior of one or more mothers with cubs 3 to 12 months old of the Seronera and Masai prides at a large kill. In 57% of the instances, the cubs were with their mother when prey was killed (table 21). If they were not there, one or more lionesses fetched them on 23% of the occasions. A piece from a large kill was seldom (3%) carried to the cubs; the only time this occurred, a male appropriated it. At other times (18%) mothers showed no concern over the absence of their cubs. Young that are present when a large kill has been made always obtain some meat. But if they are absent, their mother eats before fetching them to the kill, sometimes from several kilometers away, with the result that little or nothing is left when they arrive.

On the other hand, males sometimes inadvertently provide cubs with meat. The following quotation from Edey (1968) exemplifies a general opinion regarding the behavior of the male at a kill: "One of the first to arrive, if he is close enough at hand to realize that a kill has been made, will be the dominant male. Up he trots, and any others who may have arrived before him give way. He will eat when he wants and, depending on his disposition, may allow others to share the kill while he is still on it. However, more often than not he prefers to dine alone." Such behavior is infrequent. I observed the actions of the Masai and Seronera pride males in detail at 34 wildebeest, zebra, and topi kills. On only one of these occasions was the whole carcass taken over by a male and in that instance he had killed the animal himself. In

53% of the instances one or more males fed with the lionesses without an attempt to get the lion's share (plate 19). In 44% of the instances a male at first ate with the others, but when only a portion of the kill was left—usually head, neck, and ribcage—he suddenly drove the lionesses off. The sight of males alone by such remains has probably given rise to the notion that they always appropriate kills. Whereas a male prevents lionesses from approaching the kill, he permits cubs to feed even though he may try to discourage them at first with a slap (plate 21). Thus when cubs arrive late at a kill or when lionesses have a portion which they refuse to give up, cubs may share meat with the males.

The situation was different when lions preyed on Thomson's gazelle. Although cubs were present half of the time when a kill was made (table 21), they failed to get a portion in 51% of the instances, either because the lionesses tore the carcass apart and refused to share or because a male grabbed the kill and retreated with it. Cubs seldom attempted to follow a male with a carcass, but on 6 out of 8 occasions when they did so he permitted them to share after he had eaten a substantial portion. There was seldom time to fetch cubs to the kill, and lionesses did so less often than when the kill was large ($p = <.01$). On three occasions a lioness hid a gazelle under a bush and then led the cubs to the kill, but on one of these occasions the cubs lost sight of their mother when she entered the thicket and she finally ate the kill herself. Of the 7 gazelle that a lioness carried back to her cubs, one from a distance of 2.5 km, they obtained meat from only 3 of them: twice other lions took the kill before the cubs could eat; once a lioness carried meat to the cubs then refused to let them have it; and on the fourth occasion a lioness carried a live fawn to 10-month-old cubs, gave it to them for 30 seconds, then suddenly took it back and ate it. On one occasion, a lioness caught a gazelle, left it in a dry river course, and brought her cubs to the site. Failing to note that she had descended the embankment, the cubs waited 50 m away. Ten minutes later they apparently heard her eating and ran up, but she grabbed the carcass and fled. Finally, after a considerable amount of mutual aggression, she permitted them to eat—a good illustration of the conflict between self-indulgence and parental care. Twice a lioness killed a gazelle, hid it, and walked toward her cubs. But when she met another lioness she hurriedly returned to her kill and ate it; similarly, a lioness immediately ate the fawn she was carrying to her cubs when she saw several other members of the pride.

One lioness permitted her own cubs to share her kill but not others. Another lioness gave a gazelle to her starving 12-month-old cubs even though she had not eaten herself. A third lioness caught two gazelle and after eating one of them permitted her own 15-month-old cub to feed, while slapping at any others that came near. Such sharing was rare, however, and for one reason or another cubs failed to obtain meat from at least 75% of all gazelle kills so that some starved when no large prey was available.

Cubs Acquiring Their Own Food

Although cubs as young as 2½ months of age watch movements of prey alertly and stalk and rush spurfowl and similar animals, they do not participate in the actual hunt until about 11 months old. One cub of that age joined the lionesses in stalking a wildebeest herd. Once I came upon a zebra stallion standing in a pool with water up to his belly. His bloody head indicated that he had been attacked by, but escaped from, at least one of the three lionesses and seven cubs, 14½ to 18 months old, that stood along the river bank. Two of the younger cubs jumped into the water and swam to the zebra. One clambered on his back, the other on his head, and after some thrashing he fell on his side and drowned. Cubs 15 to 16 months old commonly participated in hunts for gazelle and on several occasions caught one. I have no evidence that cubs of that age can catch adult zebra and wildebeest unaided. Cubs less than 2 years old would no doubt find it difficult to subdue large prey unless assisted by adults. Thus, cubs are fully dependent on the adults for food until at least 16 months old. Even after that they need the help of lionesses unless they can subsist largely on small prey or live in areas where meat is readily scavenged.

Following by Cubs

The choice of following a lioness or of remaining behind remains with the cubs. Small ones often retreat to a thicket without signal from the lioness and wait there for her return. If they follow the female, which they often do in response to her grunts, they may find it difficult to maintain her pace. She usually walks more slowly than usual if the cubs are small, but after they reach the age of 5 to 7 months she often ignores their miaows as they lag. Cubs of that age sometimes become separated and remain alone for a day or two. Those I observed always rejoined, but if the pride moves suddenly to another part of its range such lost ones probably die of starvation or are killed by predators.

Starving cubs too weak to walk are abandoned: lionesses make no attempt to nudge them to their feet, carry them, or otherwise help them. However, lionesses seem to be aware of the fact that their own cubs are not present at a kill even when others are there. On one occasion a lioness left a kill on which 5 other lionesses and 12 cubs fed and brought her missing young to it. Only lionesses which have young of their own lead cubs to a kill or from one site to another. I noted only one exception to this: a lioness which had lost her litter two months previously left a kill and brought 12 cubs to it.

As noted earlier, the cubs' best chance of obtaining a meal is to follow the group. Until young are about 4½ months old, and occasionally until they are up to 7 months old, they are oblivious to the distinct postures of hunting adults. They miaow, run ahead of stalking lions, play, and so forth. Cubs older than that do not vocalize and often wait or move forward only slowly and far behind stalking adults. The following example of the behavior of two 1-year-old and two 5-month-old cubs in response to three hunting lionesses was typical. The group moves out at 1815. Ten minutes later it spots some zebra on the skyline and watches them for 30 minutes. At 1855 the lionesses spread out and advance in a front 270 m wide. As they do so the cubs drop behind, at first walking slowly then sitting on a termite hill, watching. Suddenly, at 2005, they bound ahead and after a .6 km run reach the lionesses feeding on a zebra they have caught.

Cubs continue to associate closely with their mother until she has another litter, usually when they are almost two years old. A mother and her cubs are generally in the same group, but as they grow older the amount of contact decreases even though all continue to belong to the same pride. Observations of one cub illustrate this behavior:

Age of cub/months	*No. observations of cub with a group*	*% in which mother was in same group as cub*
12–18	34	88
18–24	30	80
24–30	28	54
30–36	27	60

This lioness conceived when her cub was 23 months old. Thereafter I was unable to discern any special relationship between them. A lioness of the Seronera pride became pregnant when her two cubs were 15½ months old. When the cubs were 17 months old she severed her social ties with them and they then joined another lioness which had two cubs of similar age. In general, when the cubs are 17 to 18 months old the lionesses cease to care for them in that they do not lead them to kills or bring food to them.

PLAY

Although the cause and function of play remain obscure, observers usually have few difficulties in recognizing a playing animal by its repetitive and exaggerated movements, the incompleteness of the actions, and the erratic sequence of some patterns (Loizos, 1967). Play obviously lacks immediate survival value. Both cubs and adults play vigorously at times, providing exceptionally fine opportunities for studying this behavior in a large social mammal. But with the exception of some notes published by Schenkel (1966b), little has been written about playing lions. The following observations are based primarily on the Masai and Seronera prides.

Time and Duration of Play Periods

Lions may play at any time, but the frequency, time, and duration vary considerably. On some days they do not play and on others there may be several active periods. Play showed two daytime peaks, one at dawn and the other before dusk; 30% of 156 daytime play periods occurred between 0600 and 0700, 17% between 0700 and 0800, 6% between 0800 and 1600, 15% between 1600 and 1700, 17% between 1700 and 1800, and 15% between 1800 and 1900. The morning play sessions usually began after 0600 and with rare exceptions were over by 0800. The average duration of play periods was 22 minutes, with 55% of them 15 minutes or less and none longer than 90 minutes. Cubs did not play with equal vigor the whole time, as each rested and was active in spurts. Play was perfunctory between 0800 and 1600. The late afternoon play periods were, on the whole, longer than the morning ones, averaging 42 minutes in length; 16% of them lasted more than 90 minutes and one continued for 165 minutes. In one 120-minute-period, for example, cubs played vigorously for 60 minutes, sporadically for the next 45 minutes, and vigorously once more for the final 15 minutes. Some play occurs during most hours of the night, as shown by figure 29, which is based on continuous observation of several groups for three to four consecutive days. There were 9 to 13 cubs in each of the three groups and all were less than 1 year old and in good condition. Lions have the reputation of being playful, yet the animals in these three groups spent an average of only 1.5%, 1.6%, and 6.0% of their days in play.

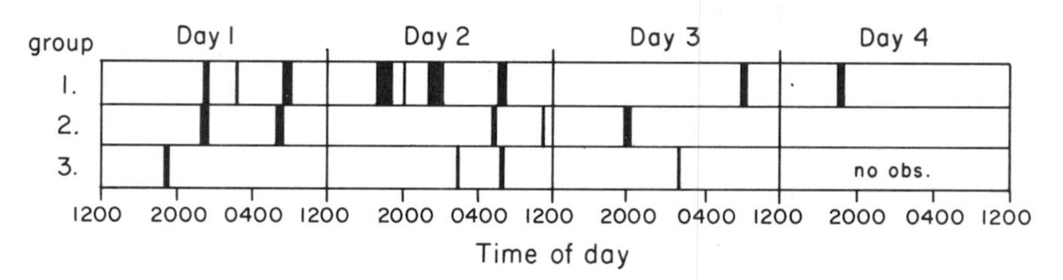

Fig. 29. The time, frequency, and duration of play periods by small cubs and lionesses in three groups that were observed continuously for 3 or 4 days.

Types of Play

The type of play used by a cub varies with its maturational development. Since I seldom saw cubs play when they were less than 6 weeks old, I was unable to note the first appearance of various patterns. One cub which we raised from an age of about 3 weeks gives some indication of the increasing complexity of his play, although the precise age at which his patterns first appeared may not be representative because of the poor physical condition he was in when I found him.

Approx. age in days	
25	He tries to trot for the first time.
29	He bats at our feet with his paws; he lies on his back, hooks my hand with his claws, and pulls it to his mouth; he backs up, pawing and growling at a piece of paper in his mouth.
35	He tries to bound for the first time. His most common form of play is to hold our ankles with his paws.
42	We pull a piece of paper tied to a string; he walks slowly, then suddenly rushes, hits it with a paw, and pulls it to his mouth as soon as he hooks it.
68	He stalks Kay carefully from a hiding place and rushes. From now on such behavior is common. He is now more careful than he was about scratching us while playing.

By the age of 2 months all play patterns are more or less fully developed. To give some idea of the types of play in their social context, I quote from my field notes on how a 6-month-old male cub played for 10 minutes within its group, which consisted of a male, 3 lionesses, and 7 cubs.

> The cub paws a twig, then chews it. When another cub passes, he lunges and bites it in the lower back. It turns and swats, then walks

> away. The cub sits. Suddenly he stalks a cub and rushes. The one attacked rolls over with a snarl and both grapple. The cub desists in its attack and bites at a tuft of grass instead. He then flops on his side. After lying briefly on his back and waving his feet, he rolls over and watches other cubs play. One of these ambles closer. He crouches behind some grass, then rushes and swats and in the same motion turns to another cub and nips it in the flank. The other cub whirls and hits him with a paw. He leaves. Two cubs wrestle, and he grabs one of these with his paws. One clouts him in the face. He lies. A cub trots by. He lunges and both rear up on their hindlegs. The other cub rolls over, the attacker lying on him, biting the throat. He releases his hold and sits. Another cub comes from behind and hits him on the head; he turns, snarling, and slaps. He lies. Suddenly he grasps a twig with both paws and bites it, shaking his head from side to side.

Solitary play. A cub occasionally crouches, stalks, then rushes a meter or two. Twice a cub strutted by itself. Both adults and young bound with exaggerated leaps. Sometimes a cub rolls over and kicks its legs or bites at one of them. Most play, however, involves some object. Cubs commonly paw at sticks, pieces of bark, or tufts of grass and then grasp them between the paws and roll over with them clasped to the chest; they pull at vegetation with their teeth and shake it from side to side. Other objects used in this fashion included an elephant dropping, two tortoises, and a milk can. Schenkel (1966b) watched cubs paw at and splash water. Once I gave an ostrich egg to several large cubs. One lay on its side and pawed it gently; then another held the egg between its forepaws and leaped backwards while trying to retain a grip on it. Fallen or leaning trees are often used in play as an animal bounds up and down the trunk; occasionally it jumps into the low fork of a tree or scrambles a meter or two up the stem (see table 16). On one occasion several cubs climbed one at a time up a small sapling and held on as it bent slowly and deposited them on the ground; two lionesses then did it too. Play by one animal often attracts others which then behave similarly.

Small objects are often carried in the mouth, sometimes as far as .5 km, if the group is moving. These include grass, dried feces, bones, and sticks. If the stick is too large to be transported easily, cubs drag it in the typical posture used by adults to move large prey. When one animal carries an object, this often stimulates others to chase it. This form of solitary play typically becomes a social activity.

Small cubs sometimes stalk birds and rush at them, vainly slapping

R.KEANE

the air as they escape. Such birds included francolins, spurfowl, guinea-fowl, Egyptian geese, vultures, and a lark. Two cubs once stalked a white-tailed mongoose. Such hunts contain all the adult components except that the rushes often consist of playful bounds.

On several occasions a cub played with the carcass of a wildebeest or zebra: it clawed the hide, clambered up on the body only to tumble down it, pulled at a leg or head, tore at an ear or nose with its teeth, made mock rushes, and in general behaved exuberantly. Once a sub-adult female caught a Thomson's gazelle. She mouthed the carcass, pawed it gently, occasionally licked it or shook it lightly until after 20 minutes of this she began to eat. On another occasion, a lioness gently pawed a wildebeest calf after she had killed it, and another time a subadult male bounded around a zebra he had found dead, kicking his hind legs into the air and lashing his tail, and then rolled on the carcass. In the Serengeti National Park Monthly Report for November, 1963, is an observation describing how a lioness played cat-and-mouse with a warthog young she had caught.

Social play. Social play usually involves two animals although three or more may join briefly in such activities as chasing each other. Several gestures are used to convey a readiness to play; these include an approach with exaggerated bounds, lowering the front part of the body (fig. 30), rolling on the back, or direct contact in the form of a nip or a push with the head. The play face is often assumed in such situations. Because play patterns are so variable, I found it difficult to classify them into discrete categories, but four main types of social play were recognized—chasing, wrestling, pawing, and stalking and rushing. These were used both by cubs and adults except that the latter also exhibited some of these patterns in other contexts, such as hunting.

1. Chasing. One animal bounds after another, swats at its legs, and tries to grasp its rump with the forepaws. Occasionally it runs beside the other, puts a forepaw on its shoulder, and pulls it down (fig. 31). In such a situation the fleeing animal sometimes collapses in an exaggerated manner by falling on its side with a thump. Wrestling and pawing often follow a chase, but some end without further contact

---

Fig. 30. Adult lions playing with cubs. Top—a male invites an approaching cub to play by lowering his forequarters; middle—a male clouts a cub lightly on the head; bottom—a lioness and a large cub slap each other playfully using positions typical of fighting lions.

RKEANE

when the fleeing animal turns and faces its pursuer. Chasing is especially prevalent when the group travels.

2. Wrestling. While one cub lies on its back, another is on top of it, mauling the throat, pulling a leg (fig. 32). Contact is often brief, but occasionally two animals wrestle for several minutes, rolling over and over as they paw and bite. This type of play is common when groups rest.

3. Pawing. Animals sit or stand facing each other and swat with a forepaw (fig. 32); occasionally they rear up on their hind legs and slap; or they lie on their sides and paw each other gently.

4. Stalking and rushing. On seeing another animal, a lion may crouch and wait and then advance in the stalking or crouching walk. On about half of the occasions no further actions occur, but on the other half the lion rushes as far as 10 m. Usually contact is brief, a simple token touch on the rump with a paw, but sometimes it ends in a pawing or wrestling bout.

Play between cubs. Cubs are far more playful than subadults and adults, not only in the frequency of their interactions but also in their duration and intensity. The 7 lionesses and 13 cubs of the Seronera pride were more playful from March to June, 1969, than any other group observed. Of 1,716 animals involved in play, 18% were lionesses, half as many as expected but still a higher percentage than usual. For instance, of 196 playing animals seen on a typical day in a pride, 190 were small cubs and 6 were lionesses. Cubs use some forms of play more often than others. The types of interactions were noted among cubs of the Seronera pride when they were 3 to 5½ months and 5 to 10½ months old. About 60% of the play of the smallest cubs consisted of wrestling (table 22). I confirmed this figure by 490 observations on cubs 2 to 6 months old of the Masai pride: 68% of their play was wrestling. In the second half of their first year, cubs wrestle significantly less often and chase and stalk more, whereas the frequency of pawing remains the same. Large cubs a year or more old wrestled infrequently, in part because those I watched were often starving and in part because

---

Fig. 31. Body positions of playing cubs compared to those of adults attacking buffalo. Top and upper middle—after a playful chase a cub lunges at the back of another cub; a lioness clutches the back of an escaping buffalo (after a photograph by Beyers, 1964). Lower middle and bottom—a large cub playfully straddles another cub and pulls it down; a lioness rears up beside a buffalo and bites him in the shoulder in an attempt to pull him down.

Fig. 32. Social play among cubs. Top—two cubs paw and mouth each other; bottom—two cubs wrestle.

this form of play tends to become less prevalent with increasing age, a trend evident from table 22.

Cubs of all ages play together even though one may weigh several times as much as its opponent. I observed 437 play interactions in groups which contained cubs of two age groups divided about equally between quite small cubs and others at least twice their age. Two large cubs played together in 21% of the instances, a large cub and a small cub in 26% of the instances, and two small ones in 53% of the instances. Although, in general, vigor of attack and retaliation are related, the larger of the two animals contains its strength when the size difference is great.

Play between cub and adult. The size difference between a lioness and a cub is great and vigorous contact play between them is not possible. Wrestling, for instance, is infrequent (table 22), and in some forms of play the lioness remains a fairly passive member. When she lies, cubs climb on her, slide down her side, drape themselves across her neck, swat her, and tug at her paw. When she walks, they grab her legs and jump up against her head, side, and rump (plate 22). Of 121 such leaps noted, 59% were against the rump or thigh, one of the few social contacts in lions which are not predominantly directed at the head. At times the adult lets itself fall playfully under the impact of a jumping cub. Cubs frequently attempt to capture the flicking black tassle at the end of a lion's tail (plate 23). Some observers (Schenkel, 1966b) assume that a lioness twitches her tail as a playful gesture to the cub. I noted no definite instance of this. Lions often whip up their tail when bothered by flies or for some other reason, and a cub may then pounce on the tassle, hold it between its paws, bite it, and shake it. The lioness pulls the tail out of the way, but the cub grabs it again, and she ultimately may rear up with bared teeth, obviously annoyed.

A lioness usually ignores playing cubs, but when they become too bothersome she either moves to another site, cuffs them lightly, or faces them with bared teeth. When one cub chased after a lioness, she uttered a sharp growl and the young desisted immediately. Once two cubs had a tug-of-war with the tail of a lioness until she rose with a snarl and they scampered off.

Lionesses initiate many encounters with cubs and indeed sometimes play after the cubs have lost interest. Pawing is a common form of play, with the lioness often using rapid, hard, downward thrusts to which the cub may respond with bared teeth (fig. 30). At other times she touches the cub lightly here and there while it lies on its back and waves its legs. Occasionally a lioness leads with one paw and when the cub turns toward it quickly swats with the other, then lunges in and mauls the cub with her teeth. She may hold the cub down with her paws until it struggles free, and sometimes she covers the youngster with her whole body while it squirms.

Lionesses often chase cubs and simply run over them or knock them down with a sweep of the paw and afterwards mouth them. Of 38 chases tallied in table 22 a female chased a cub 30 times and the reverse occurred 8 times. Once a chased cub hid behind a tree and dodged from side to side while a lioness attempted to hit it.

The possession of an object initiates some interactions between females and cubs, just as it does between cubs. One cub, for example, had a piece of bark when a lioness pounced and grabbed it. But the cub held one end and growled and slapped at her until she released it. On another occasion, a lioness picked up a twig and ran with it, chased by two cubs. She dropped it, and when the cub fled with it she pursued and retrieved it after a brief scramble.

Play between males and cubs is infrequent. Sometimes a cub jumps up against a male, rakes its claws in his mane, rolls against him and paws his face, or bites into his tail. Of 30 such encounters, the male did not respond 6 times, he moved to another rest site 3 times, and he reciprocated with play 9 times, usually by pawing the cub or mouthing it briefly (fig. 30); on 12 occasions he rebuffed the cub with a growl, snap, or slap. On one occasion two pride males gently pawed each other, a rare event; when a cub bounded up and participated, one male drove it off with a snarl.

Play between adults. I observed males 3 years old and older play less than a dozen times, a striking quantitative difference from the lionesses, which readily play into old age. Chasing, stalking, and pawing, in that order, are the most common forms of play among adults (table 22). Prolonged wrestling does not occur. All interactions are brief and precise, with such actions as slaps consisting of short, forceful strokes. Although body contacts were sometimes violent, play never turned into a fight. Once three lionesses teased a fourth one by cuffing her from all sides. Suddenly she coughed and the others desisted.

Derivation of Play Patterns

Holzapfel (1956), Loizos (1967) and others have pointed out that play patterns originated for purposes other than play. This is particularly striking in the lion, which has borrowed patterns from most aspects of its behavior, exaggerated a few of them, but has not developed new ones. The derivation of the pattern is obvious in some instances, such as when a cub struts, a female attempts to carry a cub like a newborn, a cub drags a stick like an adult would a carcass, or an animal rolls on its back when attacked, as if in defensive threat. However, when a cub chases another cub, slaps at its legs, pulls it down, and bites its throat are these patterns derived from intraspecific aggression or from hunting prey? As described earlier, a hunting lion assumes the alert face and one involved in agonistic interactions either the tense open-mouth face or the bared-teeth face in conjunction with growls or snarls. Playing lions have relaxed or alert faces or their mouths are open in the characteristic relaxed open-mouth face. They are silent except for an occasional hum. Only when hurt or when discouraging further contact are the teeth bared and snarls and other vocalizations emitted. From this it might be assumed that crouching, stalking, rushing, chasing, slapping, biting, and so forth are derived from hunting rather than from agonistic patterns. It is, however, important to remember that play consists of specific behavior without the corresponding emotional state. The only way to distinguish a lion attacking its prey from one attacking another lion is in the facial expressions and vocalization. Without these, I was unable to find out from which pattern the play derived, and indeed it may be impossible to do so by observation alone.

Function of Play

As Schenkel (1966b) has pointed out, lions play only when free from environmental and physiological pressures. They seldom play in the heat of the day, and when gorged with meat their actions are notably restrained; cubs usually play in the secure presence of a lioness, not when left unattended. Hunger had a striking influence on play. The 13 small cubs of the Seronera pride played often between November, 1968, and mid-May, 1969, when food was plentiful. By late May, after most prey had left the area, they seldom played, and by June 15 all play had stopped. In 3½ days of continuous observation in late June not a single instance of play was observed. There was no play until on September 14 the seven survivors chased each other briefly. Two weeks previously zebra had moved into the area, and the cubs had fed well.

If the premise is accepted that play has survival value, or it would not be found in so many mammals, then it must have some function. It has often been pointed out that play is a form of practice prior to when the actions need to be performed. This hypothesis has much to commend it. A cub which has learned by stalking other cubs that it can approach more closely by using the cover of a bush will presumably be more successful later in capturing prey; an attacking cub which learns to bite the neck rather than the rump may learn that its opponent is thus easily incapacitated—useful knowledge later when tackling prey with sharp hooves and horns. However, as Loizos (1967) has noted, an animal does not need to play to obtain such knowledge. And it is by no means clear that information derived from play actually benefits the animal. Subadult lions and tigers are inefficient in their hunts, and observations suggest that they mainly learn the points of stalking and killing by trial and error and watching the adults. Wrestling, the most common form of play in small cubs, is rarely used by adults. The question arises as to why, if play is practice for adult behavior, so much effort is expended on a pattern that is little needed and does not involve precision. On the other hand, stalking is a most intricate and important pattern, yet it is relatively unimportant in play. Furthermore, adults do not need the experience that play has to offer. This is not to deny that play may help to integrate various inborn motor patterns into a complex functional unit and that cubs may learn to respond to various social gestures as a consequence of interacting with other members of the group. While play is not essential for such learning, it does provide an innocuous alternative to the typical adult pattern, one which is not likely to elicit serious reprisal in the case of a misunderstanding.

Play also serves to expend excess energy and this may be the reason why it does not include sedentary patterns such as social licking and various forms of marking. It may also strengthen the muscles, stimulate growth, and so forth (Leyhausen, 1965b). Recent primate research has shown that adult patterns, such as those used in copulation, may fail to develop properly if the young animal is deprived of the necessary social stimulation (Mason, 1965). Play in lion cubs contains components of almost all major adult patterns (although sex is rare), and it might be conjectured that play may be essential for the normal neural integration of at least some of these patterns and that it may keep from atrophying those which developed early but are not needed for a year or more. Playing is also a social activity, one of the lions' few overt friendly actions, and it may therefore help to strengthen bonds between group members.

6 Population Dynamics

From the standpoint of conservation and possible management of the lion no topic has more relevance than population dynamics, yet it was an aspect of the study for which it was difficult to obtain unbiased quantitative information. To estimate accurately the size of the lion population in 25,500 sq km is a project in itself. Three years of work was clearly not long enough to elucidate such topics as birth patterns and mortality rates, much less to find out general trends in the population. Some of the conclusions in this chapter are therefore tentative rather than final.

Population Size

G. Adamson (1964) worked in about 5,200 sq km of the Corridor and plains and estimated that about 450 lions were in that part of the park. Talbot and Stewart (1964) surveyed the 31,200 sq km Serengeti and Mara regions from the air and calculated that 300–400 lions were in the former area and 250–300 in the latter. I tried various methods of censusing to confirm or dispute these figures but discarded most of these methods for one reason or another. The amount of prey available, the intensity of persecution by man, and other factors influenced the density of the lion population to such an extent that the results of sample counts in a few small areas would make an extrapolation for the whole ecological unit meaningless. It was not feasible to census or count lions intensively by car over large areas. To use a low-flying airplane for such a purpose was too expensive and, at any rate, would have provided only a rough estimate, because lions are readily overlooked in riverine forests and other dense vegetation. I attempted to census lions by playing roars at night over a loudspeaker with the hope that animals in the vicinity would answer, but responses were too erratic for the technique to be of value. Population estimates can be derived by tagging individuals and then applying the Lincoln index in which the population (N) is related to the number of animals marked (M) in the same way as the total number later seen (n) is related to the tagged ones resighted (m), giving the equation $N = Mn/m$. However, resightings of my tagged individuals were so biased due to their moving outside of the study area, that the index had limited application. Finally, lions were censused by (1) counting all prides in a large block of woodlands, and (2) estimating the number of nomads that frequented the plains.

Residents

There were 19 prides totalling about 325 animals, or one lion per 6.3 sq km in about 2,050 sq km of woodlands and woodlands-plains border (fig. 6). The size of some prides was known precisely and that of others approximately. Of course the number of lions in a pride fluctuated somewhat over the years, and some prides near the borders of the study area moved out of it for varying periods and others moved in. It was assumed that these factors averaged themselves out.

The most westerly 1,300 sq km of the park consist of a narrow corridor hemmed in by cultivation. Poaching there is heavy and many lions get caught in snares (Turner, pers. comm.). Sachs (pers. comm.) found lions scarce and estimated their number at fewer than 40 in the area north of the Grumeti river. Bell (pers. comm.), who studied wildlife intensively for three years in this western portion of the park, considered lions rare on the Ndabaka plains and thought that perhaps 40 to 50 frequented the Dutwa plains area. His estimate for the western 1,300 sq km of the park was 100 lions (including a few nomads), or one lion per 13 sq km.

The rest of the Corridor comprises about 2,250 sq km of woodlands, mainly around the Ndoha plains and the Duma and Simiyu rivers. Judging by the number of lions seen around the Ndoha plains and the woodlands border area south of the Moru kopjes, lions were as abundant there as in the study area. Lions seemed to be less abundant near the park boundary, both because of poaching and lack of prey. My guess is that about 250 lions used that area, bringing the total for the Corridor to 675 resident lions, or one lion per 8.3 sq km.

In driving through the 5,500 sq km of woodlands of the Northern Extension, I found lions as readily as in the Corridor, and prides appeared to be of similar size. In three days of searching for lions in about 200 sq km of terrain just south of the Mara River, twenty different lions were encountered in spite of the density of the vegetation and the shyness of many animals. The southeastern portion of the Northern Extension lacks the prey concentrations usually found in the other parts, and presumably there are fewer lions too. An area of about 350 sq km in size was added to the park in 1968 after a period of heavy poaching. Assuming an average density of one resident lion per 8.3 sq km, as in the Corridor, the Northern Extension has 662 lions.

The combined Corridor and Northern Extension estimates come to 1,337 resident lions. A few prides inhabit the plains area, such as those at the Simba kopjes and around Lake Lagaja, raising the total a little. But precise figures convey a spurious accuracy. A figure of 1250–1500 lions seems more reasonable.

The woodlands area outside of the park but within the ecological unit comprises about 9,200 sq km. My contact with these areas was limited to occasional visits by car and plane. The southern portion of the Masai Mara Game Reserve, about 800 sq km in extent, with a large resident population of prey, probably has a lion population comparable to that of the Serengeti Park. In the Loliondo Controlled Area, wild prey is less abundant than within the park but cattle contribute considerably to the biomass (Watson et al., 1969). Lions are found throughout, but I do not know in what numbers. In the Ikorongo Controlled Area to the north and west of the park, a considerable amount of wildlife survives in spite of human settlements, and the area holds a fair number of lions, judging by those that are shot each year. The Maswa area south of the park has few lions, probably because of poaching. While driving 200 km in two days through the area in September, 1969, only one shy lion was seen as compared to 37 lions encountered in 110 km of driving near Nyamuma plains during two days in July, 1969. In most of the woodlands outside the park there are probably, on the average, only one-fourth to one-third as many residents as within the park, or some 254 to 336 lions. Another 100 or so lions are in the southern part of the Masai Mara Game Reserve, bringing the total to about 350–450 lions. The ecological unit has thus about 1,600–2,000 resident lions.

Nomads

I attempted to assess the number of lions on the plains by using the Lincoln index. During the first few months of the rainy season the numbers of tagged individuals appearing in the study area were counted to give an indication of the total marked sample. In subsequent months the index was applied on the basis of tagged to untagged animals seen. While the use of the index in this situation may not be fully justified, I believe that the resulting figures are of the correct order of magnitude. Table 23 shows the size of the lion population as based on the index during months of lion abundance on the plains. In 1969 the rains almost failed, and the low figure for February, the only month when there were enough lions to sample, was not typical of other years. In 1968 the lions and their prey were so widely scattered during the prolonged rains that some nomads probably did not enter the study area, making the index figures too low. In 1967, however, the migratory herds first massed at the edge of the woodlands and later moved to the study area, and I think that the figure for that year is the most accurate one.

Some nomads remained in the woodlands during the rains, but unfortunately I had no way of estimating how many did so. Perhaps as many as 400 lions, or one-sixth to one-fifth of the total population, are nomadic.

Assuming that half the nomads spend most of the year within the park and half outside of it, the total park lion population is 1,450 to 1,700 lions or one lion per 7.9–9.1 sq km. For the ecological unit as a whole the population is 2,000 to 2,400 lions or one lion per 10.6–12.7 sq km (4.1–4.8 sq km). However, during the dry season most lions desert the 5,200 sq km of the plains, raising the number in the woodlands to one lion per 8.4–10.2 sq km. I would like to reiterate that these figures are not precise, but I believe that they are of the right order of magnitude. It is also worth mentioning that these are average figures for a huge area. Local conditions may be strikingly different. An intensive search in about 200 sq km during July, 1969, near the Nyamuma plains revealed at least 66 different lions, or one lion per 3 sq km, at a time when both prey and predators were concentrated near the Mbalageti River. Another count within an area of 340 sq km around Mukoma plain, Lake Magadi, and the upper Mbalageti River during November, 1968, when many wildebeest were there, showed at least 99 lions, or one lion per 3.4 sq km.

The densest permanent lion population—one lion per 2.6 sq km—occurs in the Lake Manyara National Park (see table 72). Ngorongoro Crater and Nairobi National Park have one lion per 3.7 and 4.6 sq km respectively. Kruger National Park contains one lion per 17 sq km (Pienaar, 1969). The Masai Steppe of Tanzania has an extremely sparse lion population—one animal per 306 sq km (Lamprey, 1964).

Population Composition

I classified 6,152 lions, excluding those in the main study prides, according to sex and approximate age, about half of them in the woodlands and half in the plains.

While tallying nomadic lions in the plains, it soon became apparent that some animals, particularly those localized in their movements, were seen again and again, whereas others were encountered only once or twice. For example, three lionesses with eight cubs were seen often in 1968. This biased the results because cubs were scarce on the plains. I compared several monthly samples with the known number of different individuals observed (table 24). There were generally more males in the sample than in the count because several territorial nomads were seen repeatedly. A striking aspect of table 24 is the great variation in the ratios from month to month even though the number of animals seen was fairly large and the data were collected in the same area. The differences, due to the erratic wanderings of nomads, emphasize the difficulties of obtaining an accurate sample in a short time. The ratio of males to 100 females varied from 96 to 109 in the two sample averages, and it was 88 in the count, indicating an approximately equal sex ratio. The ratio of cubs to females averaged to about one cub per $2\frac{1}{2}$ to 3 lionesses, including subadults.

Along the woodlands-plains border the ratio of males to 100 females, both nomad and resident, was about 82 in both the sample and in a count taken around the upper Mbalageti River in November, 1968 (table 25). Nomadic males frequently concentrated in the area while moving to and from the plains, and in early 1967 many remained there for several months during a dry spell. The ratio of cubs to females was nearly double that found on the plains. The sample ratio was more characteristic than the count because the latter was made at a time when an unusually high number of small cubs were in the area.

In the central part of the Corridor the ratio of males to 100 females was 68 in the sample and 55 for two counts, one made along 50 km of the Mbalageti River in September, 1967, the other in 250 sq km of the Musabi area in November, 1968 (table 25). The number of males in relation to the number of females had dropped considerably when compared to the figures from the plains, suggesting that many of the nomads were not in the area. There were about 2 cubs to 3 lionesses as was the case along the woodlands-plains border.

The woodlands population consisted of 40 to 50% adults with females outnumbering the males 2:1 (table 26). G. Adamson (1964), who tallied 428 lions in the woodlands and plains, obtained an adult (3+ yrs) ratio of 3:1 in favor of females. About 20 to 25% of the population was subadult, about equally divided between males and females. The lack of a difference in the sex ratio of subadults, coupled with a similar death rate within the study area later in life, indicates that the disparate sex ratio of adults there is due largely to emigration by adult males. Large cubs seemed to outnumber small cubs slightly (about 15% to 12%), but the latter are often hidden, biasing the sample in favor of large ones.

Adults constituted 57% of the population in the plains, almost equally divided between males and females, with the increase over the woodlands due to the larger number of males. Subadult males were almost twice as common as subadult females, because all young males become nomadic whereas only some females do so. Seventeen percent of the population consisted of cubs, a considerable difference from the 27% in the woodlands. The low figure was due not to the number of cubs produced, for the percentage of small cubs was similar in the plains and woodlands, but to the smaller number of large cubs present. Of 965 adult lionesses seen on the plains, only 5% were accompanied by one or more large cubs as compared to 23% of 184 females checked in the Musabi-Ndoha area.

The population composition of lions in other areas is similar to that in the Serengeti, although precise comparisons sometimes are difficult to make without knowing the criteria that were used to age the animals. In the Lake Manyara National Park, which had a population of 35 lions, about 10% were adult males (4+ yrs old) and 25 to 34% were adult females (3+ yrs old), the number varying somewhat from year to year as subadults grew up to become adults. "In general, adult and subadult females outnumbered the males by a ratio of at least 2:1. Cubs, two or fewer years old, comprised 20 to 25% of the population" (Makacha and Schaller, 1969). Of 24 lions observed by Schenkel (1966a) in the Nairobi Park, 5 (21%) were males, 13 (54%) were females, and 6 (25%) were cubs. Of 1,033 lions tallied in the Kruger National Park, 27% were adult males, 39% adult females, and 12% cubs; a further 21% were designated as young, presumably large cubs and subadults (Anon., 1960). Mitchell et al. (1965) reported a male-female ratio of 1:1.7 in the Kafue National Park.

Captive lions show an equal sex ratio at birth. For example, of 137 births listed in Jarvis (1966), 67 were males and 70 were females. Among 261 cubs born in the National Zoological Gardens of South

Africa, there were 131 males and 130 females (Brand, 1963). An attempt was made to sex every new litter, but in this I was not always successful because small cubs ran into thickets or high grass at the approach of the car. The sex ratio was equal among cubs a year or less old, both in the woodlands and in the plains (table 27). In the woodlands, cubs aged 12 to 18 months showed a more disparate sex ratio than young ones but the difference was not statistically significant. Cubs older than 18 months were often not with their siblings, or, if many were together, it was not possible to find out how many litters were involved. However, the fact that subadults had an equal sex ratio suggested that this age class did too. Robinette et al. (1961) reported a sex ratio of 96:100 among 229 small puma young, and of 80:100 among 252 large young, suggesting a slightly higher death rate among males than females in these cats.

REPRODUCTION

Much of the information on reproduction, particularly on age at sexual maturity and birth intervals, was obtained from the Seronera and Masai prides, which I saw often enough to be certain that litters were not born and lost without my being aware of it.

Age at Sexual Maturity

Captive lions may become cyclic at the age of 24 to 28 months (Cooper, 1942). "Young female lions usually become cyclic when about 3 years old, while males appear to require several months longer to achieve sexual maturity" (Crandall, 1964). Of several tame but free-living lions raised by G. Adamson (pers. comm.; 1968), a brother and sister mated when 2 years 9 months old, and the female gave birth 110 days later. The Serengeti lions reached sexual maturity later than these captive or tame animals. Five lionesses of the Seronera pride were known to have been born in March or April, 1963. Two conceived in September, 1966, when about 3 years 6 months old, one of these having also mated at the age of 3 years 5 months; a third conceived at the age of 3 years 8 months; and a fourth at 4 years 2 months. The fifth lioness had not to my knowledge given birth by the age of 6 years 5 months, possibly because she had been injured seriously in the abdomen as a subadult. In the Masai pride, three of the four daughters of female L came into estrus, presumably for the first time, in September, 1969, when 3 years 5½ to 6 months old. Three other young females in the two prides had not mated by the age of 3 years 4 to 6 months, nor had two females aged 3 years 1 month. I encountered no courting lionesses that were less than 3 years old, although a male once mounted a 31-month-old subadult briefly. The evidence indicates that most lionesses in the Serengeti had their first estrous period or conceived between the ages of $3\frac{1}{2}$ to $4\frac{1}{4}$ years. Two lionesses in the Lake Manyara National Park were also seen to mate for the first time when about 3 years 6 months old.

One male observed by Adamson mated successfully at the age of 2 years 9 months, as mentioned earlier, but to my knowledge no such young animals did so in the Serengeti. A few males estimated to be $3\frac{1}{2}$ to 4 years old courted with adult lionesses, and this, together with the fact that the males had a spurt of growth in body size and mane length at that age, suggested that they became fully mature sexually then. One male in the Lake Manyara National Park mated for the first time at the age of about $3\frac{2}{3}$ years.

Birth Rate

Lions have a high reproductive potential. Females are polyestrous, with periods occurring throughout the year. Cooper (1942) gave the average length of 51 gestation periods as 109.7 days with a variation of 100 to 114 days, and Sadleir (1966) found that four gestation periods, taken from the second day of estrus, were 111, 112, 117, and 119 days, respectively. If a lioness lost her cubs she sometimes mated again within a few days. One lioness whose cub was killed by another lioness was courting 7 days later. Two to three weeks after the disappearance of their cubs, two other lionesses mated without conceiving. A litter after 4 to 5 months is thus possible if newborn young are lost, yet lionesses seldom lived up to this potential.

In the Masai and Seronera prides the interval between the death of one litter and the birth of the new one was $4–4\frac{1}{2}$ months in 3 out of 13 instances, indicating that the lionesses conceived 2 to 4 weeks after their young died (fig. 33). The interval was $6\frac{1}{2}–7$ months in 2 instances, $8–9\frac{1}{2}$ months in 4 instances, $12\frac{1}{2}$ months once, and $15–16\frac{1}{2}$ months in 3 instances, for an average of about 9 months, double the expected interval. These figures do not include lionesses which lost a litter and failed to have another during the study. Female R, for example, lost her first litter in October, 1967, and two years later still had not had a second one. Female M, an elderly animal, had no cubs in 1966 and 1967 but produced a litter in mid-1968. Female L lost her litter in

October, 1968, and mated in September, 1969, just before the study terminated.

If one cub out of a litter survived, the birth interval was in all instances more than 18 months. Six lionesses in the Seronera and Masai prides raised one litter and subsequently had cubs again after 18½, 22, 22, 25, 26, and 26 months respectively. One lioness mated unsuccessfully when her cub was 27 months old. Although one female conceived when her previous litter was only 15 months old, others usually did so after their young were 19 to 23 months old.

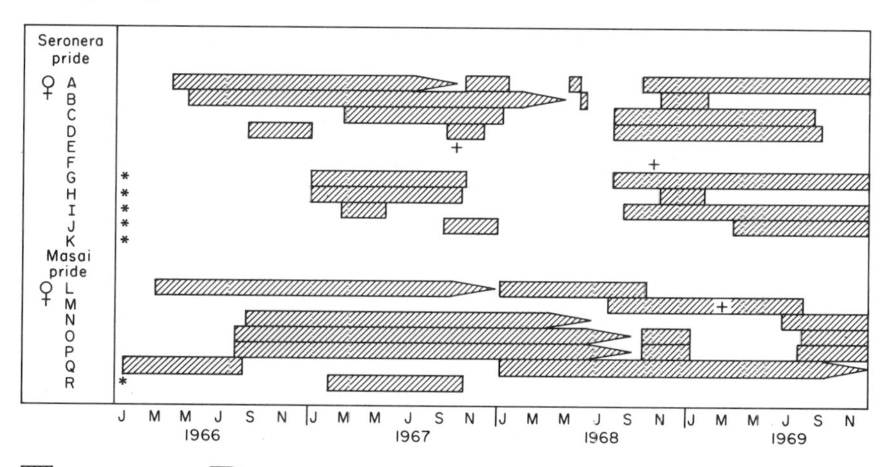

Fig. 33. The reproductive history of each lioness in the Seronera and Masai prides for four years, showing birth and death months of litters. An asterisk indicates that the lioness was subadult when the study began, a cross that she died.

Three (17%) of the 18 lionesses in the two prides produced no young during the study: one was a young adult and the others were old. Steyn (1951) noted that lionesses reach the limit of their breeding age by the fifteenth year, and two of the Seronera lionesses were at least that old. One lioness in the Nairobi National Park had no cubs between 1962 and 1967, when she died at an estimated age of 22 years (Foster and Coe, 1968), lending some support to the supposition that old lionesses cease to bear young. On the other hand, female M was also old, yet she suddenly had a litter after at least two barren years. One lioness with heavily worn teeth died in the Manyara National Park leaving cubs behind, and twice I saw aged nomads with small young. It seems likely that at least some lionesses conceive until they die of old age.

The number of litters produced by the twelve lionesses of the Seronera and Masai prides which were adult when the study began varied from 0 to 4 and the number of cubs from 0 to 9 (table 28). Lionesses failed to come into estrus, or at least they did not conceive, while the cubs were suckling and for nearly a year after they had ceased to do so. However, if all cubs died before or after they were weaned, lionesses tended to come into heat soon afterwards, indicating that the mere presence of dependent young inhibited the resumption of a regular cycle. Assuming no cub mortality, an average litter size of 2.3 (see table 27), and a birth interval of 24 months, each adult lioness should be able to raise about 1.2 cubs per year. The 18 lionesses of the Seronera and Masai prides were observed from June, 1966, to September, 1969. When I began the study, four of these lionesses had given birth recently, four were pregnant, six were subadult, and two never had cubs subsequently, making it unlikely that any important event in their reproductive history was missed in early 1966 by not observing them; the two remaining lionesses could have had cubs and lost them, but there was no evidence for it. Bertram provided me with information about the prides for the period from October to December, 1969. Thus, the births and deaths of cubs in these two prides was recorded for a period of four years. Taking into account the year in which the subadults were not reproductively active and the deaths of 3 adults, the 18 lionesses contributed a total of 62 years during which each could have raised an average of 1.2 young, or a total of 74. Actually they raised 10 young to the age of 2 years and 7 to just over one year of age, about 23% of the expected total. Only female L approached her potential.

While the Loliondo pride and some others were somewhat more successful at raising cubs during the study than the Seronera and Masai prides, others were less so. The Simba pride, for example, reared no cubs to an age of one year during three years of observation, and one female in that pride lost 3 litters in succession before 2 out of 3 cubs in her fourth litter survived to the age of a few months.

Blondie, a lioness in the Nairobi National Park, followed a reproductive pattern similar to that of the Serengeti lionesses (Guggisberg, 1961; Schenkel, 1966a). She had two cubs in July, 1955, but both died. After mating in August, she had four cubs in December; these she raised. In February, 1957, when the cubs were 14 months old, she mated without producing offspring. Between late 1957 and June, 1958, she gave birth twice and abandoned both litters. Another litter was born and abandoned in March, 1959. She courted later that month and had a cub in July which disappeared. Between 1960 and 1962 she successfully reared two cubs, her final litter. She died in 1967. Over a period of 12 years she had at least seven litters of which she raised two, a total of six cubs.

One lioness of the Chemchem pride of the Lake Manyara Park was observed frequently between June, 1967, and July, 1969. She had four cubs in December, 1967, and retained two of them until the study ended. Another female, in the Mahali pa Nyati pride, had no cubs in June, 1967, and still none two years later. Most other lionesses in the park left their prides at intervals, leaving gaps in our notes on their reproductive history.

The Estrous Cycle

According to Asdell (1964) "heat may last for a week and it recurs at intervals of 3 weeks." However, published records from captive lions show that both the estrus period and the time between periods is more variable than this statement indicates. Cooper (1942) found that estrus lasted 7 to 14 days and the interval between 44 periods varied from 1 to 16 months. Fourteen periods of estrus averaged 7.2 days in length with a range of 4 to 16 days, and the average interval between the midpoints of 9 periods was 55.4 days (Sadleir, 1966).

Periods of estrus in the Serengeti lions varied from less than one day to 22 days, but it was not possible to derive an average because a courting animal was infrequently seen for more than two consecutive days. I rarely knew when the period began or when it ended. Many periods lasted at most a day or two. On 23 occasions the male and female were observed together for 3 to 5 days. The seven longest known periods lasted 8, 9, 13, 14, 16, 16, and 22 days. The duration of six successive periods of one lioness was 14, 9, 4, 4, 1, and 5 days, in that order, showing a large amount of individual variation. The intervals between estrus periods were also highly variable. Female Q of the Masai pride was in estrus in January, March, June, August, and September, 1967, but some periods may have been overlooked. Five consecutive intervals in one lioness, measured from the approximate midpoint of one cycle to the next one, were 65, 11, 16, 36, and 14 days. Two consecutive intervals in another lioness were 40 and 26 days, and in a third one 12 and 25 days. The average of 17 intervals was 34.5 days, with a range of 11 to 65 days, a figure which is, however, a minimum. Some lionesses did not seem to come into estrus for months and even years, as was the case with female M of the Masai pride in 1966 and 1967. Such animals are excluded from the computations.

Mating Success

When a male remained close to a female as she rested and moved, I assumed that she was sexually attractive to him whether or not they copulated during the period of observation. Sexual activity was seen in the Seronera and Masai females on 40 occasions between June, 1966, and May, 1969, and 8 (20%) of the contacts led to conception and birth. In contrast, of 239 lion "breedings" recorded in captivity by Cooper (1942), 38% resulted in births. A total of 28 litters were conceived during the three years, and, since the copulations leading to 8 of these were seen, it can be assumed that only about 28% of all sexual activity was observed. Thus about 143 instances occurred in the two prides during 36 months, or 4 per month.

The 80% of sexual activity which did not result in births could be traced to several causes. On 5 occasions (13%) the lioness was already pregnant, having conceived 3 to 8 weeks previously. One female, for example, conceived around mid-May. A month later a male remained close to her for a day but no copulation was observed. However, she mated repeatedly 3 weeks later on July 8 and 9. Another lioness conceived on about October 20. Males attended her between November 12 and 19 and she copulated on at least two of the days.

Lactating females with small cubs courted on 2 (5%) occasions. The association of one female with a male lasted only a few hours and probably no mating occurred. Another lioness left her 3-month-old cubs with other females and copulated many times during four days.

At least 12 (30%) of the sexual contacts appeared to be typical of the kind leading to conception, including courtship behavior and frequent copulation, yet there were no births. I saw no evidence of miscarriages. In the house cat, and probably also in the lion, ovulation is induced by coitus (Asdell, 1964). If for some reason fertilization does not occur, the animal may become pseudopregnant; its abdomen swells, its nipples enlarge, and, in general, it gives the appearance of being pregnant until past the expected parturition date. Pseudopregnancy was reported in captive lions (Cooper, 1942), but I did not observe it in free-living ones. Females N and R, for instance, both copulated repeatedly in November, 1968, but neither became pseudopregnant or gave birth.

The remaining sexual contacts (32%) were brief, lasting a few hours to a day. Mating was sometimes sporadic, or none occurred either because the male seemed disinterested or because the lioness did not accept him. It is possible that the female was in false estrus, that she failed to come fully into heat. In a few instances courtship behavior may have had no relation to estrus at all. A lioness sometimes appeased a potentially aggressive male by presenting herself sexually to him. For a few hours such a pair can give the appearance of courting.

Because of unsuccessful sexual contacts and other factors only a few lionesses were accompanied by cubs. If the percentage of lionesses with cubs in the Seronera and Masai prides is taken arbitrarily at 6-month intervals, starting with June, 1966, and ending with June, 1969, the figures are 22, 39, 55, 47, 35, 70, and 53, in that order and with an average of 45. Similarly only 41% of 299 mature female puma had young at the time of capture (Robinette et al., 1961), and only about a third of the tigresses appeared to have young in a given period (Schaller, 1967).

Litter Size

The number of cubs in a litter range from 1 to 6 and once, exceptionally, 7 (Crandall, 1964). Some averages of litters born in captivity were 2.5 (Cooper, 1942) and 3.1 (Brand, 1963). The 34 litters born in the Masai and Seronera prides consisted of 5 with one cub, 17 with 2 cubs, 9 with 3 cubs, 2 with 4 cubs, and one with 5 cubs, for an average of 2.3. Sixty-three litters in the woodlands were of similar average size when 12 months or less old (table 27). Average litter size in the plains was somewhat smaller than in the woodlands, ranging from 1.7 to 1.9 cubs each. These figures do not represent litter size at birth, much less before parturition, because cubs are rarely observed before the age of two months. In the puma, for instance, Robinette et al. (1961) found that the average number of young in 66 prenatal litters was 3.4 but in 258 postnatal ones it was 2.9. Some of the lionesses which were accompanied by only one cub may have lost several others. Average litter size of cubs 12 to 18 months old was 1.9, somewhat less than in the younger litters.

Litter size is sometimes said to be influenced by food abundance (Stevenson-Hamilton, 1954). Of 15 litters conceived by nomads on the plains at a time of year when prey was abundant, average size was 2.1, little different than, for example, the figure for the Seronera and Masai prides which had a periodic food scarcity. The number of cubs in 10 litters in the Lake Manyara National Park, which has a large population of resident prey, averaged 2.2. Litter sizes reported for captives—which receive a daily portion of meat—are similar to those found generally in the woodlands of the Serengeti, especially if it is remembered that averages for free-living animals are somewhat low. This suggests that litter size is not affected much by variations in food abundance.

Birth Season

Published accounts agree that cubs may be born during any month of the year (Pease, 1914; Anon, 1960), but some naturalists have felt that a birth peak exists. Most births in Kruger Park are said to occur from March to July (Stevenson-Hamilton, 1954) and those in East Africa from November to March (Percival, 1924) and around July (Denis, 1964). I estimated the birth month of each litter on the basis of size and behavior of the cubs as compared to those of known age in the main study prides. To decrease the chance of error, only litters aged 6 months old or less were used. Figure 34 shows the birth months of 34 litters from the Seronera and Masai prides and 86 from the rest of the park. In the former, there were two birth peaks, one in January, the other in August–September; the latter sample shows an April peak. The August–September peak was due to 4 litters having been born to the Seronera pride in 1966 and 4 to the Masai pride in 1968. In contrast, 4 or possibly 5 litters were born to the Seronera pride in March–April, 1963, and had my study been conducted during that period, the curve in figure 34 might have been quite different. The pronounced April peak was influenced by the fact that two sets of 3 litters and one of 5 litters were born to different prides in that month.

For litters to be born in a pride at the same time, several lionesses must lack cubs and must come into estrus simultaneously. The first condition is commonly met with because the death rate of small cubs is high, lionesses often fail to conceive for several months after the disappearance of their litter, and, if several litters of the same age grow up, the lionesses would all be ready for more young concurrently. When one lioness is in estrus, others in the pride often come into that condition too. For example, 5 out of 9 females of the Magadi pride copulated in late September, 1966, and 6 out of 8 Mbalageti pride females also did so that month. Had all these lionesses conceived, a December or January birth peak would be evident in figure 34. Groups of nomadic lionesses also come into estrus together: 4 out of a group of 7 copulated in December, 4 out of 5 in a second group mated in January, and all

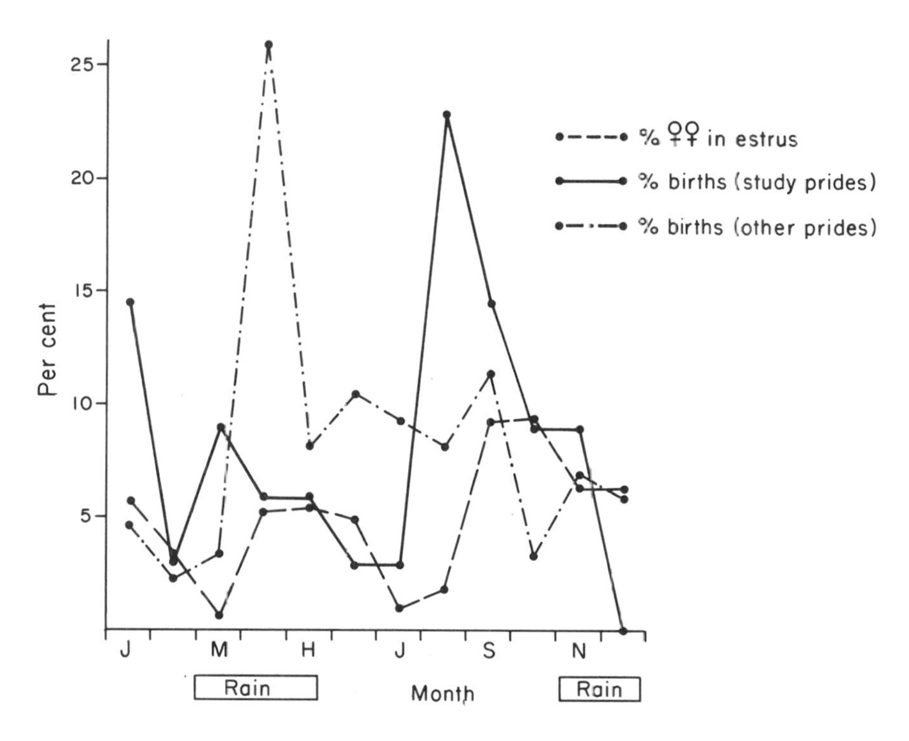

Fig. 34. The percentage of litters born and percentage of lionesses in estrus each month.

three of another group did so in April. Such synchronization is striking, and it may be that the presence of one estrous lioness stimulates others to behave similarly. Among gray langur of India, Sugiyama (1967) found that many females became sexually active after new males took over a group. He concluded that "social stimuli play a direct part in activating sexual activity."

To determine if my observations of sexual activity in lionesses are correlated with the April birth peak, the average percent of females in estrus each month is indicated in figure 34. The sample is based on sightings of 2,712 lionesses, 3 years old and older, and on 140 records of sexual activity, excluding the Seronera and Masai prides. Estrus was least prevalent during February, March, July, and August. A December or January activity peak that correlates with the birth peak three to four months later is not evident. The percent of estrous lionesses each month varied considerably from year to year, and a curve based on a three-year average may be somewhat misleading. For example, 10% of the lionesses were in estrus in June, 1966, 5.7% in June, 1967, and 2.3% in June, 1968. I have too few birth data to analyze the results on a yearly basis. In fact, it may be a mistake to expect an obvious correlation. As noted earlier, about 80% of all sexual contacts did not result in births, and on the basis of casual encounters it is impossible to evaluate the effectiveness of most sexual contacts.

While synchronous mating of lionesses in a pride no doubt affects the magnitude of birth peaks, the environmental variables of rainfall and availability of food possibly influence conceptions and births too. To find out if they do so, the number of sexual contacts seen each month in the Seronera and Masai prides and the number of litters born to them were tabulated between June, 1966, and August, 1969. Rainfall records were obtained at Seronera (table 1). Prey abundance was evaluated in two ways: the number of days when zebra and wildebeest were available to lions around Seronera each month, regardless of their abundance; and the average relative number of zebra and wildebeest present each month. The latter figure was obtained by using a relative abundance scale (0 = none to 4 = very abundant). I checked prey numbers each day and averaged the resulting figures for the month. Table 29 presents the cross-correlations of all variables, using the Kendall rank correlation coefficient, tau (Hays, 1963). Of the 21 cross-correlations possible among the 7 variables, two are trivial because they are mathematically related (wildebeest and zebra days versus relative wildebeest and zebra abundance). There is duplication between two variables, the relative number of wildebeest and zebra and the average number of days that these species were available. For purposes of assessing the significance of the data, the latter will be ignored. Ten combinations of five variables remain: rainfall, litters born, sex contacts seen, relative zebra abundance, and relative wildebeest abundance. Of these, the correlations between sex contacts and relative prey abundance are significant at the .05 level, indicating that the availability of food influences estrus in lionesses. The sex contacts seen are not significantly correlated with births 3 to 4 months later (at 3 months the correlation coefficient is .107, at 4 months .142, and the one-tailed probabilities are .1788 and .0778, respectively), results which are to be expected on the basis of only one conception per five sexual contacts, as discussed earlier.

The evidence indicates that the timing of birth peaks is to some extent fortuitous, the result of environmental conditions which may, for instance, act on cub mortality. If cub survival in a particular area is poor, perhaps because of prey scarcity, many lionesses are then ready to conceive, and peaks are created, or at least accentuated, through synchronous breeding. The availability of food also affects the frequency of sexual activity and by inference the birth peaks, an important consideration in the Serengeti where the erratic wanderings of wildebeest and zebra continuously change the levels of local prey abundance. Birth peaks would thus be expected to vary from area to area and year to year.

MORTALITY

Mortality and its causes are difficult to measure. Known animals disappear and may be dead, or they may merely have wandered away. Vultures strip the meat off carcasses and hyenas scatter the bones so that often the cause of death can not be determined. Cubs in particular disappear without a trace and their fate can only be surmised. The topic of mortality is particularly interesting in the lion, for as an adult the cat has no mammalian enemies other than man and members of its own species.

Causes of Mortality

Most lions died from disease, starvation, abandonment, old age, or as a result of violent contacts with their own or other species. These categories were not mutually exclusive; a sick animal sometimes died of starvation because it could not hunt. One male was apparently attacked by a swarm of bees, for his face was so swollen and lumpy for several days that he was barely able to see, an encounter which could have fatal consequences. Pienaar (1969) mentioned floods, grass fires, attack by ants, and bites by poisonous snakes as mortality factors but no evidence of these were noted in the Serengeti. Table 30 lists the major causes of death of cubs in the Seronera area. About a quarter died violently, bitten to death by other lions or other species of predators, about a quarter died of starvation, and half simply disappeared.

Twenty-two subadult and adult lions died from the following causes: snared or shot—9 (41%); disease—4 (18%); fight with other lion—5 (23%); probably injured by prey—2 (9%); probably old age—2 (9%). Three-fourths of the lions in this sample had violent deaths through contact with man and as a result of fights with each other.

Skulls of dead lions were collected and an attempt was made to age them. B. Mitchell of the Nature Conservancy kindly sectioned several teeth but the cement layers, which are useful indicators of age in some temperate carnivores, were not distinct enough for counting. The skulls were then assigned to age classes based on the relative amount of wear on the teeth. The diameter of the wear surface of the conical second premolar in the upper jaw was assumed to become larger with age. In young animals it was sharp, in old ones worn flat. It was also noted that the root of the canine projects less from the socket in subadults (2–3 mm) than in young adults (5–8 mm) and prime adults (over 9 mm, except in one skull), but in the older classes the distance showed no correlation with the tooth wear classes. Animals died at all ages, but 41% of the deaths were among prime adults, as is evident from table 31, which is based on an examination of 26 skulls and on the skins of 3 subadult males caught in snares. Few (10%) of the lions reached old age. Of 35 subadult and adult lions which I examined or which were reported to me, 21 were males and 14 females, a difference not statistically significant.

Disease. I collected ecto- and endoparasites and whenever possible took blood samples to determine from what diseases, if any, the cats might suffer. Four sick lions—one adult and three cubs—were killed for autopsy after they were so ill that they could only move with difficulty. Two males which I found freshly dead and a female which died from an overdose of drugs during a tagging operation provided additional parasitological material. R. Sachs, who did most of the autopsies and identified the parasites, generously provided me with his findings.

Ticks infested most lions, particularly around the ears and neck and in the groin. A sample of ticks from 18 lions, identified by Hoogstraal (pers. comm.) included *Rhipicephalus carnivoralis*, *R. simus*, and *Haemaphysalis leachii*. The last species is known to be a natural vector of *Babesia* which causes tick fever. Yeoman and Walker (1967) listed an additional eight species of ticks known to infest lions in various parts of Tanzania. The possible disease relationships of these species are unknown, with the exception of *Rhipicephalus sanguineus*, which transmits *Babesia* and is also the intermediate host of the protozoan *Hepatozoon canis*. Hippoboscid flies (*Hippobosca* sp.), which are common on all lions, appear to be mainly a nuisance to the animal.

Between April and June, 1962, an outbreak of biting flies (*Stomoxys calcitrans*) occurred in Ngorongoro Crater and severely affected the lions (Fosbrooke, 1963). The cats climbed trees or crawled into hyena burrows to evade the blood-sucking flies. Most became emaciated and their bodies were covered with bloody bare patches. The lion population dropped from about 70 to 15 as a result of death and emigration. The same species of fly is found in the Serengeti but was not abundant enough to cause harm.

Five lions checked by Dinnik and Sachs (1969) were infested with *Taenia gonyamai*; two of the five were also infested with *T. bubesei*. Another *Taenia* differing from the other two and possibly representing a new species was found in three lions. The larval stages of *T. gonyamai*, and probably of other tapeworms as well, occur as cysticerci in the muscles of antelopes (Sachs, 1969). When lions eat the meat of these antelopes they also ingest the cysticerci, which then develop into adults in the intestines of the lions. All adults had trematodes attached to the stomach lining and one had several nematodes in the intestine; neither parasite has been identified as yet. V. Haffner et al. (1969) found numerous *Neolinguatula nuttali* (family Linguatulidae) in the nasal passages of two lions.

Sachs et al. (1967) checked blood slides from 32 Serengeti lions and found that 22 (69%) were positive for trypanosomes. Trypanosomes of the *T. congolense* group had already been reported from lions (Baker, 1960), and when we injected the blood of two infected animals into white rats these became positive for trypanosomes of the *T. brucei* group in 4 to 6 days. Of 11 lions checked by Baker (1968), 7 were positive for *T. brucei* and 2 for *T. congolense*. Normally, trypanosomes become infective in the mouthparts of the tsetse fly (*Glossina* sp.) and are transmitted when it bites an animal, but they may also penetrate lesions in the mouth of a carnivore as it eats an infected herbivore (Baker, 1968). The fact that many healthy lions were infected with trypanosomes indicated that the parasite had no pathogenic effect under normal conditions, but could have if the resistance of the animal was in some way lowered. For instance, one 9-month-old cub was on the verge of starvation when it obtained a large meal. Yet its condition did not improve and soon it was unable to stand. Both types of trypanosomes were found in its blood, and it is possible that these contributed to its decline. Wildlife is generally immune to trypanosomiasis, but it is interesting to note that game from tsetse-free areas may succumb rapidly to the experimental inoculation of trypanosomes (McDiarmid, 1962). One hand-reared lion became infected with trypanosomes and showed various symptoms of the disease, including a high fever, after he was returned to the wild, but he recovered when treated (G. Adamson, 1968). Possibly cubs born and raised for several months on the plains, where tsetse flies do not occur and where lions prey mainly on wildebeest and zebra which are not heavily infected, are more susceptible to the disease than those which are born in the woodlands where flies are abundant.

Four out of seven blood slides checked by Young (pers. comm.) were positive for *Babesia*, probably *B. felis*, potentially a deadly parasite which is transmitted by ticks and acts like malaria in that it breaks down the blood cells, causing anemia. Adults are probably immune to it, unless their resistance is first lowered by other means, but cubs may be susceptible before they acquire immunity and after their female antibodies are worn off by the age of about 3 months (Young, pers. comm.).

Krampitz et al. (1968) examined the blood of 56 Serengeti lions and found that 27 (48%) were infected with the protozoan *Hepatozoon* of an undescribed species. The parasite occurs usually in ticks, and carnivores become infected by ingesting these. The protozoan travels from the digestive tract to the liver of the host where it forms a schizont and divides. When the schizont is liberated, the cell is ruined. A heavy infection could impair body functions.

Thirteen blood samples were tested for brucellosis (*Brucella abortus*), a bacterial disease causing spontaneous abortion, by using the tube agglutination test with serum dilutions of several strengths (Sachs et al., 1968). If the results are interpreted as in cattle, a titre of 1:40 and greater can be considered positive and a titre of 1:20 to 1:40 doubtful. Eleven sera were negative, one had a titre of 1:10, and one of 1:20.

One adult male in emaciated condition, weighing only 81 kg, died shortly after we found him. He was infected with trypanosomes (*T. brucei*) and his kidneys were inflamed with a glomerular nephritis caused by an unidentified agent. No disease could be identified in an adult lioness that was too weak to stand. Both animals obviously suffered from a debilitating illness which prevented them from hunting and ultimately caused death from starvation.

During the study, thirteen lions were seen which had emaciated hindquarters but otherwise seemed in good condition. Ten of these were adult males, two were subadult males, and one was a female. The hind legs of three of these individuals were almost paralyzed.

Pienaar (1961) found two lions dead of anthrax in Kruger Park. The lions were probably infected through a wound in the mouth while eating a diseased animal. Anthrax was present in the Serengeti, but I have no evidence that lions were affected by it.

Two lions had umbilical hernias. One had a nervous twitch of the upper lip and another of the head. A young adult lioness had a strange, heavy, rolling gait and was unable to recline without collapsing.

I have no evidence that the various intestinal and blood parasites actually caused the death of lions, for both healthy and sick animals were infected with them, and those that were sick sometimes contained several pathogens, any, all, or none of which could have been responsible for the condition. For example, one female cub, 22 months old, in good physical condition, swayed and staggered as if under the influence of alcohol. After observing her for 24 hours, during which she did not improve, we collected her. According to Young (pers. comm.) she was infected with trypanosomes (both *T. congolense* and *T. brucei*) and with *Babesia*, and was anemic, but he could not determine what caused her symptoms.

Violence. Many lions died violently as a result of coming directly or indirectly into contact with man. The number of animals shot or snared in and around the park was considerable, but difficult to estimate, and to the total should be added small cubs that starved after their mothers died. Turner told me that he confiscated the skins of five lions in one poacher's camp in 1957. Four male lions were found in snares during the period of study. In September, 1969, I accompanied a park patrol for two days. By a snare line we found a skinned male lion hidden beneath some grass so as not to attract vultures, and in a poacher's camp was the skin of a lioness. Shortly after I left the park, Turner wrote to me: "I've just got information of a car that was stopped by police up at Mugumu and searched and 10 lion skins were found in it." Although some lions are shot illegally for their skins or because they interfere with the poachers by scavenging the captured animals, most die inadvertently when they enter snares that are set for hoofed animals. Because of this, Turner told me, a heavily poached area characteristically has few large predators left.

Between 4 and 6 males may be shot annually in each of the three game-controlled areas surrounding the park, but professional hunters and their clients also account for many more. "In mid 1959 heavy shooting of male lion started in the Ikoma area. In December, 1960, I got it stopped. . . . At least 88 male lion were shot in this period. About the end of 1966 lion hunting again started in this area, and by the time it was again stopped in September, 1967, at least 50 males had been shot, possibly more, quite illegally as the area had never been opened again since closure in 1960" (Turner, letter, 1967). Many of the males taken from this relatively small area were probably nomads; in fact, my male No. 57 was shot there. Males were probably drawn out of the park as persistent shooting vacated territories and lowered the population of nomads.

Other predators sometimes kill lions, particularly cubs not attended by adults. A leopard bit a small cub of the Seronera pride to death, and visitors watched several hyenas tear apart a cub of the Masai pride. One evening I saw a 14-month-old cub with suppurating wounds surrounded by three hyenas while its mother sat several hundred meters away. The following morning the lioness was alone. "The undigested nails of young lions" have been found in the droppings of brown hyena in the Kruger National Park (Pienaar, 1969). Hyenas may kill adult lions, too, according to Turner, who gave me the following extract from his field notes: "On the night of February 25th, 1961, at about 9 P.M. very loud hyena howling and grunting broke out on the camp site ridge and continued on and off throughout the night. On investigating next morning we found the few remains of one of our fine resident male lions. This old lion had been very sick from a wound in his flank and we had been keeping an eye on him." Wild dogs have been reported to kill lions in the Kruger National Park (Pienaar, 1969) as well as in the Kafue National Park. "7.1.63. Eight wild dogs were worrying an adult maned lion near the camp. The lion chased the dogs, but they kept coming back. Later they crossed the Kasompe and were heard fighting till late in the night. The next day it was found that the dogs had killed and eaten the lion" (Hanks, pers. comm.).

Lions were aggressive toward other lions, both toward pride members on a kill and toward strangers, with the result that wounded animals with claw and bite wounds were common. Of 92 resident lions in the Seronera area in April, 1969, six were blind in one eye, four had the terminal part of the tail missing, and one had lost an ear. Deep cuts in the nose, patches of torn skin, ripped lips, tattered ears, swollen paws, all attested to violent encounters. Most wounds were not serious, but some became infected or incapacitated the animal. One lioness, bitten in the face and neck, died of septicemia. In a dispute at a kill on September 8 a lioness slapped a male, hooking one claw into his forefoot. The wound became infected, his leg swelled until he could only limp slowly. His foot was unusable until September 22. The leg of one lioness withered after she was bitten in the flank, and she was entirely dependent on the rest of the pride for food until nine months later when she finally recovered. A nomadic male, severely bitten in a leg and clawed in the scrotum on January 1, was by January 4 almost unable to move. He withdrew into a kopje and remained there without eating until, on January 21, his leg had healed. Much of his mane had fallen out, probably as a result of his scrotum injury. Lions have great recuperative power and even large wounds tend to heal. One male which had been bitten through the nasal bones was fully recovered 1½ months later.

Lions showed little restraint when biting each other and some wounds caused rapid death. Several cubs were bitten to death. One of the Seronera males died after a fight. One nomadic male was killed presumably by another male in a brief but violent battle: the nomad had been bitten through the nape, breaking his neck, and there were tooth punctures in his throat, yet the surroundings showed no signs of a prolonged struggle. A subadult male was killed in a fight with several other lions. On March 16, 1969, G. Dove watched a lioness walk, then crawl toward a male on a kill near Lake Lagaja. Suddenly he attacked her, and, after a brief flurry, she collapsed, quivered, and lay still. She had been bitten through the back of the neck. The male continued to eat and finally reclined beside her. Schenkel (1966a) watched a male pursue a lioness, bite her in the lower back and shake her, thereby breaking her vertebral column in two places. The lioness was paralyzed in her hind quarters and died about 20 minutes later. Pienaar (1969) and Guggisberg (1961) described further instances of lions killing each other.

Buffalo and elephant occasionally kill lion cubs. Three cubs were gored and trampled by buffalo and one was stepped on by an elephant in the Lake Manyara National Park in 1967 and 1968. On one of these occasions, a herd of buffalo chased a lioness with three small cubs. The mother and two cubs escaped by climbing into a tree but the third was trampled to death (Makacha and Schaller, 1969). Bere (1966) wrote that "a lioness in Murchison Falls Park killed an elephant calf. Discovering this, the elephant followed the lioness into a patch of long grass where her own cub was hidden. Immediately she saw the cub the elephant killed it."

Careless and inexperienced lions may be crippled or killed by the horns or hooves of the prey. One lioness with a smashed jaw in the Ngorongoro Crater and a male with a similar injury in the Serengeti possibly were kicked by zebra. I found a lioness dead with a broken leg and one alive with a horn wound in her side, both no doubt the result of a mishap while hunting. Sable, roan, kudu, and buffalo have all been known to gore lions (Pienaar, 1969). Beyers' (1964) photographs of a lioness attacking and in turn being attacked by buffalo illustrate how a slight misjudgment could well end fatally for the cat. Mangani (1962) found a lion killed by buffalo, and I came across the old skeletons of a male lion and a bull buffalo side by side. Goddard (1967) observed a subadult male attacking an adult rhinoceros: "He bit her just above the hock, attempted to hang on, and clawed her thigh. The female wheeled around with incredible speed and gored him twice in the centre of the ribs, using the anterior horn with quick stabbing thrusts. The lion rolled over, completely winded. The rhinoceros then gored the lion once in the centre of the neck, followed by another thrust through the base of the mandible, killing him instantly." Porcupine quills in the paws and face incapacitate lions on occasion and crocodiles may pull a lion into water (Pienaar, 1969).

Starvation. Healthy adult lions probably never starve in the Serengeti, but cubs frequently do. Starving cubs were encountered in several parts of the park including the plains, the upper Mbalageti River, along the Mara River, near the Nyamuma plains, and around Seronera. In the last-named area, 28% of the cubs died of starvation (table 30), most of them 6 to 12 months old, an age when they were too large to obtain much milk but too small to hunt for themselves or compete successfully at small kills. Cubs lost condition rapidly when not fed well. Some born in August, 1968, were in excellent condition through May 22, 1969. By June 20 they were lean, with their ribs and hip bones prominent, and between late July and early September several died. However, cubs revive quickly after a few meals. Ten days of ample food were sufficient to transform the survivors of the above mentioned group from staggering skeletons to frisky youngsters whose recent deprivation was not evident. In the previous year, three out of a group of five cubs died of starvation in mid-October. By November 2 the remaining two cubs had fully recovered.

Abandonment. Probably more cubs died as a result of having been abandoned by their mothers than through any other cause, but I have no data to support this assumption. Frequently a lioness had small cubs, healthy active ones, then suddenly the whole litter disappeared without a trace. Of the 17 cubs, 0 to 6 months old, shown in table 30 as having died of unknown causes, most were in this category. One lioness in the Nairobi National Park was known to have abandoned several litters (Guggisberg, 1961). One elderly lioness of the Seronera pride abandoned her 3-week-old cub beside a wildebeest carcass after the cub was nearly dead from starvation. On one occasion, I watched a lioness with three cubs, about 2 months old, at a waterhole in a ravine. After drinking she left with two cubs, but the third one, unaware of her departure, puttered around. Suddenly it became aware that it was alone. It ran miaowing up and down the ravine, but the wind blew strongly in the wrong direction and its mother failed to hear it. Steadily she walked away. She never returned and subsequently lost her other cubs as well.

Old age. A few lions were old, with canines worn to stumps and most incisors missing, their pelage unkempt and body lean. The tread of such an animal was heavier than that of a prime animal, its head was held less erect, and its face had a drawn look. Four (12%) of the resident females around Seronera were well past their prime in mid-1967 (table 5). Two of these died presumably of old age. Their condition deteriorated slowly even though they were able to hunt and could obtain food at the kills of other pride members.

Mortality Rate

I tried to note all births and deaths of cubs in the Seronera and Masai prides, but a few young may have died during the first two months after birth without my being aware of it. Some of the adults also died or disappeared and were presumed dead. These two prides provided the most detailed information on mortality patterns, but other data based on an occasional dead lion and missing litter contributed to the observations made around Seronera.

Cubs. Stevenson-Hamilton (1954) noted that about half of all cubs die in Kruger Park, and Guggisberg (1961) gave a similar figure for the Nairobi Park. Of 79 cubs known to have been born in the Seronera and Masai prides during the four years of observations, 53 (67%) died. Among the 26 survivors, 10 became subadults, 7 were over a year old at the end of the study, with good chances of survival, and 9 were less than a year old, making it likely that several of them would die before the age of two years (table 28). Thus, the mortality figure of 67% is a minimum.

Cubs in some other prides also had a high death rate, judging by those that were known to have disappeared. For example, of 9 small cubs that were present in the Mukoma pride late in 1967 only one survived, and of the 10 cubs in the Plains pride in mid-1966 one reached subadulthood. In contrast, out of a total of 22 cubs born in the Loliondo pride in 1967 14 (64%) survived, and in the Mbalageti pride 7 out of 8 cubs born in about November, 1966, grew up. Several factors, such as prey abundance and the number of litters born at the same time, affect the survival of cubs within each pride separately, but, in general, I would estimate that about half of the young die from one cause or another.

The Mahali pa Nyati and Chemchem prides in the Lake Manyara Park had 6 litters totalling 15 cubs between June, 1967, and July, 1969. Seven (47%) of the cubs died in spite of the fact that the lions in the park were not subjected to periodic food shortages as were those in the Serengeti.

Subadults and adults. Of 23 male and female lions 2 years old and older, resident in the Masai and Seronera prides in June, 1966, one male was killed in a fight, another male probably died of wounds incurred in a fight, one female died of old age, a second one, also old, was last seen in poor condition and unquestionably died, and a third female suddenly disappeared with one of her two cubs and possibly was killed by other lions. At the end of 1969 eighteen of the original members were left. One male of the Seronera pride also vanished, but I presumed that he was evicted rather than killed by other males. Assuming that no lions died in the few months preceding the study, the death rate in this sample was 22% in four years or about 5.5% per year. Lamprey (1962) estimated a 6% annual replacement rate in the Tarangire National Park.

During the four years, the Seronera and Masai prides raised ten cubs to an age of at least 2 years, an increase of 44%, or 11% per year. Subtracting the percentage of the animals that died, the actual annual increase in this sample population was 5.5%. Although these figures are based on only a few animals, I believe that they are applicable to other parts of the study area. The Loliondo pride raised 14 young to subadulthood in 3 years, the Kamareshe, Mbalageti, and Mukoma prides each 7 young, the Magadi pride 6 young, the Banagi pride 4 young, the Plains pride one young, and so forth. Few subadults and adults died in these prides, judging by how consistently known individuals were seen during the study. This supports the idea that the survival rate of cubs to independence generally exceeded the death rate of adults.

The history of the Seronera pride shows, however, that pride size tends to remain fairly stable over the years primarily as a result of emigration by subadults which become nomads. Thus the calculated annual increase of 5.5% should on a long-term basis be apparent not in the prides but in the nomadic segment of the population. These nomads undoubtedly have a higher death rate than the 5.5% calculated for pride members because many nomads leave the park and are then snared and shot. Previously it was determined that 20 to 25% of the population consisted of subadults. In a stable population, 10 to 12% of the adults would have to die each year to make room for the subadults entering the adult class. With a population of 2,000 to 2,400 lions in the ecological unit, about 200 to 300 adults (and a number of subadults) would be expected to die. Perhaps half of these would be pride members, judging by the death rate in the prides around Seronera, and the other half nomads. The high mortality of nomads outside of the park may be sufficient to remove the annual surplus raised by the prides, with the result that the lion population is relatively stable.

Two prides containing a total of two males and six females at least 2 years old at the beginning of the study were observed for two years in the Lake Manyara National Park. During this period a male was shot and a female died of what appeared to be old age, a death rate of 12.5% per year in this small sample.

POPULATION TREND

The prides in the study area were increasing at a rate of at least 5.5% per year, but an unknown number of nomads were killed as they moved outside of the park and the residents there also suffered from poaching. The lions in most parts of the park presumably have been able to reach some sort of equilibrium with their environment, even in the Northern Extension, which was not protected until 1959. The Ngorongoro Crater provides a good example of how rapidly a lion population can recover to its former level after being decimated. An estimated 70 lions inhabited the Crater (Wright, 1960) before a plague of biting flies reduced their number through death and probably emigration to about 15 in 1962 (Fosbrooke, 1963). In 1969 the lions numbered about 70 again (Des Meules, pers. comm.) with most of the increase due to a high survival rate of cubs. It was my impression that the number of lions in my study area and, probably, also in the Northern Extension had become stabilized. G. Adamson (1963) commented:

> My opinion, for what it is worth, is that at the present time the southern part of the Serengeti National Park carries as many lions as it can comfortably hold. Old-timers as is their wont, compare numbers unfavourably with the past, but it must be remembered that in those days it was legitimate to feed the lions in the Serengeti. And as I myself have witnessed, as soon as a car was heard, lions would appear from all directions.

Judging by the accounts of hunters, the western part of the Corridor once had more lions than it does now. In August, 1957, D. Ker camped for three days near Kiwawira and saw 51 lions. In September, 1960, the Serengeti National Park report stated: "Lack of lions in the western part of the Park is very noticeable. In 210 miles of patrol during 8–10th December in the Mbalageti-Duma areas no lion were noted. Warden Poolman recorded none in 196 miles patrolling in the Western Corridor. It will take several years for our lion population to return to its 1957–1958 level if no other areas adjoining the park are opened to unrestricted lion hunting." The area still held few lions between 1966 and 1969, undoubtedly as a result of heavy poaching along both sides of the park boundary. Given the available prey, some parts of the ecological unit outside of the Serengeti National Park and the Masai Mara Game Reserve could support more lions than they do. Many of the lions in these areas seem to be nomads that have left the park for terrain less densely occupied by lions. Thus the park serves as a reservoir of lions with the excess spilling over and in fact being siphoned off into the surrounding areas where lions have been decimated by poaching.

7 Food Habits

Predation is exceedingly complex. On the one hand it involves the size of the predator population and the way in which predators capture their prey, including predator group size, choice of prey, and hunting methods. On the other hand, prey shows not only direct responses to a predator, such as flight, but also indirect ones which may be morphological, physiological, or behavioral. These manifest themselves in pelage color, birth season, herd size, and so forth—any responses that decrease the chance of a fatal encounter. Many, if not most, indirect responses evolved under various selection pressures, not just under those from predation (Kruuk, 1964). In general, "the race between the adaptations of the predators for capturing prey and those of the prey for escaping a predator may be viewed as a race whose finish line is constantly moved ahead of the contestants" (Klopfer, 1964). Ultimately, predation requires an integrated analysis with the use of mathematical models, such as those used by Holling (1965) and Schoener (1969), but until the whole behavioral system has been studied in detail I will have to limit myself to a description of a few of its fragments.

The Prey Population

The Serengeti lions share their habitat with a wide variety of potential prey but most species are, for one reason or another, eaten so seldom that they need not be considered in detail. Some species belong in a size category which lions typically ignore. These comprise dik-dik, hyrax, hare, and vervet monkey, to name a few at one end of the scale, and hippopotamus, rhinoceros, and elephant at the other. Porcupines and baboons have weapons to which lions respond with caution. Oryx, mountain reedbuck, bushbuck, roan, and aardvark are either so rare or restricted in distribution that lions seldom encounter them, and the bush duiker and oribi are largely confined to the northern part of the Northern Extension. Birds are seldom hunted and, with the exception of the ostrich, provide at most a snack. Fourteen species constitute the main prey of lions, and for each I present some data on numbers, social organization, and other information of relevance from the standpoint of predation.

Ecological Distribution

The prey population consists of migratory species and those that are semimigratory or resident. The migratory ones—wildebeest, zebra, Thomson's gazelle, and eland—have a wide ecological distribution with most of the animals spending the rainy season on the plains and the dry season in the woodlands. All are principally short-grass feeders which avoid thickets and other dense vegetation.

The wildebeest comprise a large migratory population and small resident ones at Kirawira and in the Mara (fig. 35). There is also a resident population in the Lolionda area (Watson et al., 1969) and one in and around Ngorongoro Crater. The migratory animals generally

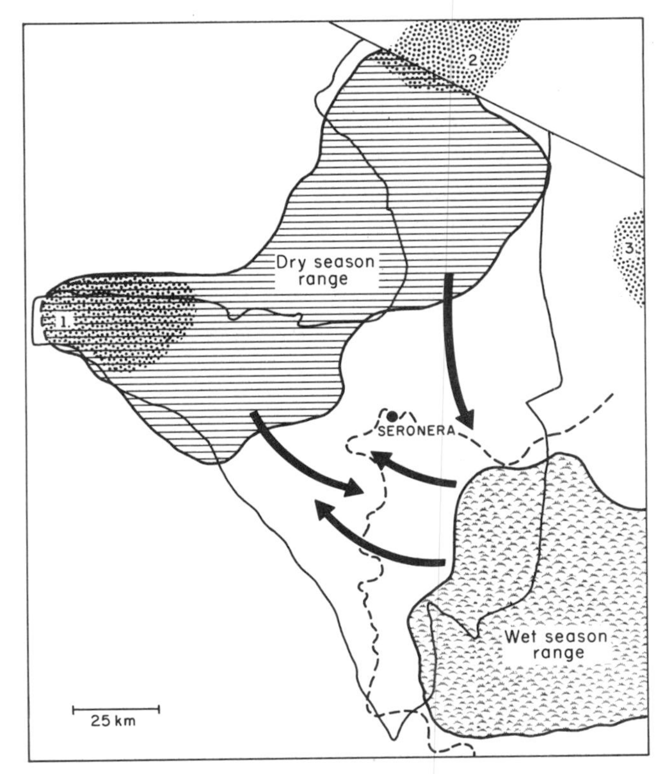

Fig. 35. The distribution of wildebeest in the Serengeti area showing the three resident populations and the movement pattern of the migratory population.

move to the plains in November or December, abandoning the woodlands so completely for several months that they are rare there at that time. The westward movement off the plains is rapid and usually takes place between May 15 and June 15. The animals stream down the Corridor, then spread to the northeast and into the northern part of the Northern Extension where they are usually found in September and October. Between 1962 and 1965 the population spent 53% of its time on the wet season range. About 51% of its total range lies within the park (Watson, 1967).

Zebra follow the same general pattern of movement to and from the plains except that the population does not travel in vast herds and a fair number of individuals remain scattered throughout the woodlands even at the height of the rains. Their migration from the plains is more gradual than that of the wildebeest, and the animals then scatter widely through the woodlands during the dry season. About 42% of the total range of zebra lies within the park, but actual use of the park by the animals in terms of occupancy is 65% (Watson, 1967). Eland resemble zebra in their movement pattern.

Most Thomson's gazelle spend the rains on the eastern short-grass plains and at the onset of the dry season follow the wildebeest and zebra westward, generally reaching the Seronera area between late May and late June. Although some gazelle move far down the Corridor and into the Northern Extension to about the Grumeti River, the main concentration remains near the edge of the woodlands from the Moru kopjes and Seronera, eastward to the Musabi plain and northward to Banagi, a preferred dry season area quite distinct from that used by the other migratory species. About one-third of the total range of the gazelle population lies outside of the park (Watson, 1967).

Grant's gazelle are widely distributed in the ecological unit and at least those on the plains are semimigratory with numbers fluctuating considerably around Seronera (see table 34). The same may be true of ostrich. Topi and hartebeest, both long-grass feeders for part of the year, are woodland animals which also occupy the plains sparsely. Both are semimigratory in that noticeable shifts in their population may occur. Topi are particularly abundant around the Ndoha plains.

Most resident species have a more restricted distribution than the migratory ones. Warthog are common around the Musabi plains and in parts of the Northern Extension. They are found in moderate numbers throughout the rest of the woodlands and are sparse on the plains. Impala are confined wholly to the woodlands where herds are widely scattered during the rains but concentrate in the vicinity of the Mbalageti, Grumeti, and other rivers during the dry season. Buffalo herds are found through much of the woodlands and some may on occasion follow a drainage line into the plains. They are long-grass feeders whose most important habitat during the dry season is the grasslands bordering rivers. Bohor reedbuck and waterbuck are usually found near watercourses, a highly restricted habitat, although a few animals of the latter species occur on ridgetops and other localities far from riverine forests. Giraffe are browsers which are found singly and in herds wherever there are trees and shrubs, sometimes even at isolated kopjes in the plains.

From the point of view of lion ecology two points in the distribution of prey are particularly important. One is the shifting of migratory herds to and from the plains, leaving that area almost devoid of wildlife during the dry season and the woodlands impoverished during the wet one; the other is the concentration of both migratory and resident prey near rivers in the woodlands during the dry times of the year.

Numbers

I reluctantly present figures on the number of prey animals in the ecological unit. Wildlife censuses have been done over large areas in several parts of Africa but with few exceptions these represent only rough estimates no matter how precise the figures seem. In the Serengeti, repeated attempts have been made to count wildebeest, buffalo, and elephant, but work on other species has been sporadic or is still in progress (see also Appendix C). Counts made prior to my study (Grzimek and Grzimek, 1960; Talbot and Stewart, 1964; Watson, 1967) may or may not have relevance because population numbers may have fluctuated. Some species, particularly wildebeest, can be counted quite accurately by using aerial photographic methods when the animals are massed on the plains; others may be censused either by making total counts or by taking a stratified sample (see Watson et al., 1969), but for a few species, such as warthog and reedbuck, accurate censuses are difficult to conduct. During the rains I tried to count all individuals of certain species from the air on several woodland plots, covering the whole area in a definite pattern and photographing large herds so as to be able to tally individuals later. While this provided me with some figures of resident prey numbers in the park, the information can not be extrapolated to other parts of the ecological unit because wildlife generally is less abundant there.

Migratory species. The Serengeti is justly famous for its vast migratory herds, but until recent years the number of animals actually involved has been based largely on guesswork. "At least ten million head of zebra and wildebeest covered the veldt for miles in front of us," wrote Johnson (1929), no doubt an exaggeration. Yet even after several years of intensive research, only an approximate figure can be given for most species.

Wildebeest. The resident populations of wildebeest are small and relatively unimportant to the lion in the ecological unit either because they exist peripheral to it or because few lions inhabit the area. The Kirawira population numbered about 5,500 in 1966 (Bell, 1967), the Mara one about 18,000 in 1961 (Talbot and Stewart, 1964), the Ngorongoro one about 14,000 in 1965 (Estes, 1966) and the Loliondo one about 5,900 in 1968 (Watson et al., 1969). There is no evidence that these populations have fluctuated markedly in the past few years. The migratory population, however, has increased more or less steadily during the 1960s, presumably because a succession of fairly wet years provided ample forage and because rinderpest, which contributed heavily to mortality in the past (Talbot and Talbot, 1963), has not been a major cause of death since 1962:

Date	*Total*	*Source*
May, 1961	221,699	Talbot and Stewart, 1964
May, 1963	322,000	Watson, 1967
May, 1965	381,875	Watson, 1967
June, 1966	334,425	Watson, 1967
May, 1967	400,000	Watson, 1969
March, 1970	500,000+	Kruuk, pers. comm.

For purposes of this report a figure of 400,000 migratory and 10,000 resident wildebeest is used.

Zebra. Talbot and Stewart (1964) derived a figure of 151,006 zebra for the Serengeti and 20,867 for the Mara. Samples taken by Watson (1967) in 1966 suggested a population of 280,000. In September, 1969, the Serengeti Research Institute sampled zebra throughout the ecological unit and obtained an estimate of 130,400 $\pm$ 26,700 (Kruuk, pers. comm.). I have no way of evaluating the accuracy of the various estimates and therefore use a figure of 150,000. Bell (1967) tallied 12,000 zebra in the western Corridor during the rains in 1966, and scattered herds occur throughout the woodlands as well. An estimated 17,500 zebra, or 12% of the total population, do not migrate to the plains during the rains.

Thomson's gazelle. Gazelle are difficult to census accurately from the air because young ones are readily overlooked and adults blend into the surroundings unless light conditions are good. Talbot and Stewart (1964) estimated 480,000 to 800,000 gazelle, but Watson (1967) placed their number at 243,000 (including Grant's gazelle). Censuses made by the Serengeti Research Institute in September, 1969, revealed 125,300 ± 31,200 gazelle (Kruuk, pers. comm.). In this report I use the figure of 180,000 of which 10,000 remain in the woodlands during the rains, 5,000 stay on the plains during the dry season, and the rest are migratory.[1]

Eland. Talbot and Stewart (1964) estimated 4,900 to 7,300 eland, and my guess is about 7,000.

Semimigratory and resident species. I counted resident prey from the air on five sample plots in January and February, 1968. One plot was near Seronera, two were in the Corridor, and two were in the Northern Extension. In addition, P. Jarman made a ground count of resident species, except buffalo, at Kogatende near the Mara River in September, 1969; earlier, in March, 1968, Sinclair had made a rainy season count of wildebeest, zebra, and buffalo in the same area. I have combined these two counts to give an indication of the prey that lions have available there during the rains. The six plots comprise 9.3% of the woodlands area of the park. The results are extrapolated to include all park woodlands (table 35), but not the ecological unit as a whole.

Buffalo. The buffalo population has been increasing steadily since 1961 when Talbot and Stewart (1964) tallied 15,898 animals, a count which may have been low. Watson (1967) noted 39,000 in 1965, and by 1969 there were 52,000 (Sinclair, pers. comm.). I use the figure of 50,000 buffalo as the average for the period of my study.

Topi and hartebeest. Counts indicate a topi population of about 18,000 individuals in the park, but the western Corridor, where no sample plot was located, has a higher than average density. Bell (1967) counted 13,000 there in a census covering 2,860 sq km inside and outside of the park. On the other hand, the eastern part of the Northern Extension has a lower than average population density. There are probably about 27,000 topi in the ecological unit as a whole. Although hartebeest outnumber topi in some areas (see table 34) they are gener-

1. H. Lamprey told me recently (November, 1971) that a census in 1971 gave a figure of about 750,000 gazelle, results strikingly at variance with the prior count.

ally less abundant and an estimate of 18,000 is probably of the correct order of magnitude.

Others. Impala are the most abundant resident prey in the ecological unit. There are at least 45,000 in the park, judging by my sample counts, and to these I have arbitrarily added 20,000 for the rest of the ecological unit for a total of 65,000. Grant's gazelle are seemingly more abundant on the plains than in the woodlands and my guess is a total population of 10,000. Giraffe are surprisingly difficult to spot from the air when they stand still among trees, and my count may be somewhat low; there may be 8,000 in the ecological unit. Warthog cannot be censused adequately from the air and my samples give misleading results. There may be as many as 15,000 in the ecological unit. My estimate for waterbuck is 3,000 individuals, for ostrich 5,000, and I have added 10,000 animals to represent roan, reedbuck, bushbuck, and several other species.

Assuming that these figures are reasonably correct, then 958,000 large prey animals occupy the 25,500 sq km of the ecological unit, or 37 animals per square kilometer, almost half (16.1) of them wildebeest, followed by Thomson's gazelle (7.1), zebra (5.8), impala (2.5), buffalo (2.0), and topi (1.1). However, such averages do not convey actual conditions. Some 74% of the animals are on, or move to, the plains and are unavailable to resident lions for part of the year. Resident prey comprise about 240,500 individuals. Of these, most zebra, wildebeest, and Thomson's gazelle crowd into the western Corridor. Table 35 shows what is actually available to lions in the park woodlands during the rains. The amount of prey varies from as few as 8 animals per sq km near Seronera to as many as 33 in parts of the Northern Extension. Most prey consists of buffalo, impala, topi, and hartebeest, in that order of numerical importance; only one wildebeest was seen in 1,035 sq km. If all warthog, reedbuck, and others were included in these figures the total for each area would be raised somewhat. However, the resident species are not distributed evenly which may cause extensive areas to be almost devoid of prey. Buffalo, for example, characteristically travel in large herds, and mixed aggregations of zebra, impala, topi, and Grant's gazelle occur scattered here and there.

In February, 1968, Watson et al. (1969) censused the 6,734 km Loliondo Controlled Area, the western portion of which is part of the ecological unit. Their density figures for some resident species are similar to those found in parts of the park: hartebeest, .50 per sq km, impala 2.59, giraffe .39.

Biomass

Ideally the biomass of a species should be derived by multiplying the average weight of each age class and sex by the number of animals in these categories, but since counts are not precise and the population composition of each species is not known, I use three-quarters of the weight of an average adult female times the number of animals to obtain the biomass figure for a species. In this way any overestimation caused by attributing to young animals the weight of an adult is compensated for by the underestimation of the weight of large individuals, particularly of males.

The total prey biomass in the ecological unit is 107,666,000 kg (236,865,200 lbs) or 4,222 kg per sq km (24,150 lbs per sq mile). Only 37% of this biomass is available to resident lions during the rains and just over half of the biomass consists of buffalo (table 32). The biomass of resident prey in the six sample plots varied from 998 to 7,234 kg per sq km; buffalo constituted 64% of it and giraffe 13%. Both of these species are so large that lions are diffident about attacking them alone. Thus only about 23% of the resident biomass in much of the woodlands is readily available to any lion, a point to which I will return later.

Bell (1967) made a count of large mammals in 2,860 sq km of woodlands in the western Corridor where resident zebra and wildebeest are common. Excluding hartebeest, warthog, ostrich, and a few others, he tallied 57,000 animals with a biomass of 3,110 kg per sq km about 42% of which consisted of buffalo, and 6% of giraffe.

The Masai and Seronera pride areas, consisting of long-grass plains and wooded river courses, have a low resident prey biomass. As table 34 shows, this biomass fluctuates considerably as Grant's gazelle, hartebeest, and others move in and out of the area. There are no resident buffalo herds, and no migatory species remain in the area during the rains. At that time only 2 to 4 animals with an average biomass of 100 to 300 kg could be available per square kilometer.

I made a prey count on the plains between Simba and Gol kopjes in an area of 130 sq km on October 5, 1966, at the height of the dry season. The following animals were seen: 7 zebra, 197 Thomson's gazelle, 263 Grant's gazelle, 2 topi, and 45 ostrich—an average of 4 animals per sq km and a biomass of 131 kg per sq km. The census area included a waterhole near which many animals were congregated; some other parts of the plains had a lower biomass.

Direct comparisons between the Serengeti and other parks are difficult to make because so many different species weights have been used to compute biomass. Watson and Turner (1965), for example, used a figure of 500 kg for buffalo when the average animal weighs about 420 kg, at least in the Serengeti (Sinclair, pers. comm.). I have computed the biomass for four parks (table 33), using the weights as given in table 32 even though differences in the size of a species are known to exist from area to area. Several kinds of animals occur in Kruger Park but not in East Africa and for these the weights as given by Pienaar (1966) are used. The resulting biomass figures differ considerably from the published ones, of which we also availed ourselves earlier (Makacha and Schaller, 1969). For instance, Turner and Watson (1965) derived a biomass of 9,000 kg km² (excluding elephants, etc.) from a census in Lake Manyara Park in 1965. The wildlife populations had not changed appreciably by 1968, except for an increase in the number of impala, yet the biomass figure using my weights is 7,785 kg km². Until precise average weights are available for each park, computations have to be standardized in some way if comparisons are to be meaningful.

Mating and birth seasons

Prey is especially vulnerable during the mating and birth seasons: in the former because adults may be less attentive and in the latter because newborn young are physically unable to avoid predators as readily as adults. Birth season information from other areas can not be applied to the Serengeti. For example, warthog breed all year in western Uganda (Clough, 1969) but they have a sharp birth peak in the Serengeti. I kept notes on the appearance of newborn young. The results are summarized in figure 36. All species for which I have information show a more or less sharply defined birth peak, which, however, may vary in some species not only with the part of the park the animals occupy but also from season to season. Warthog and topi in the western and northern parts begin to have young a month or so earlier than those around Seronera. In 1967 I noted the first topi young of the season at Seronera on September 23; the following year a young topi was seen on August 20 and a birth peak was reached by about September 11. In general, eland, impala, Grant's gazelle, and giraffe have young throughout the year with one peak season. Zebra and buffalo also have a sharp peak and there may be 2 to 3 months of the year when births are rare. Discrete birth seasons, with most young being born within a period of two months, are found in wildebeest, topi, ostrich, and warthog.

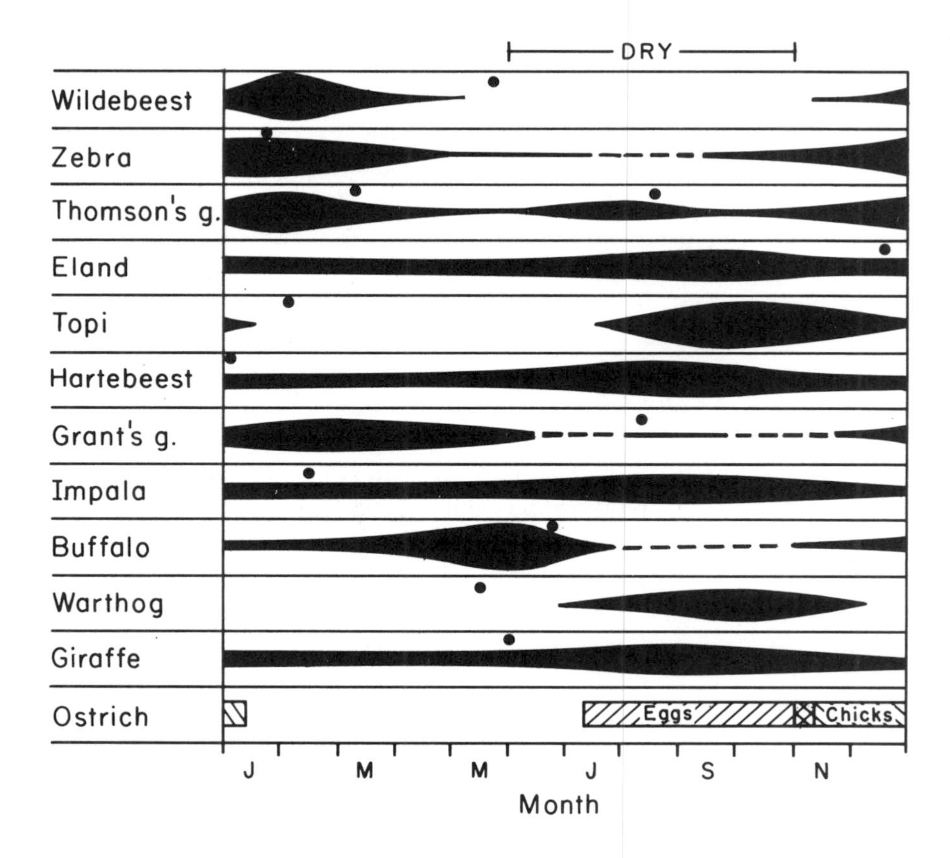

Fig. 36. The birth seasons of the major lion prey species in the Serengeti Park. The thickness of the black lines represent the relative number of young born each month, and the black dots indicate the approximate rutting peaks.

Thomson's gazelle have a major peak and a minor one and a few young are born throughout the year. I noted only one birth peak in hartebeest but it is possible that there is another one in February and March as has been reported in Nairobi Park (Gosling, 1969). Newborn waterbuck were noted in October and November and small reedbuck in September and April, but I do not have enough data to delineate a season with precision. Ostrich lay between July and November with a peak in August and September, but so many nests are destroyed by predators and fires that small chicks are not often seen. The approximate peak of mating activity is also indicated for each species in figure 36.

Social Organization

The social organization of a species is the result of selection pressures operating on mating, feeding, antipredator, and other patterns. The resulting systems vary from the solitary bushbuck and paired dik-dik to the family units of zebra, discrete herds of buffalo, and huge, amorphous masses of wildebeest. Most species form two basic kinds of herds for at least part of the year—male herds and herds consisting predominantly of females and subadults. With the exception of eland, male antelope may be territorial and thus often alone. Zebra, buffalo, giraffe, and warthog show no evidence of being territorial, but in these species too a segment of the male population tends to be solitary; females are seldom so in the first two species.

Wildebeest. The resident wildebeest in Ngorongoro Crater are divided into fairly cohesive nursery herds of females and young, loosely knit bull herds, and a network of territorial males which may retain their territories for two or more years (Estes, 1966). The migratory Serengeti population is dispersed during the dry season with herds varying in size from a few to several thousand individuals. Although the sexes are mixed in these herds, many bulls are in small groups of their own. The segregation of the sexes becomes more pronounced in December at the beginning of the birth season. Some bulls make scrapes, rub their forehead on the ground, and exhibit other forms of territorial behavior with increasing frequency in January but such activity does not reach its peak until mid-May, the height of the rut. At that time many bulls have established contiguous territories, some no more than 30 m in diameter, and each attempts to retain as many passing cows as possible. The activity of bulls is ceaseless as each circles his harem, keeping the cows tightly clumped, chases nonterritorial bulls, and challenges his neighbors. Territories may be retained for only a few hours, for when the cows move on the bull usually does too, perhaps to establish another territory farther on. Rutting ceases by late June.

Zebra. The zebra population is organized into stallion herds and family herds consisting of one male and one or more females and their young for an average total of five. The family herds remain constant in composition even when hundreds of them aggregate, except that young males join stallion herds by the age of 4½ years and young mares are abducted by stallions at the age of about 14 months. "This type of social organization is unique in being based on non-territorial coherent families and stallion groups kept together by personal bonds between members" (Klingel, 1969).

Thomson's and Grant's gazelle. The size of Thomson's gazelle herds varies from a few individuals to several hundred, with an average of about 10 to 50. Herds change constantly in composition when migrating, but when stationary tend to segregate into male and female herds and into territorial males. "As a basic scheme there is a central area with the territories surrounded by the home ranges of the bachelor herds, through both of which the herds of the ♀♀ wander in daily circuit" (Walther, 1969). Territories are about 100 to 300 m in diameter (Walther, 1968).

Grant's gazelle have a social system similar to that of the Thomson's gazelle except that on the average herds tend to be smaller. Often the territories of males are not contiguous and some may be 1 km or more in diameter (Walther, 1968).

Topi and hartebeest. Topi usually occur in small herds of some 2 to 25 individuals, but at the end of the dry season, when a local green flush appears, several hundred of them may congregate. Males are territorial and at least some of them remain so for several months of the year. A characteristic sight near Seronera is a solitary male topi standing on a termite hill, sometimes the only animal of its kind in 50 sq km or more. It was my impression that the female herds are not stable in composition, but that changes are less frequent than, for example, among wildebeest and gazelle. Hartebeest seem to behave much as topi do.

Impala. Herds consisting of females, subadults, and sometimes one or more adult males may number as many as 300 individuals but most consist of fewer than 100. Some males are territorial, but in the dry season they may abandon their territories and move to areas with green grass. "Loss of territorial status was not an irreversible event for a male, since several were seen to resume old territories or to invade new ones after a period of non-territoriality" (Jarman and Jarman, 1969).

Buffalo. The buffalo population is divided into discrete herds. These herds, each of which numbers some 100 to 600 or more individuals, consist of cows and their young and a few bulls. Such herds retain their identity and confine their wanderings to distinct home ranges. In addition, about 3,000 bulls, or 6% of the population, live singly and in small groups along seepage lines and rivers. Bulls may be expelled from a herd during the rut in July and August. Some leave on their own volition at that time, possibly because suitable habitat becomes more fragmented as the dry season advances. Many bulls rejoin herds after the onset of the rains, but those over 12 years of age remain solitary. Old animals may retain a limited range along a section of river for many months (Sinclair, 1970).

Giraffe. Giraffe are often solitary and in groups of two and three, but herds numbering 20 to 30 and up to 50 may be seen. Relations between animals are casual: "never has a large herd been comprised of the same individuals in two consecutive observations" (Foster, 1966). One male I knew at Seronera wandered over at least 60 sq km, a range he occupied throughout the year. The bond between mother and young is loose—young seldom suckle after they are a week old (Foster, 1966)—and a common sight is a small young browsing alone up to several hundred meters from an adult.

Warthog. Solitary adult boars and sows are common, as are pairs. Two to four yearlings are often together, apparently comprising litters that have separated from their mother after she had new young. A sow is often accompanied by several young and at times by a yearling or two as well. The largest group I saw comprised 4 sows and 14 young. "Warthogs are diurnal and sedentary. They spend the night in holes that seem to be the most important fixed points in their home range. The home range is neither marked nor defended against other Warthogs . . ." (Frädrich, 1965).

Others. Eland herds usually consist of several cows and subadults, and sometimes one or more large bulls as well; small bull groups and solitary individuals also occur. Most herds contain fewer than 25 individuals but one that I counted numbered 184. Eland travel ceaselessly, with some herds possibly retaining their composition for a considerable period.

Some waterbuck males are territorial, "having living and breeding territories" of about .5 to 2.5 sq km. (Kiley-Worthington, 1965). Female herds and herds of mixed sex range in size from 3 to 20, but one that I saw numbered 39 animals; one male herd consisted of 10 individuals. Such herds seem to change their composition constantly as small groups leave and join.

Bohor reedbuck occur singly and in small, loose aggregations. I have counted as many as 35 together on a patch of short green grass. Associations between adults were so casual that a specific social organization was not readily apparent.

Ostrich males and females occur singly and together throughout the year with most flocks containing fewer than 15 birds. Courtship displays were seen between March and November. Males are polygamous and several hens may lay eggs (as many as 40) in the same nest. Both sexes incubate and remain with the chicks. Several broods may join, and once I saw 75 chicks with a male and four females.

Food Habits

Lions are catholic in their tastes. The prey they kill ranges from crocodiles, guinea fowl, hares, and baboons to various antelopes, buffalo, and on occasion other lions. Between 1966 and 1969, the Serengeti lions were known to have eaten 18 kinds of mammals and 4 kinds of birds (table 36). Turner told me that prior to my study these lions once killed and ate a yearling rhinoceros, a porcupine, and a leopard cub; and Bell saw them with an aardvark carcass. The list does not include an elephant, which was scavenged by lions after park authorities shot it for control purposes, and a freshly laid ostrich egg the lions broke open after the hen fled. Nor does it mention animals which lions killed but did not eat: several lions and hyenas, one black-backed jackal, one leopard, one cheetah, one vervet monkey, one tortoise, and one white-backed vulture. Table 37 presents lion food items from four other national parks for comparison with those from the Serengeti. The Kruger Park lions fed on 38 different species, those in Manyara Park on 7, a difference due in part to the greater variety available in the former area. Other reported foods include python, catfish, locusts, fruit, and a shirt (Guggisberg, 1961). Near Nairobi Park the lions "began a concentrated attack on chickens, ducks, and turkeys of six plotholders . . . . During this period over 300 head of poultry were killed but not all of them eaten" (Cowie, 1961).

Lions obviously may eat whatever they can catch, but a glance at the percentages in tables 36 and 37 indicates that in each area fewer than five species, most of them large, contribute about three-quarters of the food. The Serengeti lions prey mainly on wildebeest, zebra, buffalo, and topi; the Kafue ones on buffalo, hartebeest, zebra, and warthog; the Kruger ones on wildebeest, impala, zebra, kudu, and waterbuck; the Nairobi ones on wildebeest, zebra, and hartebeest; and the Manyara ones on buffalo and zebra. Of 42 kills reported by Bourlière and Verschuren (1960) from Albert National Park, kob constitute 31%, young hippopotamus 19%, topi 17%, waterbuck 12%, and buffalo 7%. Indian lions prey almost exclusively on domestic cattle and buffalo because little wildlife survives in the forest where they occur.

Food habits are most conspicuously influenced by four factors.

(1) *Size of prey*. Adult elephants weigh 3,500 kg and more, hippopotami at least 1,800 kg, and giraffe and rhinoceros usually over 1,000 kg. One elephant calf which had become separated from its mother was killed by lions in Lake Manyara National Park (Douglas-Hamilton, pers. comm.), and Guggisberg (1961) and Goddard (1967) describe attacks on rhinoceros, but these species are so large and the adults may defend their young so vigorously that lions usually make no attempt to prey on them. Adult hippopotami are occasionally killed in the Gorgongoza Park (R. Ardrey, pers. comm.) and the young ones are readily taken in Albert Park (Bourlière and Verschuren, 1960). Aside from their size, their habit of staying in water makes them unavailable to lions for much of the day. Giraffe are taken in small numbers (.5–4%) in all parks in which they occur. However, when giraffe in the Timbavati Nature Reserve were weak from starvation, 18% of the lions' prey items consisted of this species (Hirst, 1969). Thus prey that weighs over 1,000 kg is relatively safe from lions and in fact has almost escaped the influence of predation in general.

At the other extreme, small mammals and birds are not hunted much, undoubtedly because the energy output in trying to subsist on, for example, hares or dik-dik is not commensurate with the output. Lions do capture gazelle fawns weighing one kilogram or two but only incidentally. In the absence of other prey, prides around Seronera may subsist for long periods on Thomson's gazelle weighing 15 to 20 kg. According to Guggisberg (1961), bushpig and warthog are the main prey of lions in areas of Kenya and Tanzania where large species have been eliminated. The usual prey size thus ranges from about 15 to 1,000 kg.

Bourlière (1963) wrote: "Carnivores actually prey upon herbivores of about the same weight as themselves; they also appear to avoid animals that are much lighter in weight and smaller than themselves. Only those predators like the lion, the cheetah and the Cape hunting dog, that hunt their prey in organized groups, may succeed in overcoming animals much larger than themselves." An adult lioness averages about 120 kg in weight and a male about 170 kg; alone, both kill with little difficulty animals such as adult zebra, waterbuck, and wildebeest weighing as much as 280 kg. Topi and hartebeest weigh as much as 150 kg. Adult buffalo, some 425 to 850 kg in weight, may be attacked by solitary lions but usually several cats do so cooperatively. Of the lion's important prey in the parks that have been studied, only impala, warthog, and Thomson's gazelle average much less in weight than a lion. Bourlière's generalization restated to apply to the Serengeti lions would read: solitary lions tend to prey on herbivores weighing between 50 and 300 kg, one half to twice their own weight, but animals weighing over 1,000 kg may be killed by groups hunting together; small prey is captured either incidentally or out of necessity, because little else is available.

Prey preference with respect to size is readily apparent when lions have a choice of which species to hunt. During the dry season around Seronera, prides tend to ignore buffalo and giraffe whereas they stalk Thomson's gazelle persistently. But this prey is below optimum size for lions, and cubs often starve during the period when gazelle are the principal quarry. Gazelle in turn are disdained if topi, wildebeest, zebra, or others of that size are available and vulnerable. I have seen lions stalk through gazelle herds to reach such prey. Hanks et al. (1969) suggested that lions in Kafue Park find waterbuck meat unpalatable. No preference for a particular kind of ungulate meat was noted in Serengeti lions.

(2) *Availability*. Prey availability has a profound influence on all aspects of the lion's biology in the Serengeti. Once it was thought that lions follow the moving prey. "Some lions are resident year long in game concentration areas near water, but the majority follow the prey species on their seasonal movement within the region" (Talbot and Stewart, 1964). As this study has shown, only the small nomadic segment of the lion population follows the migratory herds. Most wildebeest, zebra, and Thomson's gazelle—about 62% of the prey biomass—are unavailable to most lions for part of the year. Since movements of prey are largely influenced by erratic weather, availability of the migratory herds is usually a matter of chance. I recorded the presence and absence of prey within the foci of activity of the Seronera and Masai prides, an area of about 250 sq km (fig. 37). A species was listed as present if it occurred within some part of the area

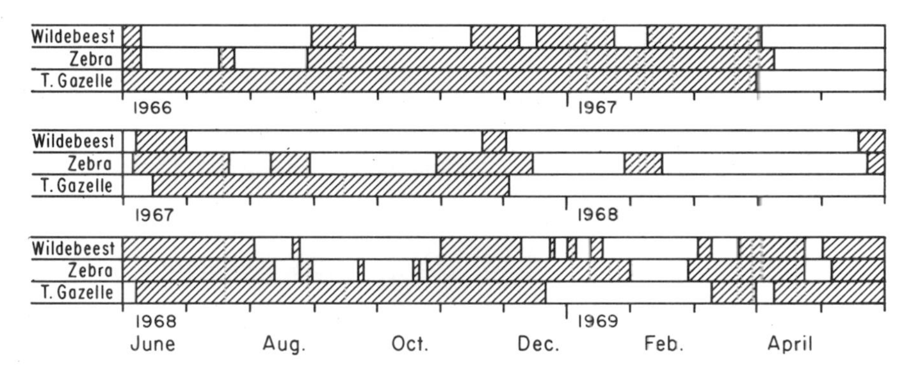

Fig. 37. The presence (crosshatching) or absence (white) of migratory prey around Seronera, showing the erratic availability of these species to the resident lions.

even though the lions were often not diligent enough to find it. The vagaries of weather produced striking yearly variations in availability. In 1967–68, a wet year, no wildebeest were at Seronera from early December to late May, but in 1968–69 they were there commonly. Over a period of three years, wildebeest were available to the lions an average of 35% (13–52%) of the time, zebra 63% (44–71%) and Thomson's gazelle 68% (47–83%). The percentages for wildebeest and zebra apply to large areas of the woodlands as well, but Thomson's gazelle are so sparse and localized in most parts that they provide little food.

The differences in availability of migratory species around Seronera are also reflected by the kill record (fig. 38). Zebra were captured during every month with peaks in September and November; no wildebeest were taken in February and October but many fell prey between April and June. Most gazelle were killed in July and August even though they were often present at other times as well: lions preferred large prey if it was available. Many gazelle were captured in June, 1966, too, but in subsequent years other prey was present.

Although resident prey in the woodlands varies in its density, and seasonal shifts in abundance may occur, the lions there always have buffalo, topi, impala, and others available. However, those lions that move to the plains may be almost wholly dependent on migratory species for several months, a fact apparent from the kill record (table 36).

(3) *Density*. The more abundant a preferred species is in an area the more likely it is to fall prey. Lions are conservative in their movements, which, together with the tendency of all pride members to be in the

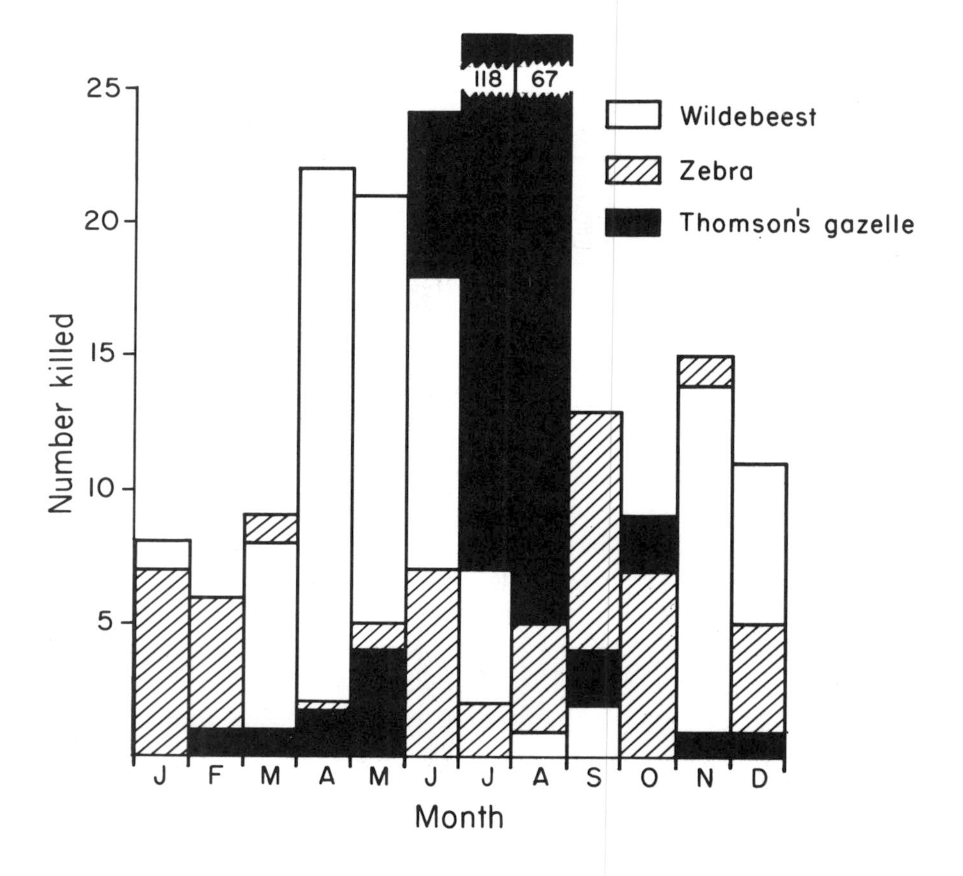

Fig. 38. The number of wildebeest, zebra, and Thomson's gazelle known to have been killed by the Seronera and Masai prides each month.

same locality, may leave some parts of their range unoccupied, and small, scattered herds there may escape the effects of predation temporarily. Furthermore, the probability that a lion will encounter a rare animal is low. On the other hand, both residents and nomads are attracted by prey concentrations, and their greater than average food intake at such times attests to the increased vulnerability of animals when they are abundant. With densities changing constantly as a result of movements of migratory and semimigratory prey, and even as a result of local movements by such residents as buffalo and impala, there is a continuous adjustment by lions, with prides confining their shifts to pride areas but nomads roaming over a large part of the ecological unit. As the density of a particular species in an area decreases through emigration, vulnerability decreases too but only up to a point—the point at which lions merely begin to work harder to meet their daily requirements.

In certain situations a species remains highly vulnerable to predation regardless of its density as long as a few animals are available and lions prefer them. This is evident late in the dry season when an occasional zebra family, often the only one for many square kilometers, approaches a waterhole at Seronera and in the process meets lions which usually hunt there at that time of the year. Indeed preference alone may induce a lion to prey proportionately far more heavily on a species than would be expected from its density. Again zebra are a case in point. Early in the dry season, when only a few zebra but many gazelle remain on the plains, the former are selected, judging by the kill record.

(4) *Scavenging.* In addition to killing their own prey, lions readily scavenge food from other predators and eat animals that have died from disease and other causes. Of the 1,180 food items that the lions ate, 75% were killed by them, 16% were scavenged, and 9% were of unknown origin. Frequently lions ate scraps with several hyenas nearby, but I was unable to determine which predator had killed the animal. In a few instances prey had probably died of disease but the clues that could be used to confirm this had been obliterated. Sometimes a feeding lion was spotted from a plane or observers reported kills to me without being certain that they had not been scavenged. All such questionable carcasses are placed in the "unknown" category.

If the scavenged food is analyzed separately for the woodlands (including the plains border), the Masai and Seronera prides, and the plains, a striking difference becomes apparent (table 38). Whereas lions in the woodlands killed 78% of their food items and the Seronera ones 88%, those in the plains captured only 47% of them. The high incidence of scavenging on the plains is attributable to three main causes: hyenas, from which lions scavenge much of their food, are more abundant in the plains than in the woodlands; herds tend to be concentrated, making animals that die from disease fairly localized; and in the treeless expanse lions can see vultures descending to a carcass several kilometers away.

Lions obtained their scavenged meat from one of eight sources (table 39). If a hyena or other predator found an animal that had died of disease but a lion then appropriated it, the lion is listed as having scavenged from that predator regardless of how the animal died. However, jackals are considered only if they had killed the prey themselves. Any animal that died of causes other than predation and drowning, including malnutrition, old age, and childbirth deaths, is placed in the disease category.

Lions took 42% of their scavenged meat from hyenas, a figure which may be low, for hyenas hunt mainly at night in contrast to cheetah and wild dogs which do so in the daytime. Most of this meat consists of wildebeest and zebra, in a ratio of 4:1, whereas lions kill wildebeest and zebra themselves at a ratio of 3:2. Relatively few gazelle are scavenged even though these comprise some 30% of the hyena's kills, mainly because hyenas either eat the carcass rapidly or flee with it when a lion approaches. In Ngorongoro Crater, lions parasitize hyenas even more than they do in the Serengeti: they took over 31% of the wildebeest killed by hyenas and 26% of the zebra (Kruuk, 1972). Cheetah provided lions with 12% of the scavenged meat, mainly Thomson's gazelle around Seronera during the dry season. Wild dogs (8%), leopard (5%), and jackal (2%) also provide lions with an occasional meal.

Disease and malnutrition killed many animals. Lions obtained 16% of their scavenged food from such sources, often alerted to their availability by the behavior of vultures. Once, two lionesses sat on a ridge at 0815 about .4 km from an adult zebra that had died in a pool. At 0830 a lappet-faced vulture descended to the carcass. One lioness watched it alertly, and, when a second vulture came, she walked toward the birds. A third bird landed and the lioness trotted up and found the zebra. On another occasion, three subadult males were in tall grass when four hooded vultures wheeled low about 130 m away. The lions saw this, walked toward the birds, and discovered a dead zebra.

I found a total of 75 zebra freshly dead from disease and other such causes. Lions appropriated 15 (20%) of these before other predators did so or vultures stripped all the meat. On four occasions I spotted a carcass first and lions suddenly arrived without my having been aware that any were in the vicinity. Of 40 wildebeest (classes II–X) that had recently died, only 2 (5%) were discovered and eaten by lions, a figure much lower than that for zebra possibly because many wildebeest often die in a small area and the lions there are then satiated either from scavenging or killing sick animals. For example, many wildebeest died around Kogatende in the 1967 dry season. We found 46 carcasses there with the skeletons still articulated and encased by dry skin, indicating that no large predators had fed on them. Buffalo, too, die of disease and lions occasionally find the bodies. Between September and November, 1967, I searched for kills along the Mbalageti River and found eight freshly dead buffalo. Three had been killed by lions, one had died and had been scavenged by lions, two died but were not found by lions, and at two sites the history was obscure.

Every year many wildebeest drown when herds cross rivers. Lions scavenge only a few of these. For example, 62 wildebeest drowned in one pool near Seronera in June, 1967, seven more died there in June, 1968, and 83 in April 1969, yet none were eaten by lions.

In Lake Manyara Park, lions scavenged 3 out of 100 prey items—two drowned wildebeest and one impala taken from a python.

The effect of vultures on the food habits of lions needs special comment. Of the six species of vultures, the white-backed is most abundant (64%), followed by Rüppell's (13%), Lappet-faced (12%), hooded (8%), white-headed (3%), and Egyptian (.7%), according to Kruuk (1967). However, the relative abundance of each species varies with the locality. Hooded vultures are particularly common around Seronera, whereas Rüppell's griffon remain primarily in the eastern plains near their nesting cliffs. Because many white-backed vultures follow the migrating wildebeest, concentrations may occur in parts of the park which at other times support only a sparse resident population. By virtue of their abundance and their preference for viscera and meat directly off the carcass, white-backed vultures and, to a lesser extent, Rüppell's griffon are the most important scavengers. With many birds gorging on a zebra or wildebeest, little meat may be left for a predator after an hour. Much of their food consists of animals that have died of disease and malnutrition, rather than of kills, which predators usually consume themselves. Such carcasses could provide lions with an appreciable amount of meat if they were permitted to remain free of vultures, and this is turn would reduce the need of predators to kill as often as they do. The other kinds of vultures have little effect on the availability of meat for lions to scavenge. Lappet-faced and white-headed vultures feed mainly on small items, such as gazelle fawns, or they tear meat and skin from bones that have been discarded by other vultures or predators; hooded and Egyptian vultures tend to collect the scraps of a carcass. Any kill in the open is kept under constant surveillance by vultures in daytime. A few birds may wait nearby, but often none are in sight until lions leave the vicinity of a carcass. At that point, white-backed and Rüppell's vultures plummet from the sky, the movement and, possibly, noise alerting vultures for many kilometers in all directions, with the result that a few minutes later the remains are covered with a writhing, hissing mass of birds. Lions may be deprived of a second meal from a kill if for some reason they fail to guard it.

Lions, like hyenas and wild dogs, are vulture-watchers. If several birds fly past and land nearby or if they descend suddenly from the sky, a lion may trot over and investigate the site. On about one-third of the occasions there is nothing to eat. One male, for example, hurried toward several vultures 1 km away only to find, after much sniffing and searching, a solitary rib of a gazelle. At other times, vultures have finished the meat or a hyena has snatched and departed with the remains. Often, however, a meal is to be had. Of the 193 scavenged items listed in table 39, at least 11% were found by lions only after they had responded to the presence of vultures. In this respect vultures benefit lions, but I would estimate that, on the whole, lions lose more food from vultures than they gain.

Evaluation of Food Habit Data

The changing pattern of food availability and other factors make it difficult to obtain an unbiased kill sample, and the question must be asked how well the percentages in table 36 actually represent the proportion of each species eaten. First, it should be emphasized again that food habits vary from pride to pride. Data covering a large region like the Corridor can give no more than a rough approximation. Second, most woodlands kills were collected between June and December; during the rains, when the grass is high and the vultures have largely left the area, it is difficult to find carcasses with lions still in attendance. Yet it is the season during which lions prey primarily on the resident species, and these are therefore not fully represented. Percentages may also be skewed in favor of wildebeest because I found pleasure in being around the huge herds. Third, small prey is always underrepresented in kill collections.

I believe that the data from the plains are fairly accurate. Wildebeest and zebra are of similar size and generally available so there is little bias in locating carcasses. Thomson's gazelle are not represented fully, but with so much other prey available lions hunt them there infrequently. The other species are not abundant and their contribution to the lions' diet is small.

I know of no other area in the park where gazelle are hunted as much during the dry season as around Seronera, a fact clearly shown in the kill sample even though the species is underrepresented. This provides a good example of how data from a small area may not be typical of the region as a whole. Topi, kongoni, and Grant's gazelle are killed infrequently, except during the rains when little else is available. All resident species except buffalo, giraffe, and impala are undoubtedly captured somewhat more often than indicated.

The woodlands sample reflects primarily the lions' dry season food habits, not those for the whole year. I searched for kills intensively for several days in September, 1967, and in July, 1969, along the Mbalageti River and found kills of 2 wildebeest, 9 zebra, 3 topi, 3 Thomson's gazelle, 2 buffalo, and 1 impala, a sample which may be quite representative for that area in the dry season. During the wet season, lions there have to subsist almost wholly on resident prey, as pointed out earlier.

In a list of kills a Thomson's gazelle and a wildebeest rate equally even though the latter provides ten times as much meat. On the average at least twice as many lions feed on a wildebeest or zebra as on a gazelle (table 40) and the amount of meat they obtain is greater. At Seronera, an average of 2 subadult and adult lions fed on an adult gazelle, each obtaining about 7 kg of meat. On the other hand, the edible portions of a wildebeest gave each of seven large lions an average portion of 16.5 kg, which, incidentally, is one good reason why the lion prefers to hunt large prey.

A list of food items also fails to differentiate between prey that has been killed and prey that has been scavenged. Only about half as many lions eat on a scavenged item as on one they have killed for themselves (table 40). Wildebeest and zebra that have died of disease or malnutrition sometimes provide much meat, as do the remains of prey that has been killed by wild dogs. On the other hand, cheetah and leopard, which seldom capture animals larger than gazelle, furnish lions with little food. Unless lions arrive at a hyena kill within a few minutes after it has been made, only scraps tend to be left. The relative amount of meat that lions appropriated from 46 wildebeest and zebra killed by hyenas was as follows: much meat—enough to gorge at least two lions—12 times (26%); a moderate amount, 12 times (26%); and just a few kilograms of scraps, 22 times (48%). In Ngorongoro Crater, where the predator population is denser than in the Serengeti, Kruuk (1972) found that when lions scavenged from hyenas they obtained much meat in 63% of the instances, little in 27%, nothing in 8%, and an unknown amount in 2%.

The number of lions which actually feed on a particular carcass gives a more accurate impression of food habits than does a mere list of kills (table 41). Occasionally most lions had left the kill site or precise information was not collected by an informant; in such cases I applied the average number of lions known to feed on such a carcass. Table 41, which is based on 6,785 lion meals, shows that wildebeest rank first, followed by zebra, Thomson's gazelle, and buffalo in that order. Scavenged meat provides only 8% of the meals.

Even these figures convey only a superficial impression of the actual importance of the prey species to lions. To obtain "a true indication of the real food preferences of a particular predator irrespective of the density of its various prey species" in Kruger Park, Pienaar (1969) calculated a "preference rating" by dividing the number of a species into the frequency with which it was killed. Thus, waterbuck, which was fifth in the list of kills, received the highest rating when compared to the number of animals available; but impala, by far the most abundant hoofed animal in the park, rated tenth in spite of the fact that it was second among kills. The term "preference rating" is unfortunate, for a kill does not only signify a predator's preference but also the prey's availability and vulnerability. While such a rating system may be useful in areas where wildlife is resident and where an unbiased kill sample has been obtained, it has limited application for the Serengeti with its constantly moving prey population. However, an interesting point can be made with respect to zebra and wildebeest.

Zebra comprise about 25% of the total migratory wildebeest and zebra population, yet 36% of the kills of these two species on the plains consisted of zebra; at Seronera and in the woodlands the figures were 42% (see table 36). The importance of the zebra becomes even more apparent when multiple killings are considered, those in which lions capture more than one animal during a cooperative rush. Thirty-four percent of the wildebeest kills were multiple (table 42), most of them of two wildebeest at a time, but on three occasions four wildebeest were captured together and twice five of them were taken together. Of the zebra killings, on the other hand, only 10% were multiples ones. Not included in these percentages are three instances when zebra and wildebeest were caught at the same time. If food items are tallied on the basis of the number of successful hunts rather than on the number of animals killed, wildebeest and zebra would contribute almost equally. The greater proportion of zebra over wildebeest in the kill sample is largely due to the greater availability of this species throughout the ecological unit at all times of the year, thereby making it the most important prey animal in spite of the fact that it ranks third in number and second in biomass.

Given the biased nature of the kill record from the woodlands, the contribution of the various prey species to the lion's diet can only be estimated—for example, by assigning to each one a relative rank based on abundance, availability, vulnerability, and preference. These factors can be considered on an annual basis or only for the critical time of the year, the rains, when the migratory herds are on the plains. The seven most important prey species from the point of view of biomass (excluding giraffe) are listed in table 43. By virtue of their abundance and vulnerability when available, zebra and wildebeest rank at the top on an annual basis. Buffalo are difficult to place. They provide much meat but solitary lions may find them invulnerable, whereas topi, the species next in importance, is readily killed. The rating of these two species is interchangeable depending on the size of the lion group that is hunting them. When only the critical time of year is considered, buffalo and topi occupy the most important positions while zebra and wildebeest rank low. These ratings apply to the woodlands as a whole and may, of course, vary with local conditions.

If the various species are rated on an annual basis for the percentage of food they provide to lions throughout the ecological unit, my guess would be as follows: wildebeest 20–25%, zebra 30%, Thomson's gazelle 2.5%, buffalo 15%, and resident and other species not mentioned so far 27.5–32.5%. The figure for gazelle may at first seem low, but it should be kept in mind that lions prefer large prey; they kill gazelle either incidentally or, around Seronera, out of necessity for a brief period each year.

Feeding on Primates

Baboons are abundant in the park, yet the lions I watched either ignored them or merely trotted toward them as they scampered into the trees. T. Moore once observed three lionesses as they vainly attempted to reach a juvenile baboon in an acacia sapling. Eyssen (1949) watched as "a lioness tore into a troop of baboons right alongside the road and killed two." In contrast, the Manyara Park lions commonly kill baboons (see table 37), a difference which may be a matter of habit. Lions seldom responded to vervet monkeys, but on one occasion a lioness had just caught an infant when I arrived. Its mother screeched in the tree above while the lioness mouthed it and finally left it. The infant lay on its side whimpering until a jackal snatched it up and ate it.

Turner told me that in 1961 a male lion pulled a visitor out of his tent by the head. The man died of his injuries, the only known such fatality in the park in recent years. Around Manyara Park, however, lions have started to add man to their diet. Several villagers have been injured, and in July, 1969, a man was caught at night and almost wholly eaten. Two more killings occurred in the ensuing months, and in May, 1970, Makacha wrote me: "Satima the young male among the Mahali pa Nyati pride has been shot when he was found eating another killed person near the Park Headquarters." These lions have ample food in the form of buffalo and impala available.

Feeding on Carnivores

Lions generally disdain other predators as food, even when they are hungry. Of ten hyenas killed by lions only one was partially eaten by them (Kruuk, 1972). In addition, the Seronera pride once wholly devoured a hyena. A jackal, cheetah, and a leopard were killed but not eaten.

Cannibalism has been occasionally reported among lions (Guggisberg, 1961), and it occurred twice during this study. Once a lioness ate the cub of another lioness which had intruded into her territory. On the second occasion two males ventured into a neighboring pride area where they found and killed a litter of three cubs of which they ate one and possibly two. When the mother of the cubs returned she ate the third one. Yet in most instances dead lions were not eaten even though other lions knew of their presence: cubs dead from starvation, cubs killed by unknown assailants, adults killed in fights. The factors that trigger cannibalistic behavior remain unknown.

Feeding on Porcupines

The porcupine is protected by quills of up to 30 cm long and lions rarely kill it. One pride merely watched a porcupine waddle by; on another occasion several lionesses harassed one in a thicket and one female emerged from the encounter with a quill in her muzzle which she pawed out after repeated attempts. One night a lioness followed a porcupine for at least a half hour. She several times raised her paw as if to slap it, but the porcupine merely rattled its quills and sidled toward her, causing a quick retreat.

Feeding on Reptiles

Lions chewed on the carapace of a live leopard tortoise on three occasions without cracking it open. On one occasion, two lionesses saw a monitor lizard, 3 feet long, basking on the bank of a streambed. One lioness approached it slowly from behind. Suddenly the lizard puffed up its body and, when the lioness came close, lashed its tail back and forth once. The lioness leaped backwards with a woof, tried another approach with the same result, and, after circling the reptile cautiously, molested it no more. When a lioness met a spitting cobra "she held her head low and repeatedly sprang forward and recoiled with a 'woof' . . . After doing this several times she withdrew, followed by a litter of several small cubs" (Chief Warden's Report, June, 1965).

Feeding on Vegetation

Lions sometimes spend as long as 10 minutes biting off the tips of green grass blades, behavior also common in tigers. The reasons for eating grass are obscure. Murie (1964) found that several wolf droppings contained grass and round worms, and he suggested that the vegetation may act as a scour. Lions in desert areas are said to eat melons to obtain water (Pease, 1914).

PREDATION IN RELATION TO SEX, AGE, AND HEALTH OF PREY

Although an attempt was made to find out the sex, age, and health of animals killed by lions, it was in most instances not possible to obtain information on all three points. Sometimes a kill could be sexed even though it was not feasible to scavenge the head for later examination of the teeth to determine its age, but in most instances the health of an individual remained unknown.

Sex of Kills

My samples of sexed kills are adequate only for wildebeest, zebra, Thomson's gazelle, and buffalo, and these are analyzed here.

Wildebeest. Watson (1967) found an equal sex ratio at birth and noted, as did Talbot and Talbot (1963), that the adult ratio was approximately equal too. More male than female yearlings were killed but not significantly so. A total of 193 adults of classes V to X were sexed (table 45 and 10 unaged ones), and of these 129 were males and 64 were females, a highly significant difference ($p = <.001$). This suggests a strong selection for males by lions, assuming that the sample is not biased. The sexes of wildebeest tend to be segregated; as Talbot and Talbot (1963) observed, "on a year-round basis, 67 percent of the males noted were separated from the females." An influx of males into an area could skew the sample without selection having taken place on the part of the lions. For example, late in 1968 the lions killed 10 adult and 4 subadult wildebeest around Seronera, none of them female, but a perusal of the herds showed that these consisted mostly of males. On the other hand, the following April the kill record in the same area consisted of 5 males, 6 females, and 15 subadults.

However, I think that males in general are more vulnerable to predation than females. Single males or small groups of them are more widely scattered than female herds, are often found near river thickets, and solitary individuals may remain in an area long after the main herds have passed through—all traits that make males more available and vulnerable to lions. In addition, during the rut males behave erratically and incautiously. Some gallop in and out of high patches of grass and through ravines, and they circle the females they have collected, thereby placing themselves at the periphery where a stalking lion would reach them first. Of the kills collected during the rut, 45 were males and 19 were females ($p = < .05$), but I was unable to find out what percentage of the males were territorial at the time they were killed. Estes (1966) speculated that lions prey "largely on territorial males in Ngorongoro Crater." Although solitary individuals tend to be more vulnerable to predation than those in herds, lions find it difficult to approach a stationary animal in open terrain, which territorial males tend to occupy. The supposition thus needs verification. Males may also be more susceptible to disease and malnutrition than females, and this could increase their vulnerability. I found 29 freshly dead adults of which 19 were males. Considering the preponderance of males among kills, and possibly among diseased animals, the supposed 1:1 sex ratio is surprising. Either these factors have little influence on the population or another mortality factor selects for females, providing that the ratio is accurate.

Of 506 wildebeest kills sexed in Kruger Park, 56% were males (Pienaar, 1969), an almost significant difference ($\chi^2 = 3.56$); in the nearby Timbavati Reserve 55% of 69 kills were males (Hirst, 1969).

Zebra. Klingel (1967) observed a 1:1 sex ratio among foals but was unable to obtain a figure for the adult segment of the population. Skoog (pers. comm.) collected a large sample of adults but it was biased in favor of males, which tended to lag behind the fleeing herds and were then shot. Assuming a 1:1 ratio among adults and taking into account the cohesive groups to which most zebra belong, the kill record should be equal unless certain behavioral traits make one sex more vulnerable than the other. For example, females characteristically lead a herd to water and males bring up the rear of a moving herd, patterns which could readily bring both sexes into contact with a stalking lion. Table 47 shows that males and females are killed equally often in classes IX to XVII but that significantly more males are killed in the four oldest classes ($p = <.05$). This could mean that old males are more vulnerable than old females or that there are more males in the population. Two pieces of information point to the latter alternative. As is evident from table 47 more female than male zebra ($p = <.05$) died of disease or malnutrition in age classes XIV and XV. The shot sample also shows a higher mortality rate among females than among males between the ages of 5 and 15 years. Taken from the approximate mid-point of one class to the mid-point of the next one, males showed a 17% mortality between classes XIV–XV and XVI–XVII and a 52% mortality between XVI–XVII and XVIII–XIX. For females the respective figures were 30% and 65%, the difference between the two sexes being possibly due to disease or malnutrition. Thus lions may not select old males but merely kill what is available. In Kruger Park the lions killed zebra of both sexes equally often (Pienaar, 1969) but in Kafue Park the sample consisted of 3 males and 18 females (Mitchell et al., 1965).

Thomson's gazelle. The short conical horns of male gazelle fawns become visible by the time the animal is 5 months old, and by the age of 9 months (class IV) the horns are at least 8 cm long. A total of 179 gazelle of classes IV to X were sexed (including 42 that were not aged) and these comprised 94 males and 85 females (table 49). To find out what percentage of the gazelle population around Seronera during the dry season consisted of adult males (VI to X), I made a sample count each year. Of 3,969 animals tallied, 18% to 24%, with an average of 22%, were adult males.[2] Adult females were difficult to distinguish

2. H. Hvidberg-Hansen and A. de Vos (Reproduction, population, and herd structure in two Thomson's gazelle populations. Mammalia. 35:1–16 [1971]) found that 31% of the Serengeti gazelle population consisted of adult males. Perhaps a disproportionately small number of males visit Seronera, where I did my censusing. It is also possible that the results of these authors are biased in favor of males because their censuses were often made along roads where males tend to congregate.

from large subadults but it was my impression that they were slightly more abundant than males, constituting perhaps 28% of the population. In some seasons more adult males were caught, in others more females. The percentages in the total kill sample were: 1966—49% male, 20% female; 1967—26% male, 36% female; 1968—30% male, 34% female; 1969—49% male, 23% female. In two of the seasons the sexes caught were in about the expected proportions but in the other two over twice the expected number of males fell prey. This difference could be related to a combination of ecological conditions and gazelle behavior. The 1966 and 1969 seasons were exceptionally dry and the plains were burned. The gazelle massed near the waterholes to drink. When disturbed by lions, the females and subadults fled immediately; reacting more slowly, the males tarried and were caught. Nonterritorial males also tend to feed along river courses and thickets, as pointed out by Walther (1969), making it possible for lions to stalk undetected. On the other hand, in 1967 and 1968, gazelle only came to water occasionally and males concentrated less near the rivers because not much grass had been burned there. During these two seasons the males were no more vulnerable to lion predation than the females.

Buffalo. Table 50 shows that among subadult and young adult buffalo, 2 to 7 years old, the sexes are killed with about equal frequency but that among the older animals considerably more males than females are taken. Sinclair (pers. comm.) found that few males over 10 years and none over 12 years are in mixed herds; they are usually solitary or in small bull herds. Such individulas are undoubtedly easier for lions to attack and kill than those in herds, and, being widely scattered, they are also more available. The selection for old bulls is greater than the figures indicate because there are proportionally more young animals in the living population.

In Lake Manyara Park the ratio of males to females killed by lions was even more unbalanced than in the Serengeti. Of 50 large but unaged animals sexed by Makacha (pers. comm.), 44 (88%) were males. A sample of kills from Kruger Park comprised 193 males (67%) and 93 females (Pienaar, 1969), and one from Kafue Park consisted of 56 males (61%) and 36 females (Mitchell et al., 1965), showing that in these two areas the males also bore the brunt of predation.

Others. Among the adults of other species killed by lions were 12 male and 9 female topi, a hartebeest of each sex, 3 male and 4 female eland,

1 male and 4 female Grant's gazelle, a reedbuck of each sex, 2 male and 1 female impala, a female waterbuck, 3 giraffe of each sex, and a male and 6 female warthog. On the basis of a few sample counts, I would judge that female topi are more abundant than males and the kill sample, though small, therefore suggests that the latter are particularly vulnerable to predation. No deductions can be drawn for the other species on the basis of these samples.

Age of Kills

Klingel and Klingel (1966) published an absolute age scale for zebra, and Sinclair generously gave me one for buffalo. For the other species only relative scales can be used. However, from the point of view of assessing predator selectivity, absolute age is less important than the relative amount of tooth wear. An animal which has its teeth worn to the gums is likely to be in poor condition and thus susceptible to predation regardless of its age.

Wildebeest. The first molar erupts at 6 months of age, the second at about 15 months, and the infundibulum of the third molar divides at about 3 years in the mandible and 3½ years in the maxilla (Talbot and Talbot, 1963; Watson, 1967). The tooth eruption sequence together with the sharp birth peak make it possible to determine the age of kills quite precisely up to 42 months (classes I–V). Watson (1967) presented a relative age scale for adult wildebeest, but I found that his tooth wear criteria were not mutually exclusive and therefore prepared a scale of my own (table 44).

Of 262 wildebeest kills aged, 5% belonged to class I and 8% to class II. Watson (1967) sampled wildebeest in April, May, or June between 1964 and 1966 and found that 9% to 18% (av. 13%) of the population consisted of calves 3 to 4 months old (class II); from 1967 to 1969 the average percentage was 12% (Sinclair, 1970). From this one might assume that proportionally fewer young than expected are killed by lions. However, small animals are not represented fully in the kill sample. Among the 21 killings that were actually observed, 14% of the animals were in class I and 19% in class II as compared to 4% and 7%, respectively, in the sample for which the capture was not seen. Thus, slightly more than the expected number of young animals were taken.

Mortality between classes II and III was low, judging from Watson's (1967) sample counts, and since wildebeest of that age were not likely to bias the sample much because of their size, I think that only the expected number of this class are caught by lions. The percentage of yearlings in the population ranged from 7% to 17% (av. 13%) in the four yearly May counts made by Watson (1967); it was 10% in 1969 (Sinclair, 1970) as compared to 8% in the kill sample.

Sinclair (1970) calculated the expected number of wildebeest in each age class of the living population from the life tables as presented by Watson (1967). As table 46 shows, lions captured the expected number of adult males among young and middle-aged animals, but they caught proportionately more old males (classes IX and X) than were present in the population. Young adult females fell prey as often as expected but prime ones (classes VI–VIII) much less so. A significantly large number of females of class IX were killed, but those in class X were preyed upon in about the same proportion as they occurred in the population.

Of 510 wildebeest kills aged in Kruger Park by Pienaar (1969), 83% were designated as adult. In the nearby Timbavati Reserve, 73% of 76 wildebeest were adult (Hirst, 1969).

Zebra. I aged 174 zebra kills by using the sequence of tooth eruption and the progressive wear of the incisors as outlined by Klingel and Klingel (1966). Nine percent of the kills were young of classes I to IV. However, of 15 killings observed, 33% belonged to these classes, indicating that in this species, as in wildebeest, young animals are represented only about one-third as often as they should be. Skoog (pers. comm.) felt that his sample of shot animals of classes I to VIII was biased by being too low, and as a result the kill sample of young, yearlings, and subadults cannot be readily compared with it. Klingel (1969) found that only 31% of mares in the Serengeti were accompanied by foals in October, pointing to a high mortality of young in the first nine months of life. The 33% young in the observed killing sample may represent a somewhat larger proportion than is present in the population; the yearling kill may be lower than expected; and the subadult kill seems to be of the same proportion as the occurrence of the subadults in the population.

A comparison of the shot adult sample with the kill data in table 47 shows that lions captured as many zebra as expected in all classes except the two oldest ones, in which proportionally many more old zebra were killed ($p = <.001$). Apparently old animals are caught because they react more slowly and run less swiftly than the others, although I would not have expected such a striking difference on the basis of observing lions hunt.

In Kruger Park, 80% of the 340 kills aged by Pienaar (1969) were adult. Twenty-seven kills from the Kafue Park consisted of 30% young, no yearlings, 11% two-year-olds and young adults, 37% prime animals, and 22% old ones with heavily worn teeth, figures rather similar to those from the Serengeti.

Thomson's gazelle. I aged 204 of the 258 kills using the criteria outlined in table 48. Wildebeest and gazelle require separate scales because their tooth wear-pattern differs in many respects. For example, the incisors of wildebeest are often worn to the gums, an infrequent condition among gazelle; conversely, the third molar of a gazelle is often heavily worn but that of a wildebeest seldom so. The actual killing of 112 of the 204 gazelle was observed, and their ages are shown separately in table 49. Small fawns (class I) constitute a much higher percentage of the observed sample than the one in which the killing was not seen. In the other classes the results are generally of the same order of magnitude. Gazelle are very productive, with some females having young twice a year (Brooks, 1961); this, together with several rough counts I made, suggests that at least one-third of the population is a year or less old. About 40% of the kill sample consists of animals less than 9 months old. A few more are 9 to 12 months old, indicating that lions capture proportionally about as many young as expected. Males with horns 20 cm long and over but not yet fully grown were presumed to be 1 to 2 years old. Judging by counts made around Seronera, about 15% of the population consisted of yearlings (part of class IV and class V). The kill sample corresponds to this figure.

A. DeVos and H. Hvidberg-Hanson shot 27 gazelle belonging to classes IV to X and kindly let me check the teeth. On one occasion, hyenas killed over 100 gazelle in the Mukoma Plain on a dark, stormy night, and of these H. Kruuk collected 31 of classes IV to X. Table 78 compares the lion kills with these two population samples on the assumption that both died fairly randomly. The lion kills correspond so closely to the shot sample that there is no evidence for selective predation.

Buffalo. Lions killed buffalo of all ages but selected differently from the male and female segments of the population (table 50). Fewer males than expected were taken in the young age groups whereas males 12 years old and older were killed in proportionately greater number than they occurred in the population. Such old bulls are usually solitary or in small bull herds (Sinclair, 1970). On the other hand, the number of female kills in each category resembles the expected distribution except in the oldest categories, in which lions killed proportionately more animals. Sinclair (1970) pointed out that if the male sample was reduced to that of the female and both were weighted according to the age distribution in the living population, significantly more males than females over 12 years of age were killed ($p < .001$), an indication that life in a large herd provides females with protection. Mitchell et al. (1965) aged 61 buffalo in the Kafue Park of which 20% were less than a year old, 21% were young prime animals, and 34% were in their prime, to mention only the three highest percentages. Of 292 buffalo aged in Kruger Park, 84% were adult.

Others. My aged sample for the prey species not discussed so far is small, and, as table 51 indicates, shows only that animals of all ages, including many prime ones, are captured by lions. In addition to these, 12 warthog were aged, consisting of 2 small young, 2 large young, 2 yearlings, and 6 adults with moderately worn teeth. Among other mammals killed was a large pangolin, presumably adult, a young hare, a young bushbuck, and an adult female waterbuck.

Health of Kills

It is difficult, indeed usually impossible, to find out the health of an individual that has been captured by a lion. Even a thorough autopsy would in many instances reveal only that it harbors a wide variety of parasites and suffers from various physical defects which may or may not have made it vulnerable to predation. Brooks (1961), Talbot and Talbot (1963), Sachs and Sachs (1968), and others have shown that one or more potentially debilitating organisms afflict most animals, but baselines to distinguish innocuous from dangerous levels of infection have not been determined.

The bone marrow is commonly used to ascertain physical condition in North American deer (Hornocker, 1970). A healthy animal has fatty marrow but a starved one with its body resources depleted has translucent, gelatinous marrow. This technique cannot be uncritically applied to tropical animals. Wild ungulates in East Africa generally have a fat content below 3% (Ledger et al., 1967), much lower than that of temperate ones, so that any stressful situation may deplete the marrow without necessarily signifying that the animal is in the ultimate stages of starvation. I checked the marrow of 13 adult wildebeest during the 1968 rut. In males the marrow was generally more depleted than in females, a suggestive indication in spite of the small sample. To find out if lions prey selectively on animals with marrow in poor condition it would be necessary to collect samples from the living population at intervals throughout the year, for, as Hornocker (1970) has shown, animals with depleted marrow may not be killed proportionately more often by a predator than those in good condition. Such extensive collecting was not feasible in the park. Furthermore, depleted marrow signifies a slow-acting agent, whether disease or starvation; virulent diseases kill so rapidly that the marrow is not affected. For instance, wildebeest suffering from mange had marrow in poor condition when they finally died, but those that merely collapsed of unknown causes did not.

The health of individuals often reflects the condition of the range because animals suffering from malnutrition are more susceptible to various diseases than those which are well fed. This usually makes it difficult to determine if an illness was primarily or secondarily responsible for the poor condition of an animal.

Wildebeest. Talbot and Talbot (1963) described a wide variety of endoparasites from wildebeest but they "had no direct evidence, however, of these parasites being a primary cause of sickness or mortality." Schiemann (pers. comm.) found that the larvae of a dipteran fly (*Gedoelstia* sp.) occasionally penetrate into the brain, where their presence may cause an uncoordinated gait and other abberrations. Sporadic instances of anthrax have been reported, and Sachs et al. (1968) found a fairly high incidence of brucellosis. Animals with hoof abscesses were seen several times. In early June, 1967, I found 13 dead or dying wildebeest near Naabi Hill, all healthy looking except that some staggered, fell, and were unable to rise again. Many wildebeest died near the Mara River in September, 1967. Of 46 that were aged, 24% belonged to class III and 43% to classes IX and X, indicating that most were young or old. Of 126 wildebeest that died in the same area a year later, Sinclair (pers. comm.) found that class III comprised 36% and the two oldest classes about 24%. I mention these instances not only to convey some idea of the magnitude of mortality from disease, malnutrition, and other causes among wildebeest but also to indicate the profusion of carcasses available to lions at certain times.

An obvious disease, not encountered, interestingly, by Talbot and Talbot (1963), is sarcoptic mange, which covers animals of all ages with a gray crust. I saw only one such wildebeest dead between June, 1966, and May, 1968, but saw twenty of them between June, 1968, and May, 1969, suggesting an increased incidence of the disease. In addition, lions killed a few such animals each year of the study, two the first year, four the second, and three the third. At least 13 (4.5%) of the 284 wildebeest that fell prey to lions were in poor condition from one cause or another. One wildebeest female with a swollen foreleg, for example, stood beneath the same tree for four days and on the fifth a lioness killed her. Her bone marrow was in good condition. Lions captured wildebeest 22 times while being observed and 5 (23%) of the animals were sick: three had mange, one staggered, and one was unable to rise to its feet. Possibly, more animals that are sick are captured in daytime, when most observations were made, than at night, but the incidence of sick animals in the diet is nevertheless substantial.

Between June, 1968, and May, 1969, the bone marrow of wildebeest that died of known causes was checked systematically (fig. 39). Animals that died of disease and malnutrition generally had depleted marrow, as did many that were killed by hyenas. In contrast, of 13 wildebeest

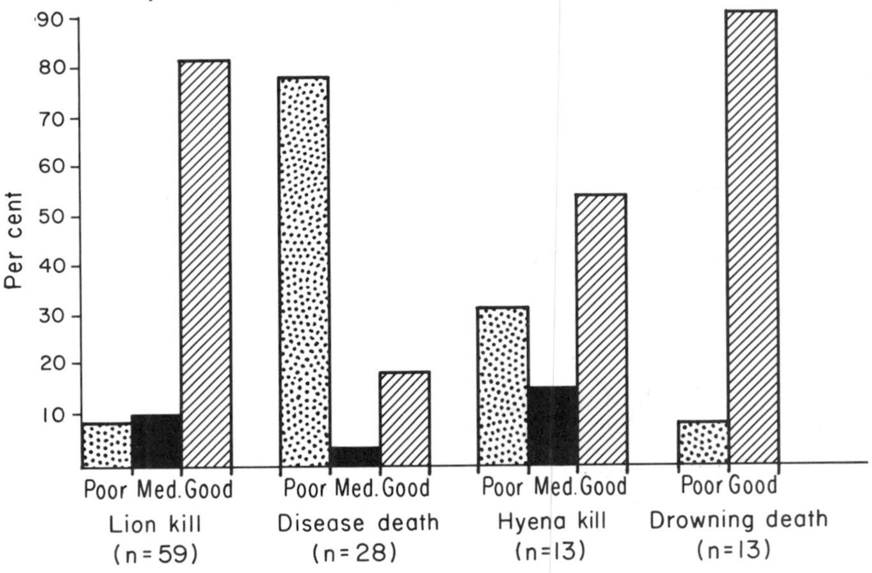

Fig. 39. The condition of the bone marrow of 113 wildebeest (classes III to X) comparing lion kills, hyena kills, disease deaths, and drowning deaths, June, 1968, to May, 1969.

that drowned while crossing a river in April, only one had marrow in poor condition. Nearly 20% of the lion kills showed signs of depletion, which together with the previous observations tends to suggest that about one-fifth of the wildebeest captured by lions are physically below par. Hornocker (1970) noted with respect to puma predation that "poor animals were not selected but were taken proportionally to their occurrence in the population." Lions do select individuals in poor condition, and it seems unlikely that the 20% of such animals in the kill sample represents the actual proportion in the population.

Zebra. Zebra dead from disease were quite common (see table 47), but except for isolated instances of anthrax their cause of death remained unknown. Most seemed to be in good physical condition, even those that fell and died within two hours. The bone marrow of zebra typically consisted of a red mushy material, making it difficult to evaluate its condition, but the fat content of the animal was usually not depleted, judging by deposits in the eye sockets. Most dead animals were found on the plains at a time of year when ample green forage was available which suggests that disease rather than malnutrition was the primary cause of death of many zebra in the sample. Several zebra with limps and serious wounds, some of these a result of having been mauled by lions, were also noted. I was present when lions killed 17 zebra. Of these 4 (23%) were in poor condition. On one occasion a male lion chased a small herd from a waterhole, and one thin animal lagged after 60 m and was caught. The three other sick individuals were solitary.

Thomson's gazelle. Brooks (1961) listed seventeen species of internal parasites that infect Thomson's gazelle and noted that the cause of death of one animal may have been due to a heavy infestation of lungworm, *Protostrongylus*, a parasite which affects many animals (Schiemann, pers. comm.). Anthrax, enteritis, and nephritis are some other pathological conditions found in gazelle. The most conspicuous disease was sarcoptic mange caused by mites of the genus *Sarcoptes*. Heavily infected animals had a thick black crust over much of the body, particularly between the hind legs. These animals were emaciated, and their flight distance to man was reduced to as little as 10 m. I saw a total of 91 such gazelle, 21% of them in June and 44% in July, indicating a high incidence early in the dry season. Mange affects mainly adult males, and 68 of the 91 animals were of that sex ($p = <.001$). By comparison, of 25 adult gazelle found dead or dying of other diseases, 14 were males and 11 were females, a more even ratio. One female died giving birth and a male died after having been gored in a fight. Animals with broken legs were seen several times. Once a lioness stalked such an animal unsuccessfully. None of the gazelle I saw being captured was noticeably ill. I think that lions take only a small percentage of animals in poor condition.

Buffalo. Buffalo that have died of disease, malnutrition, old age, or a combination of factors are noted fairly often. The blood protozoons *Theileria*, *Trypanosoma*, and *Babesia* have been found in Serengeti buffalo, and a high percentage of calves are infected with tuberculosis and foot-and-mouth disease (Sinclair, 1970). Solitary bulls are sometimes in poor condition: they are thin, they limp, and they have sensory defects, judging by their tardy reactions to potential danger. One bull, at least 15 years old and obviously sick, was killed by lions soon after Sinclair (pers. comm.) observed it. It is probable that a substantial portion of the buffalo that fell prey to lions were physically below par. Sinclair (1970) examined the marrow of buffalo that had died of various causes and noted that animals 10 years or less old had marrow in good condition but that 13 out of 14 old ones had marrow depleted of fat. The tendency of lions to select old animals was noted earlier. Of 19 buffalo kills checked for condition, Mitchell et al. (1965) considered 47% to be poor.

Others. It was my impression that migratory species were more afflicted with disease than resident ones, with the exception of buffalo. Adult eland, a migratory animal, were found freshly dead three times, but topi and hartebeest only once each. Also dead, apparently from disease, were two impala males, one male giraffe, and one male Grant's gazelle.

8 The Hunt

"There are certain things in Nature in which beauty and utility, artistic and technical perfection, combine in some incomprehensible way, the web of a spider, the wing of a dragon-fly, the superbly streamlined body of the porpoise, and the movements of a cat" (Lorenz, 1964). And at no time is such movement more vitally beautiful than when a lion tautly snakes toward its prey. I found that fleeting hesitation between the end of the stalk and the final explosive rush a moment of almost unbearable tension, a drama in which it was impossible not to participate emotionally, knowing that the death of a being hung in the balance. Surprisingly few detailed descriptions of hunting and killing by lions have been published (see Guggisberg, 1961). This chapter summarizes my observations on this topic, one which from the lion's point of view is more important than any other in its day-to-day existence.

Behavior of Prey

The success or failure of a hunt depends as much on the response of prey to the lion as on the behavior of the cat itself. While a lion is well endowed with claws and teeth to grasp and kill prey, it lacks speed. This fact is clearly recognized by the various hoofed animals and considerably influences their responses. Howell (1944) and Demmer (1966) gave the maximum running speeds for several species as follows: zebra, 60 to 70 km/hr; hartebeest, 70 to 80; wildebeest, 80; Thomson's gazelle, 80; Grant's gazelle, 64 to 80; giraffe, 50 to 60; buffalo, 56; and warthog, 48. These figures are of the correct order of magnitude, judging by my casual observations of animals running beside the car, except that warthog can attain a speed of 55 km/hr. Other species, such as eland and topi, can also run at speeds of at least 70 km/hr. It is difficult to determine the speed of running lions. Howell (1944) listed it as 80 km/hr, a figure which is too high; Guggisberg (1961) mentioned 48 to 59 km/hr and this estimate agrees with mine. On several occasions, a lion chased zebra at full speed, and although it gained at first it was easily out-distanced as soon as the animals were able to accelerate fully. In many pursuits, a prey was obviously not fleeing at its maximum speed, yet the lion was unable to catch it. In most instances lions can capture an animal only if they can approach it so closely that it has no time to attain full running speed before being grabbed.

The bulging eyes of most hoofed animals provide extreme wide-angle vision, useful for spotting slight movement. Scent often serves to alert animals to potential danger, and sight is then used to verify it and to seek an escape route, In general, vision is the most important sense in open terrain; smell plays a prominent role when the vegetation is dense, as does hearing but to a lesser extent. Prey usually ignores roars and other calls unless the lions are very near.

As long as lions are visible to them, prey behave in a remarkably casual manner. Resting lions are usually ignored. A typical sight in the plains is some lions encircled by grazing wildebeest and zebra 100 to 125 m away. Male Thomson's gazelle tend to remain in their territory when a lion rests in it (Walther, 1969), sometimes within 25 m of the predator. The response of prey to moving lions depends on the latter's proximity, suddenness of appearance, speed of movement, number, and direction of approach. When one or more lions merely walk along, the various species respond similarly: they face the danger with neck held rigidly erect, ears cocked, and, in the case of antelope, an occasional stamp of a foreleg. Most species also snort. These visual and auditory signals alert all prey in the vicinity, with the visual signals usually being sufficient in the plains but the snorts, especially the loud whistle-like snorts of reedbuck, being important in dense vegetation. Wildebeest may stop their incessant grunting when a lion approaches, thereby creating an area of silence which is as effective a stimulus contrast as an alarm call, particularly at night. Usually prey animals merely observe the predator until it moves on, but at times they advance at a trot, then line up and watch it pass (plate 24). If the lion stops or turns in their direction, the animals wheel, flee a few meters only to halt and look once more. There is no difference in behavior toward a lean or gorged lion. The distance between lion and prey in such situations is usually between 40 and 60 m, and it is obvious that the latter feel themselves immune from attack. By behaving in this manner the animals keep the lion in sight, a useful trait in response to a predator which is dangerous primarily when hidden; the behavior may also teach young animals the distance at which a lion is safe. In this context, it is worth noting that prey usually do not line up and follow wild dogs and hyenas, both species which do not conceal themselves when hunting. Lions in turn recognize their limitations and seldom indulge in futile rushes. On one occasion zebra cautiously approached two male lions on a kill to within 23 m. Walther (1969) found that 90% of the Thomson's gazelle withdrew from an approaching lion when it was 50 to 300 m away.

The flight distance of large prey is considerably lower than that of gazelle, zebra, and others. Giraffe and buffalo may watch lions pad past them a mere 20 to 30 m away. One night a hippopotamus ambled past two male lions 25 m away as if they did not exist. On another occasion four lionesses rested beneath a bush as a browsing rhinoceros approached; when it was within 7 m the lions retreated 6 m and watched without eliciting a response. These species can defend themselves against attacking lions and their behavior reflects this.

If a lion moves either at a walk or a run toward some animals, they trot aside or flee at a gallop, the method depending on the imminence of the danger. Spronking, a bouncing gait in which all feet hit the ground at the same time, is seldom used by gazelle in response to lions and then only briefly at the beginning "in a high, but not the highest, level of excitation" (Walther, 1969). Escape rarely is precipitous for more than a short distance. Animals that have barely eluded a lion often stop within 50 m and look back.

Prey is particularly cautious about entering thickets. Herds may mill by a riverine forest for an hour or more before one individual either leads the rest into or away from it. Wildebeest sometimes stampede toward a river from as much as 1 km away. The long column of animals hits the river at a run, and if the embankment is steep and the water deep the lead animals are slowed down while those behind continue to press forward until the river turns into a lowing, churning mass of animals some of which are trampled and drowned. One such herd I observed at Seronera left seven dead behind; several hundred may drown in such circumstances. At the slightest hint of danger, animals rush away from dense vegetation and into the open, then halt and look around. When attacked, gazelle and impala typically scatter, an effective method of confusing a predator. On several occasions I saw a lioness stand undecided in high grass looking quickly back and forth at gazelle leaping around her without being able to select an individual. In similar situations zebra and wildebeest tend to bunch up unless, of course, lions run among them.

Once prey has been attacked at a particular locality, it might logically be assumed that the survivors would avoid the spot that day.

This is not the case, and migratory herds may make several attempts to drink or cross a river at the same place. On one occasion 10 gazelle entered an area of high grass at 0700. They scattered when they scented the four lionesses there but one was caught and eaten, no doubt permeating the area with the odor of blood and lions. At 0810, 53 gazelle entered the same area. Twelve of these passed one lioness before she rushed but missed. The herd fled 130 m, regrouped, and 25 minutes later 43 of the same animals retraced their steps and were attacked again.

Adult reedbuck and bushbuck and the young of gazelle, hartebeest, and eland may escape lions by crouching motionless. This method is effective, and even when a lion stumbles on such an animal its startled flight is so precipitous that the predator often fails to react in time to catch it. On several occasions, lions passed within 10 m of crouched reedbuck and one such animal lay hidden in tall grass within 6 m of three lionesses for at least three hours. Similarly, a Thomson's gazelle fawn remained crouched in the open for at least a half hour within 20 m of several lions until a male accidentally flushed and caught it.

The response of prey to lion odor varies considerably. Sometimes animals merely veer from the place at which a lion is hiding or has passed through recently. At other times they first halt and stare at the spot, then detour round it. More often than not prey seemingly ignore the odor of lions, behavior that is particularly conspicuous when herds are migrating and when water sources are localized. In the former situation, animals convey the impression that to follow the route of a previous herd is more important to them than a mere lion attack. One day several lions caught six wildebeest between 0817 and 1855 near the same river crossing as herd after herd moved into the area in spite of the lions and dead wildebeest there. During the dry season the vicinity of all water sources is permeated with lion odor, and for prey to respond with flight each time they approach a river would not be adaptive. At such times, the sense of sight is more important in detecting a predator than smell.

Main Environmental Factors Affecting The Hunt

Because of the fleetness and keen senses of prey, lions must seek every possible advantage to catch an animal. Prominent among the environmental factors which may contribute to their success are the time of day, the height and density of the vegetative cover, and wind direction.

Time of Day

Unless conditions are favorable, lions hunt mainly by night, unquestionably because they are able to stalk with greater chance of success under the cover of darkness. Lions appear to be well aware of the advantage that darkness gives them. They frequently watch prey in late afternoon but wait until dusk before hunting. Similarly, a burst of activity often follows the disappearance of a bright moon. One night the Masai pride heard zebra call about 1.5 km away. The lions did not respond, but when a dark cloud obscured the moon they walked rapidly toward the animals and stalked unsuccessfully. However, lions hunt whenever the opportunity arises. Those in the plains, where cover is sparse, generally hunt at night, but the Cub Valley group, which centered its activity around a reedbed, twice caught wildebeest there in daytime within a period of 13 days. The woodlands lions hunted usually at night when animals were dispersed but often during the day as well, when they were near the rivers. For example, the Seronera pride killed 13 wildebeest and 1 zebra between April 1 and 6, 1969, all of them during the day at the following times: 0800–0900, 1; 0900–1000, 2; 1200–1300, 2; 1300–1400, 8; 1800–1900, 1.

To find out if lions hunt mainly during certain hours of the night, members of the Seronera and Masai prides were followed for 30 whole nights and nomads for 3 whole nights, and all stalks were noted (fig. 40). During this period, lions stalked prey 35 times at all hours but with a slight peak just after dark between 1900 and 2000 and a major peak between 0200 and 0400, a period of generally high activity (see fig. 25). Many nights were spent without hunting, particularly on the plains where, for example, males Nos. 159 and 134 killed nothing in 26 full days of observation and the only pursuit of prey was an unsuccessful chase after a dik-dik on Naabi Hill.

As figure 41 shows, there is a daytime hunting peak shortly after dawn, followed by a drop in activity, especially between 1100 and 1700, and finally an increase toward dusk. In general, these hunting times reflect the activity of prey during the dry season around Seronera. Gazelle often come to water with the first light whereas zebra and wildebeest tend to arrive later.

Cover

The height and density of vegetation influence the ease with which lions can stalk without being detected. Table 52 summarizes the types

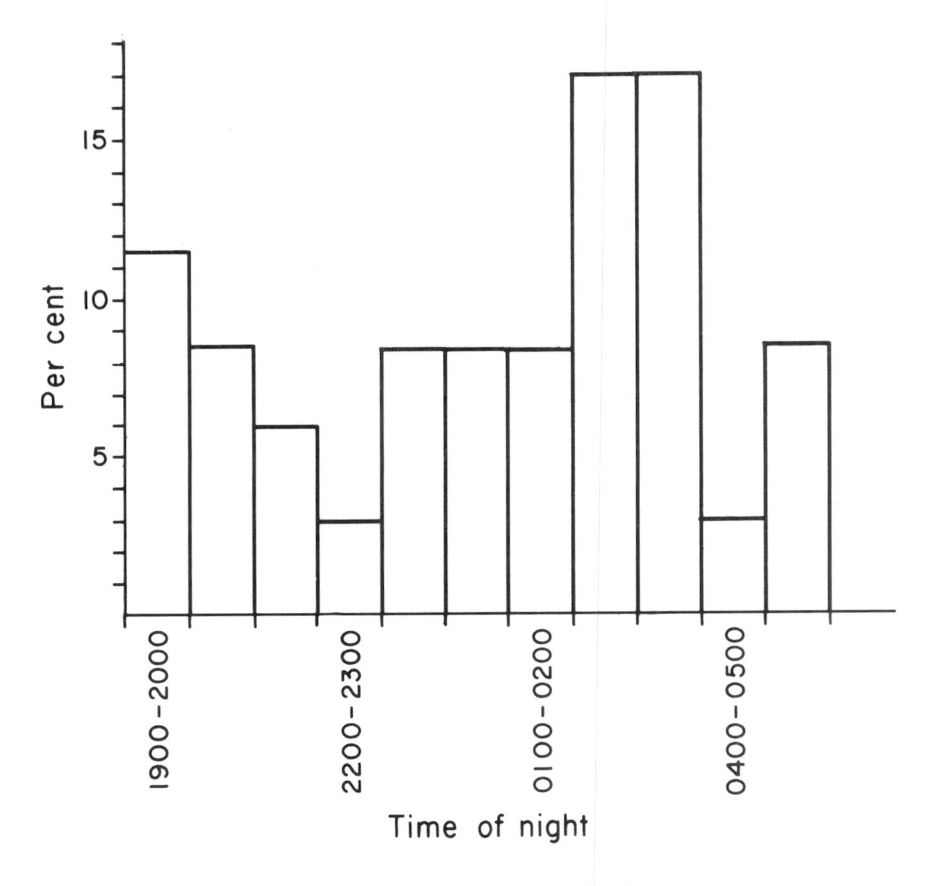

Fig. 40. The time of night when lions were observed to hunt.

of cover in which wildebeest and zebra were killed. Most data in the woodlands were collected during dry seasons, but since the two species were largely on the plains at other times the percentages are fairly representative. About three-fourths of the animals on the plains were near some cover when caught. The short-grass plains furnish lions with few hiding places except for patches of *Solanum* and *Gudigofera* and for erosion terraces. But the intermediate and tall grasslands may furnish ample concealment. The frequency with which I found kills increased as soon as prey moved into those areas. Hunting is easier in the woodlands than in the plains, even after the grass has been burned, because thickets are widespread. About 30% to 41% of the kills were made near a river where the dense vegetation and broken terrain enable lions to stalk undetected. These are high percentages considering the restricted distribution of the habitat (plate 25). The fact that about a

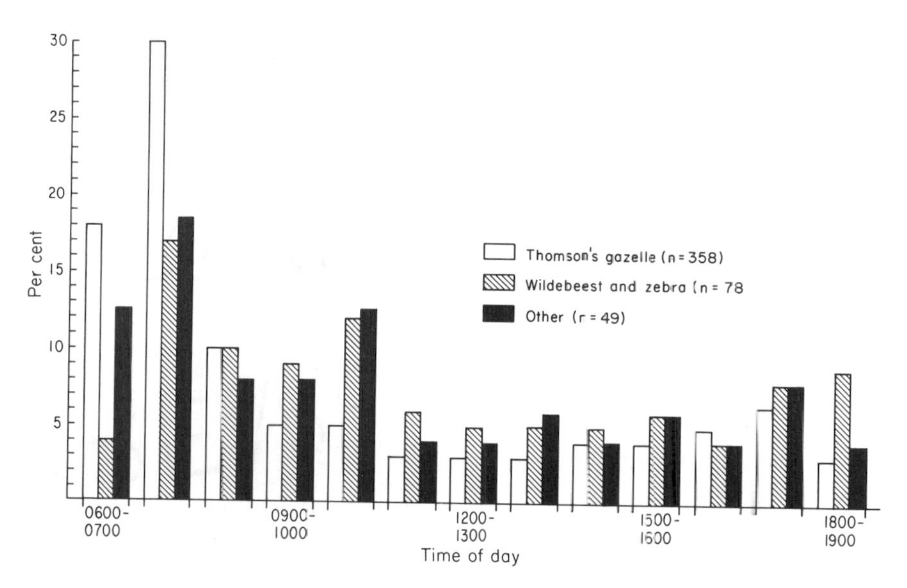

Fig. 41. The time of day when various prey species were caught by lions.

quarter of all kills in the three areas were made in sparse cover or none reflects well on the hunting ability of lions even though the percentage includes young and sick animals.

Thomson's gazelle are not much hunted on the plains. Around Seronera and in the woodlands they are pursued mainly in dense cover such as in tall grass and in the vicinity of rivers. These animals are perhaps the speediest and most agile prey species in the park. Lions have little success in catching them in the open and consequently seldom attempt to do so.

Buffalo, particularly solitary bull buffalo, concentrate their activity along river courses and therefore are killed there. Of 45 kills, 51% were within 150 m of a river, 11% were within 150 to 300 m of one, and 38% were more than 300 m away. Other species also favor riverine habitat or concentrate there during the dry season. Thus, these narrow strips of shrubs and trees provide lions with prey out of all proportion to the land area of the strips.

Wind Direction

"Of course the lion always approaches his prey against the wind" wrote Guggisberg (1961). Others (Stevenson-Hamilton, 1954; Denis, 1964) have made similar statements so frequently that it has become almost axiomatic that lions stalk in this way. Since prey may scent lions at a distance of 100 m or more, an upwind stalk would clearly benefit the cats. I observed several hundred stalks without indication that lions considered wind direction. Occasionally a lion circled prey in such a way that its movement was upwind, but this happened inadvertently during cooperative hunts or in order to take advantage of a particular piece of cover. Wind direction was noted in 300 hunts. One or more lions began to stalk upwind on 28% of the occasions, downwind on 28.3%, and with the wind from one side or the other on 43.7%.

Hunting Behavior

Lions hunt in several ways, either singly or communally. This section describes their methods of doing so.

Senses Used in Hunting

Sight, hearing, and smell, in order of decreasing importance, are the main senses used by hunting lions. At times lions spot and approach prey in the plains from 2 km or more away, but in dense vegetation they may not become aware of it until they are within 50 m or less. During the stalk they orient themselves almost wholly by sight, waiting and advancing in accordance with the behavior of their quarry. To keep prey in view, a lion often raises its head above the level of the vegetation; such movement may be detected by the prey. One individual is selected and pursued by sight alone during the final rush. Occasionally lions first become aware of prey by hearing it: gazelle walking through tall dry grass, zebra braying in the distance at night. Once several lions heard splashes in a creek; rushing to the embankment they saw a Thomson's gazelle hurriedly crossing to escape a crocodile. It eluded that danger only to run literally into the arms of the waiting lions. The cries of hyenas on a kill or the dying screams of prey also attract the cats. It would be difficult if not impossible for a lion to track a specific prey by scent among the many trails, but in certain situations the nose seems to be used. On one occasion, when a gazelle disappeared in high grass, two lionesses slowly crisscrossed the area with their noses raised as if sniffing. In spite of their keen senses, lions miss many opportunities to stalk by not being alert. During daytime in particular they may rest while a herd moves by or drinks at a waterhole without being detected by them.

Hunting Methods

Lions use several distinct hunting methods. At times they are offered an unexpected opportunity to kill—a crouched fawn, a sick animal that stumbles into the resting pride—and they take immediate advantage of it. Occasionally a swimming animal is plucked from water or a warthog is dug out of its burrow. When preparing for an ambush, a lion hides itself and waits for prey to appear. The remaining methods involve a more or less active search for prey. Lions walk slowly with frequent halts for listening and looking—waiting, it seems, for a quarry to wander into their vicinity rather than searching for one by traveling many kilometers. Once prey has been spotted, they sometimes wait until it moves into a vulnerable position. At first they may walk closer with little attempt at concealment, so that in daytime prey often flees before the lions are close enough to hunt. But if they can approach to within 100 m or so, one of several methods may be used to get still closer. The drive is made by unconcealed lions walking or running at animals in spite of the fact that these are already aware of danger. Similar to this is the run, except that it begins without the animals having seen the lions. The stalk is the most commonly used hunting method. During a stalk the lions attempt to approach undetected, usually spending less than 15 minutes to do so but occasionally as long as an hour.

I found it difficult to define a hunt. When a lion rushes at its quarry the purpose of the action is in little doubt. But when a lion merely walks in the open toward some animals which subsequently flee, is that a hunt, an investigation of possibilities, or a detour not directly associated with prey? When a lion sees gazelle and then lies down, is it hunting, waiting to hunt, or simply tired of walking? Lions see many potential prey in the course of a day, and they undoubtedly evaluate the possibility of catching any that are even slightly vulnerable. Most are given a glance, some merit a closer look, a few elicit hunting movements, and only a very few are actually pursued. Ideally each category should be quantified, as was done well by Mech (1970) for wolves hunting moose, but lions may have so many animals in view and their responses may be so ambiguous that an estimate of the number of potential prey in their sensory sphere would be a guess. Consequently a lion was considered to be hunting only if it fit certain arbitrary criteria. Any pursuit of an individual or herd at a trot or run was one hunt if the lion approached to within 60 m or closer. A stalk included an approach in a stalking posture of at least 10 m to within 60 m of prey except when animals moved toward the lion, which then crouched. This eliminated from consideration abortive stalks of a meter or two, stationary lions alert to distant prey which they made no effort to approach, and similar situations. Lions were considered to lie in ambush only if prey came to within 60 m of their hiding place and they oriented toward it; concealed lions resting at waterholes were not included even though they might have hunted had prey appeared. In effect, only the final stages of hunts are considered. Buffalo and giraffe are not included in the computations because single lions hesitate to attack them.

Lions used some methods of hunting more often than others (table 54). Ambushing was uncommon (2.7%) and driving was seen only in pursuit of gazelle (2.8%). Unexpected hunts also contributed a small amount to the total (6%). At least a few meters of stalking featured in many hunts by single lions (22%), and running did also (14%). When two or more lions hunted together, I lumped stalking and running because both methods were often used. Some 88% of all hunts consisted of runs and stalks.

Forty-eight percent of the hunts were by single individuals, 20% of them were by two animals, and the rest consisted of communal hunts with 3 to 14 lions, though more than 8 seldom hunted together (table 54). Hunting groups comprising 3 or more individuals were more common when wildebeest and zebra were stalked (51%) than when Thomson's gazelle were pursued (27%), a difference due to the fact that prides were characteristically scattered during the gazelle hunting season but together when large prey was available.

Although lions frequently hunt alone, others are usually nearby and in fact often watch the hunt. Other lions were present without contributing in 72% of the instances when a single lion hunted (table 53). At other times an effort by such lions might have hindered a hunt or at least not increased the chances of success, but often a contribution by the spectators would have been useful. Obviously hunting behavior by one animal usually fails to elicit it in the rest. It was my impression that failure to elicit communal hunting was more prevalent during the day than at night.

Lionesses hunt proportionately more often than males, a point on which my information agrees with popular supposition. In 71 hunts by single lions that were in groups of mixed sex, males took the initiative only twice (table 53). Out of a total of 1,210 lions observed stalking and running, only 3% were males; no males participated in driving or ambushing. Males, however, respond quickly to an unexpected opportunity (table 54). These figures are slightly biased in favor of females because I spent fewer hours with males but only after experience had shown that they were unlikely to hunt. The figures are also biased toward pride males which have females available to do the hunting; nomads must by necessity kill more for themselves, and various observations in this report attest to the fact that they readily do so.

Each hunting method is described in some detail below.

Unexpected Hunting. A lion may be provided with an unexpected opportunity to kill in several ways. It usually just grabs or pursues such prey without stalking. One lioness was walking along a river when two Thomson's gazelle rushed up the embankment in front of her. She lunged and, sliding sideways, hooked a gazelle in mid-air with two claws. It crashed on its side and she grabbed it with both paws while simultaneously biting it in the nape. When two warthog boars fought, a lioness immediately tried to catch one; a courting reedbuck male lost his life because he ignored some lions nearby. A bushbuck fled from my car into a group of resting lions. A zebra foal slept so deeply that it failed to hear its family move away, and it only awakened briefly after a male lion pounced on it. Lions accidentally flushed a total of 17 crouched Thomson's gazelle fawns and caught 10 of these after a chase of 150 m or less. Once an exhausted adult male gazelle trotted past a group of lions, followed closely by a hyena and then a jackal. One lioness ran after the male, slapped him once, then grasped him with both paws before biting him in the back and nape.

Wildebeest calves which have become separated from their mother may approach lions just as they do cars, people, and other moving objects. One calf trailed 20 m behind two male lions when I first saw the trio at 0645. Each time a male turned toward the calf, it halted or trotted a few meters away. One male rushed the calf 9 times in 15 minutes but it eluded the attacks easily. Finally the males lay down and the calf stood irresolutely 40 m away. When, after a few minutes, a male rose, the calf immediately came to within 15 m of him. This male charged five times vigorously, yet the calf remained nearby. At 0730, after having accompanied the males at least 2.2 km, it drifted up on a rise and saw another lost calf. It ran toward it, ignoring several lionesses in its path. One of these chased it 150 m, grasped its rump with both paws, and, after pulling it down, bit it in the throat.

A sick zebra walked alone past a resting pride and was captured, and a mangy wildebeest left its herd and unfalteringly went to a tree beneath which several lions rested. Another wildebeest was unable to rise when a male approached. The lion straddled the wildebeest, and, when it rolled over, dismounted and bit its throat.

Ambushing. One or two lionesses occasionally hide themselves and wait for prey, as two examples illustrate.

> (1) At 0835 a lioness sits in a small patch of grass by a river watching several gazelle at a distance of 50 m. At 1750 she is still there, this time crouched as five gazelle approach cautiously, alternately walking and stopping. A male gazelle descends the embankment, closely followed by a female, The lioness raises herself slowly, only her head above the grass; when the female gazelle passes at 5 m, she rushes. The gazelle leaps forward and the lioness misses but, using her momentum, swerves and bowls over the male. She hauls him by the nape up the embankment where he ceases to kick.
>
> (2) Two lionesses lie in a small thicket by a waterhole at 0950. At 1320 many gazelle drink about 30 m away. Although both lions are crouched, tensed for a rush, neither attacks. At 1345 two male gazelle drink there, and at 1435 five warthog. At 1455 a female gazelle leads another herd to water but none of the 50 animals approached the lions closely. At 1615 a solitary female gazelle passes the lionesses at a distance of 15 m. Forty-five minutes later two herds appear and one of these files past the lionesses at a distance of 10 m. Ten gazelle move by while the cats wait. Finally, when they reach the water and lower their heads to drink, one lioness rushes—but misses. At 1710 hours, both leave their hiding place.

The last example contains two points of interest: one, lions must be close to fleet-footed prey if they are to be successful, and even a rush from a distance of 10 m may fail; two, an attack is not necessarily directed at the first available animal, and the lions may permit a herd to pass before bursting among the animals from behind.

Driving. One or more lions may trot toward gazelle without attempting to conceal themselves, as if to scatter them or drive them into dense cover and in the confusion capture one. For instance, a lioness first walked then trotted 300 m toward a large gazelle herd. When it fled, she followed and 10 minutes later ran at it once more without success. On another occasion, a lioness bounded 230 m after about 15 gazelle. Twice an animal suddenly left the herd and dashed toward the lioness, but she was unable to catch it. In yet another instance, about 20 gazelle moved into a cul-de-sac lying between the junction of two creeks. Two lionesses walked toward them from 70 m away. Instead of crossing one of the creeks, the gazelle raced back toward the lions, one of which rushed, missed, pursued another gazelle and failed again. But two male gazelle tarried, and a lioness chased one of them 60 m while he seemed undecided about an escape route. Finally he entered a reedbed where the lioness caught him.

Digging. That lions may dig warthog out of their burrows was mentioned by Guggisberg (1961), Kruuk and Turner (1967), and others. I observed this once and another time found a site at which two warthog had been caught. One night, at 0035 hours, the Masai pride was moving across the plains when a lioness stopped and began to dig at the entrance of a burrow. Two lionesses joined her, one only briefly. Each pawed several times with one paw, then with the other, and occasionally with both in unison, ripping at the sod with their claws. One lioness continued to dig vigorously until at 0135 some 2.5 m of tunnel were exposed. She looked in the hole, jerked back, ducked in once more, and grabbed something—her body grew taut with the strain of pulling. She retained her hold for 8 minutes until finally she was able to pull out a screaming adult female warthog by her nape.

Grabbing prey in water. On two occasions a lioness plucked a swimming gazelle from a river. Once a gazelle apparently slipped and fell from a steep embankment into the river and a lioness helped it to land.

Running by single lions. A lion may trot or bound toward prey from as far away as 50 to 200 m, especially if a favorable hunting situation suddenly presents itself. For example, one lioness was feeding on a gazelle when she saw several others nearby. She rushed from the carcass, pursued a gazelle 40 m through high grass, and caught it, falling into a pool while doing so.

In some situations a lion ran toward prey even though there was no urgency, and it seemed to me that a preliminary stalk might have increased the chances of success. Here is a typical hunt, one of the few successful ones. Four lionesses and five cubs walk along a road in single file. One sees some 50 gazelle about 125 m away on the sloping river-bank leading to a waterhole. She runs toward them. Most flee but a few remain by the water, unable to see the lioness until she rushes over the crest of the embankment and captures a male.

In many running hunts a lion trots from 60 to 80 m toward prey which is in some way vulnerable (such as being near tall grass), without making a vigorous effort to catch an animal. The lion gives the impression of knowing that the possibilities of a kill are slight but that there is always a chance that an animal will be inattentive. For example, one lioness saw two gazelle on a riverbank above a waterhole. She trotted toward them without hope of catching one, but she nevertheless ran to the top of the bank and looked down the trail to the pool in anticipation of any lagging gazelle there. At other times a run is a preliminary to a rush except that the prey has become alerted and further pursuit is futile.

Stalking by single lions. Some hunts include only a brief stalking walk or a crouch before the rush, but in many of them lions approach prey in a classical cat pattern, slinking from bush to bush, stopping occasionally to peer at the quarry, crouching and waiting, then advancing again until close enough for a final attack (plate 28). The following incidents illustrate these methods of hunting:

(1) A lioness sees three warthog and several gazelle 80 m away in the open. She first stands, head lowered, then slowly advances in a stalking walk screened by some bushes. The gazelle see her at 60 m and flee, as do the warthogs—an adult female first, followed by a yearling female, and a tiny piglet last in line. The lioness bounds after the warthogs and after a chase of 100 m bats the piglet once, then snaps it up, holding it by the back.

(2) At 1610 a lioness sees eight kongoni .4 km away. She lies and watches them. When, at 1630, they move to a waterhole partly hidden by tufts of grass and acacia saplings, the lioness immediately trots closer in a semicrouch and from a distance of 40 m rushes at full speed. The kongoni scatter. She follows one and swipes with a forepaw but misses, and all escape only to halt 60 m away and watch her sitting by the river.

(3) A lioness sits by a tree and watches a herd of at least 40 gazelle descend the sloping river bank 100 m away. She waits and when 10 minutes later most have moved to water, she approaches slowly a few meters in a stalking walk, then trots 30 m, and finally runs. Ignoring those that run from the water, she rushes over the embankment and grabs a lagging male gazelle which she carries by the neck across the river and into the trees.

(4) Numerous zebra are crossing the Seronera River in an area of brush and reeds. Several lionesses caught a foal there at 1200 yet an hour later four zebra, including a foal several months old, approach the same place. One lioness sees them at a distance of 100 m and immediately creeps some 30 m into a clump of reeds near the trail that leads from the river. The zebra last in line halts at the crossing but the other three plod on. Two adults pass the lioness at a distance of about

9 m, but when the foal is opposite her she hurls herself forward and leaps, the only time I have seen this. Her whole body hits the zebra in the side with an audible slap and they fall, the foal screaming until she grabs hold of its throat.

(5) At 1450, a lioness sees three topi standing in the open near a river. After glancing at them briefly, she descends into a dry watercourse, moves downstream about 100 m, then lies at the top of the embankment hidden behind a screen of grass where she watches a male topi standing alone on a burned area 50 m away and broadside to her. When at 1503 he turns away from her, she creeps into the open, partly hidden from him by a fallen tree. Some distant topi see the lioness and snort. The male jerks to attention and she freezes in mid-stride. But he fails to look around, and she suddenly trots forward, body held low, and is within 10 m of him before he spots her. He flees but she pursues and after 22 m grabs his rump with her paws, throws herself to the left, and both crash to the side, the topi's legs flailing the air. She lunges for his throat and bites it until he dies two minutes later.

Perhaps the most striking aspect of a stalking lioness's behavior is the care with which she watches prey. She advances when it has its head lowered to graze or stands facing away from her, except in the case of large herds when it might be futile to wait for all animals to be inattentive. If an animal suddenly becomes alert, she halts, sometimes standing motionless with paw raised in mid-stride. She is fully aware of the advantage that cover confers, and when prey moves behind some shrubs or out of sight into a ravine, she may run closer. Yet lions often ignore the behavior of prey and the availability of cover, thereby failing in many hunts.

Because the animals react so quickly, crouched lions may miss an attempt to catch prey that is only 10 m away, but a running lion often has enough momentum to overtake a quarry before it can attain full speed. The many stalking runs of 50 m or more are probably of advantage to lions from this point of view. The average distance from which a stalking lion rushed was around 30 m whereas a crouched lion sometimes did not charge at prey that was only 15 m away. Lions judge their limitations and those of the prey fairly accurately, and chases after young animals are generally longer than those after adults. When a rush fails, the hunt is quickly abandoned in most instances, but on occasion, particularly in high grass, prey is pursued for as much as 200 m. At times a lioness bungles one hunt then follows a retreating herd several hundred meters and tries again.

Lions have little stamina—a fast run of a few hundred meters makes them pant—and they are unable to pursue an animal rapidly over long distances. The size of a lion's heart, a measure of the capacity to pump blood and hence distribute oxygen, suggests a limitation in this respect:

	Body Weight kg[1] *(excluding stomach contents)*	*Heart as % of total body weight*
Lioness	117	0.57
Male lion	193	0.46
Male leopard	36	0.36
Unsexed hyena	43	0.95

1. All cat weights were kindly given to me by R. Sachs.

The hyena has a heart considerably larger than that of the cats.

Communal hunting. Lions often hunt as a group and, according to various accounts in the literature, the strategy they use to catch prey is elaborate. "Then several creep off to get to leeward of the quarry, and only when these are in position do the others show themselves or give the animals their wind. These usually stampede downwind, to be intercepted by the lions lying in ambush" (Stevenson-Hamilton, 1954). Or they may "spread around a herd of zebra or antelope, when one may roar or grunt to scare the animals toward the others" (Roosevelt and Heller, 1922). Kruuk and Turner (1967), on the other hand, expressed doubts about the existence of such cooperative hunting and thought it "most likely that each lion goes his own way to catch and kill a victim; if one member of the pride stampedes a herd other lions may make use of this." The actual complexity of communal hunts lies between these two extreme viewpoints.

In some hunts there is no evidence of cooperation in the sense of a joint action by several individuals in the performance of a defined task. Any help the animals give each other is inadvertent: when prey flees from the lioness it sometimes approaches another which takes advantage of the situation. Frequently only one lioness rushes, and others watch as if waiting for prey to come near. Two examples illustrate such hunts.

(1) A herd of about 50 gazelle approach two lionesses sitting in tall grass. One lioness snakes toward them 60 m and crouches. Then the second also moves forward, and the gazelle, which have been advancing steadily, either see or smell her. They scatter and one runs in a fast, flat gallop directly at a crouched lioness. She reaches up and grabs it out of the air with both paws, slams it down, and bites its throat; it groans once and dies within 30 seconds.

(2) Two lionesses see two gazelle on a burn by a river 70 m away. While one lioness trots closer, the other watches. The gazelle flee at moderate speed in one direction, but when the lioness angles toward one of them, it suddenly doubles back. At that the second lioness rushes in from the side, and before the gazelle can dodge she knocks it over so violently that both slide in a cloud of dust until jarred to a halt by a sapling.

In some hunts the lions seem to cooperate but the behavior may be accidental.

(1) Four lionesses walking down a road in single file see about ten gazelle by a riverine thicket 65 m away. While three wait, one lioness continues on alone, partly screened by acacia saplings. The gazelle spot her after 40 m and trot to one side. The lioness runs toward them, causing four to double back, two of which run directly in front of the three waiting lionesses. These rush, and one captures a subadult male after 40 m.

(2) Two male wildebeest walk slowly down a shallow valley beside a reedbed. Two lionesses on the slope above see them at a distance of 250 m. One crosses the reedbed at a trot, body held low, sneaks across a barren area 15 m wide, and flattens herself in a patch of grass while the second one crouches at the edge of the reeds, facing the other lioness. The wildebeest approach slowly, downwind, but the lead animal suddenly stops 20 m from the lionesses and looks back toward his companion. At that moment a lioness bursts from the grass and the wildebeest, instead of remaining in the open, runs into the reeds pursued by both cats. His escape seems certain when after a run of 70 m he apparently steps into a hole, for his forequarters collapse and his hindlegs fly into the air. Before he can rise, a lioness hangs from his neck, her hindquarters between his forelegs as she bites at his throat, while the other straddles his back and bites at his shoulders until all topple into the reeds.

When several lions spot potential quarry they characteristically fan out and approach in a broad front, sometimes spread over 200 m of terrain. This fanning action may be well coordinated in that those at the flanks walk rapidly in their chosen direction while those in the center halt or advance slowly. By being widely spread out, lions increase their chances of coming into contact with prey which may either scatter, with one or more animals running at a hidden lioness, or veer to one side and inadvertently meet one of the hunters at the flanks. These two examples are typical.

(1) When at 1845 35 zebra file by .8 km away, seven lionesses move toward them in an irregular line. The zebra reach a thicket bordering a stream and mill there. The lionesses congregate 300 m away, look at the zebra, then fan out, at first to a width of 60 m then to 160 m. Three male lions stroll 130 m behind the lionesses. At 1905, when some lionesses are either crouched or stalking within 50 m of the zebra, these suddenly walk away. The lions follow for 5 minutes before halting.

(2) At 1845 five lionesses and a male see a herd of some 60 wildebeest 2.7 km away—just black dots moving against the yellow-grey plains. The lions walk slowly toward them. At dusk the wildebeest bunch up. The last light has faded at 1930 when the lions stop, .3 km from the herd. The lionesses fan out there and advance at a walk in a front 160 m wide, moving downwind, the male 60 m behind them. They crouch when 200 m from the herd, and I can see only an occasional head as they stalk closer; the male remains standing. Five minutes later a female on the left flank rushes and catches a wildebeest, but I am unable to see the details. Two lionesses converge on her. The herd bolts to the right and two lionesses and the male run at an angle toward it, pursuing about 100 m without success. The wildebeest is on its back while one lioness clamps its muzzle shut with her teeth, a second bites it in the lower neck, and a third in the chest. Then the male bounds up and with one bite tears open the groin.

When a lioness runs toward a fleeing animal from the side, she aims her attack ahead of the moving target. Similarly, stalking lions anticipate the direction of traveling prey by placing themselves in ambush on the track ahead. On one occasion, for example, two lionesses saw several gazelle at a distance of 120 m, a female with a small fawn in the lead. The lionesses stalked 100 m and crouched 10 m apart directly in the path of the advancing herd. But the gazelle veered aside, using a well-beaten trail, and the lionesses shifted their position too. The leading gazelle sensed danger, trotted ahead and almost into a lion, which lunged but missed.

Hunts such as these fit the definition of cooperation in that lions orient toward a common goal and their actions are similar, with each animal patterning its behavior after that of the others even to the extent of obviously looking at a neighbor during the stalk. If only one quarry is involved, all endeavor to catch it, but, if there are several, cooperation may cease as each tries to capture the nearest one. While cooperation in many hunts is clearly evident, the behavior is relatively simple in nature. However, on 29 occasions lionesses encircled prey, sometimes by detouring far to one side, behavior also described by Bridge (1951). The other lions waited during the flanking movement as if in anticipation of prey fleeing in their direction. Figure 42 illustrates the movement patterns of lionesses in such hunts. In A, B, and D the encirclement of a herd succeeded and one lion caused prey to run toward the waiting ones; in C the herd moved off before the stalk was completed; and in E one lioness rushed before two others were fully in position. During such hunts lions integrated their actions solely by observing each other's posture and movement; no sounds were used nor were facial expressions employed which, at any rate, would not have been useful at night. Encircling implies that lions are aware of the consequences of their actions in relation both to other group members and to the prey.

Multiple Killings

Lions sometimes kill more prey than they can consume in a meal. Sometimes this is done inadvertently when several lions attack a herd and each captures an animal. As many as five wildebeest, for example, may be killed in this fashion. Carcasses are usually scattered widely, the average distance between 23 such kills being 78 m (4 to 280 m). At other times, lions already have a kill but take advantage of an easy opportunity to obtain another, no matter how gorged they are. When large herds of wildebeest migrated past Seronera for several days, a group of lions killed repeatedly, leaving uneaten carcasses lying around. Cowie (1961) noted that during a drought the prides in Nairobi Park killed many starved animals without eating them. Such instances show that the lion's hunting and killing patterns may function independently of hunger.

Hunting Success

Many factors influence the success or failure of a hunt. Some of these are environmental, including time of day, brightness of moon, and wind direction, others relate directly to the lions, among them the age and sex of the hunters, the method of hunting used, and the number of lions involved; and still others involve the antipredator behavior of each species. All such physical and biological conditions which make an animal susceptible to predation can be termed its vulnerability (Craighead and Craighead, 1956).

Factors Relating to Environment

I stated earlier that lions may hunt at any time of the day, that much

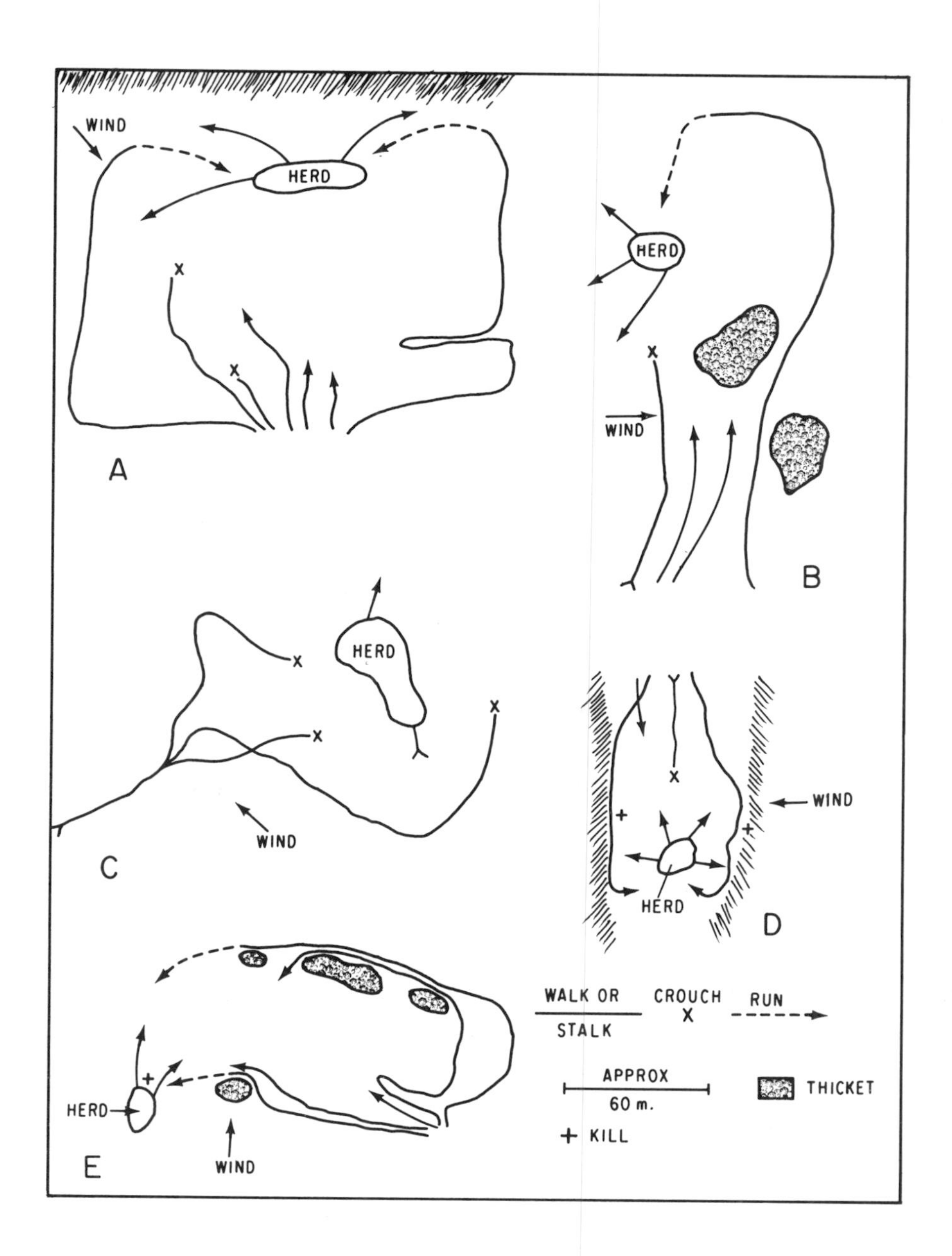

Fig. 42. Five examples of routes taken by lionesses when stalking prey cooperatively.

prey is caught near cover, and that wind direction is ignored. How much is hunting success influenced by light, cover, and wind? Generally, hunts at night are more successful than those in the day (table 55), which is not surprising, for darkness permits lions to stalk across terrain where they would be spotted immediately in daytime. Most of my observations were made on nights with a bright moon. It is probable that the difference in success between day and night hunts would be greater if only dark nights are considered. Other factors may also affect the percentages. Lions tend to hunt wildebeest and zebra alone near waterholes in the day but communally in the plains at night. When only communal hunts of these two species are considered, daytime success was 27% and nighttime 42%.

The type of habitat is of critical importance for a daytime stalk. With little or no cover, as when the plains have been burned around Seronera, lions catch few animals (table 56), and many of those that fall prey are young or sick. Of three gazelle caught in the open, two were newborn and one had a broken leg. Even if some cover is nearby, a lion is seldom able to cross the surrounding open area in time to capture prey. When a few shrubs and saplings break the visibility, a lion's chances are somewhat better with respect to gazelle but not with larger prey. As soon as the groundcover is at least .4 m high, the success rate increases for gazelle mainly because the animals are unable to see the maneuvering of lions. On the other hand, wildebeest and zebra may detect the approach of lions. By far the most vulnerable place for most animals is the vicinity of riverine thickets, not only because of the dense vegetation there, but also because prey is restricted in its use of escape routes. Only narrow trails usually lead from the water's edge, and at the first intimation of danger the animals flee along these, a habit the lions use to their advantage. The third column of table 56 shows an erratic pattern of hunting success mainly because reedbuck, which feature prominently in this small sample, behave differently than other prey.

When considering daytime hunting success in relation to wind direction, lions capture Thomson's gazelle three times more often after stalking upwind than downwind (table 57), yet they take no account of this during the hunt. My information on other species is too limited to be indicative.

Factors Relating to Lions

I have little information on how a lion's age and sex influence hunting success. Males, which hunt infrequently, are less agile than females and cannot run as fast as lionesses, but if prey appears unexpectedly they are adept at dispatching it. The cubs that grew up in the Masai and Seronera prides joined in the pursuit of gazelle by the age of 15 months. They made attempts to catch prey which adults disdained, and in communal hunts they were sometimes the first to rush. However, such behavior may benefit the group in that other lions then pursue prey that has scattered. By the age of 2½ years the hunting behavior of subadults is indistinguishable from that of adults although it is possible that the subadults' rate of success is lower.

The hunting method influences the success rate (table 58). Unexpected hunts are highly productive (61%), as are such minor ways of obtaining food as digging for warthog. Stalking by single lions, driving, and ambushing are equally successful (17–19%), but running by single lions shows a low return (8%). When two or more lions hunt together their success is 30%, an important figure not only because it is sizable but also because this method of hunting comprises nearly half of all attempts.

That groups of lions are twice as successful at catching prey as single lions is shown more fully in table 59. The data are similar for Thomson's gazelle and for wildebeest and zebra, except that there is an unexplained drop in the latter when three lions hunt together. Large lion groups do not catch prey more successfully than small ones, but there is an increase in the actual number of animals captured, something not shown in these tabulations. For example, the 18 successful Thomson's gazelle hunts that were made by groups of four or more lions hunting together included four multiple killings of 2 and 3 gazelle. While gazelle do not provide much meat, wildebeest do, and multiple killings are common in this species (see table 42).

Factors Relating to Prey

A communal stalk upwind at night through dense ground cover is the most favorable situation for a hunt, judging by data presented so far, but lions also catch animals under many other conditions. Prey decreases the probability of an encounter with a predator by a combination of morphological, physiological, and behavioral factors, some with a direct antipredator function, others only incidentally so. Most such patterns evolved under pressures from a variety of predators, not just the lion, including possibly such extinct species as the saber-toothed cats (see Cooke, 1963).

Morphological. Morphological modifications of the body that make fleetness possible, eyes that bulge laterally from the head providing wide-angle vision, long mobile pinna with which to localize sources of sounds, and large nostrils and complex nasal passages used to smell danger, all are characteristics which have undoubtedly been under strong selection pressures from predators, although other factors have also helped to shape them, with pinna, for example, being important in the expression of emotions. Most species have a brownish or grayish pelage which blends well into the environment when animals stand or lie motionless, making those in dense vegetation particularly difficult for a predator to spot. However, most species are social, and a herd cannot hide well. The constant movement of herd animals negates any cryptic value the pelage may have. The black color of buffalo and wildebeest is more conspicuous than brown, at least to the human eye, but cryptic coloration in such large social species would seem to be of limited value. The stripes of zebra are frequently cited as a good example of the value of camouflage; in addition, the suggestion has been made that the black and white pattern confuses a lion to such an extent that it may misjudge its leap (Cott, 1940). A herd of zebra is not difficult to see, especially when animals are moving, which is usually the case. At night, the time when lions do most of their hunting, the stripe pattern loses its contrast. Besides, lions seldom leap at their prey. Any striking pattern of stripes and spots is likely to be related to inter- and intraspecific communication. The toughness of the hide may also have antipredator functions in that, for example, lions find it difficult to injure giraffe and buffalo, whose skin may be over 1 cm thick.

That horns and tusks evolved primarily as fighting organs between members of the same species rather than as defensive structures against predators has been emphasized by several authors (for example, Geist, 1966). Lions are occasionally gored by the prey they attack but such instances are infrequent, except when they harass buffalo. Of the various antelope I saw being killed, only two wildebeest attempted to defend themselves. One example illustrates such behavior:

> A wildebeest bull walks alone to a waterhole through tall grass and brush. A lioness sees him, trots 60 m closer, and watches him from a distance of 30 m while hidden in some grass. A second lioness joins her. The bull turns without drinking and walks slowly away. One lioness approaches in a crouching run then charges from a distance of 12 m. He wheels, horns lowered, and lunges at the lioness. She rears back but the second lioness runs up behind the wildebeest, hooks her forepaws

> in his thighs and pulls him down. One grabs his throat, the other his nose, and he dies 5 minutes later.

Zebra have powerful hooves and the occasional lion with a broken jaw may be the result of an encounter with these. Yet those zebra I saw being killed made no effort to defend themselves either by kicking, other than some wild flailing after having been pulled down, or by biting, which is a common form of fighting among themselves. On one occasion, two zebra moved toward a river crossing when one was grabbed by a lioness. Hanging onto the zebra's neck with her paws, she bit it in the nape and attempted to pull it down, but it spread its forelegs and stood rigidly, screaming (plate 29). The other zebra merely watched from a distance of 10 m for about a minute before it left at my approach. The lioness pulled the zebra awkwardly on top of her and, when other lions arrived, she released her hold. The zebra fled and after 10 m was captured again, this time by a lioness clutching its rump. Two lionesses held its throat, and it died after 7 minutes, having made no effort to defend itself.

Physiological. Diverse physiological factors help to decrease the vulnerability of a species. Acute senses are obviously important for detecting predators. Backhaus (1959) noted that ruminants can see vertical lines better than horizontal ones, and this may be of advantage in detecting stalking predators. Grant's gazelle and several other kinds of antelope which can subsist for long periods without drinking are able to avoid waterholes where lions tend to lurk. During and immediately following birth, both female and young are highly susceptible to predation. Births are accomplished quickly and in case of danger may be delayed while the female moves to another site. Estes (1966) noted that wildebeest cows tend to give birth in the morning rather than at night when hyenas and lions are most active, and that the placenta, which could attract predators, is expelled only after the calf can run. All species have one or more birth peaks which limit the time that small young are available to predators. Among wildebeest in particular the birth season swamps the predators in a given locality with food, and even though many young are killed there is probably a smaller loss than if calves were present throughout the year.

Behavioral. Any behavior pattern which reduces an individual's chance of meeting a lion has selective value whether or not it functions primarily as an antipredator device. Reedbuck and bushbuck spend much of the day lying quietly alone in thickets where lions may pass without sensing them. This, together with their alertness and sudden startled flight when detected, makes it difficult for lions to catch them. Such a method of avoiding predators is limited to fairly small and essentially solitary species which must remain scattered to retain their invulnerability—a way of life that uses the available resources inefficiently.

Another alternative, and one used by most species, is to form herds. It may be hypothesized that danger is better detected by many animals than by one; if true, this would be a strong evolutionary incentive for herding. Table 61 shows that, among migratory species, the only ones for which I have sufficient information, solitary individuals including young and sick ones are more vulnerable than those in small herds. Animals in large herds are also vulnerable because an attacking lion often gains valuable seconds before all become aware of danger and because herds at waterholes are so densely packed that animals in front or in the center may not be able to flee until others have done so. The high frequency of multiple killings of wildebeest as compared to zebra (see table 42) may reflect a greater vulnerability of large herds over small ones, but other factors, too, are undoubtedly involved. In addition, it was my impression that small herds are more vigilant than large ones. Crofton (1958) found that a grazing sheep is oriented in such a manner that it fixes a neighbor with each eye, behavior which, if true for other hoofed animals, would insure that danger signals are rapidly communicated throughout the herd. Herding in most species can be viewed as an important antipredator device, but the type of social organization seems to be related more to other factors such as reproductive behavior. Of course, social organization determines which segment of the population is most vulnerable to predation. In many species the males constitute that segment. Often animals avoid lions by staying away from dense cover or by being particularly cautious when near it. Lions often hunt by waiting for prey to move close to them, rather than by stalking stationary animals, probably because they can then choose the most suitable place for an ambush. Migratory species, which must cross rivers and thickets, are especially susceptible to such predation, but lessen their vulnerability somewhat by moving mainly in daytime. Hoofed animals also tend to travel in single file, an action which decreases the chances of stumbling on a hidden predator. Serengeti prey comes to water only in the day, a time when lions are least active and can readily be seen, in contrast to zebra and some species in Wankie Park which drink mostly in the evening and at night

(Weir and Davison, 1965). Warthog retreat into a burrow for the night. Although some Asian pigs build large nests (Frädrich, 1965), only the warthog uses a burrow. It is probable that the habit developed primarily to keep the young safe and to avoid predators while sleeping. Warthog did not use burrows as important retreats when pursued by lions during the day: in 16 chases observed only one warthog escaped by running into a burrow. Their habit of dashing for a burrow head-first, then at the last moment turning quickly to back into it while facing the danger, no doubt deters predators from further pursuit.

The vigilance of a species is an important factor in its vulnerability. It was my impression that topi, hartebeest, and waterbuck are by far the most alert and cautious of the animals; impala and Grant's gazelle seemed intermediate in these traits. The migratory species are less vigilant than the others and it is perhaps not coincidental that they are least so when they occur in large herds—as if there were security in numbers. Wildebeest, zebra, and eland sometimes slept so heavily that I was able to drive to within 5 m of an animal without waking it. Warthog are often highly incautious, trotting to a waterhole without hesitation while gazelle stand irresolutely nearby.

Other behavior patterns also function as antipredator devices. As already mentioned, the young of some species remain crouched for a period after their birth. Hartebeest females become solitary and conceal themselves in thickets just before the birth of their young (Gosling, 1969). Wildebeest calves are able to follow their mother within 7 minutes after being born (Estes, 1966), behavior which enables them to remain with the moving herd and lessens their vulnerability. Male topi (but also females on occasion) often stand on termite hills, basically a territorial display but also effective behavior for spotting predators. Two standing zebra often place their heads on each other's backs, a position of friendliness (Klingel, 1967) which also provides them with a 360° view of their surroundings.

When actually pursued, many species have special flight patterns including, for example, the bunching up of zebra, the scattering of Thomson's gazelle, the dodging and zig-zagging of most species, and the spectacular twisting leaps of impala which to a lesser extent occur also in Grant's gazelle and eland. Wildebeest may flee in a side-stepping gallop with head lowered and twisted around to face the danger.

The effectiveness of these antipredator patterns is reflected in the success that lions have in capturing various species (table 60). An incomplete hunt in this table refers to any approach, using the criteria defined earlier, which does not terminate in a rush at full speed; a complete hunt is one in which a lion rushes or pounces. The percentage of completed hunts give some measure of the ability of a species to detect danger and avoid it, before a lion can rush, and the percentage of successful hunts reflect in part the agility and speed of prey in escaping the final attack. By these criteria, the topi is the most vigilant and least vulnerable species of those for which I have data; another resident species, the reedbuck, is second. Lack of height, speed, and caution conspire to make warthog highly susceptible to predation, judging by the percentage of both completed and unsuccessful hunts. Of the three migratory species, the zebra is most alert and Thomson's gazelle least so, the latter because it cannot see well in high grass. In spite of these differences, the success of lions in catching the three species is about equal. Gazelle are adept at dodging. I have, for example, seen one jump over the head of a lioness and another escape a rush from a distance of 4 m. The resident and semimigratory species may be, with the exception of warthog, less vulnerable to lion predation than the migratory ones.

Not mentioned so far are giraffe and buffalo, which are so large and their hooves and horns potentially so dangerous that grown animals are seldom attacked by a single lion. Buffalo may retreat into water when attacked, behavior true also for wildebeest in some circumstances (Estes and Goddard, 1967; Kruuk, 1972).

HUNTING LARGE PREY

Although solitary lions may on occasion successfully attack adult giraffe and buffalo (Pienaar, 1969), these species are usually hunted cooperatively. On many occasions one or two lionesses watched a giraffe or buffalo alertly, as if contemplating an assault, but others in the group ignored the gesture and no further effort was made. The Seronera and Masai prides frequently saw these two species, often in what to me seemed a vulnerable position, without attempting to make a kill. Yet at intervals an animal was suddenly taken. Some factor I was unable to discover triggered a sustained cooperative attack.

Giraffe

I saw lions stalk or chase giraffe on ten occasions, none of them successful. Once when a lioness rushed at a giraffe, it reared up and slashed at her with both forelegs. She jumped back and then she and another lioness chased it 200 m. S. Trevor wrote to me about an incident he observed in the Serengeti: The Seronera pride fed in its usual noisy fashion "and for some of the time at a distance of no more than 20 yards the whole scene was watched by a large bull giraffe. When the fight threatened to move right under him he moved off. Later when one of the lionesses saw him put his head down to drink she stalked him and as he moved off to feed she came right up under him and sniffed his back leg—he jumped but did not kick and ran off. After 30 yards he stopped and then went back to feed in exactly the same place 10 minutes later."

Five giraffe kills had claw marks on the rumps and lacerated throats, indicating that lions had pulled them down and throttled them in much the same manner as they do other large prey.

Buffalo

Buffalo may defend themselves against lions and even attack in return. On two occasions, the Masai pride surrounded a solitary bull while he stood facing the lions with his rump against a tree, until some 20 minutes later he walked off unmolested. When a male lion rushed at a solitary cow buffalo he had surprised in a thicket, she ran into a nearby stream and stood in the water facing him. He left. She had a large raw wound on her back, undoubtedly the result of an earlier lion attack. Wounds on the shoulders and back of buffalo are not infrequent (Sinclair, pers. comm.) and attest to the difficulty that lions have in subduing the animals.

On three occasions buffalo examined the site at which one of their species had died, driving away the lions there while doing so. For example, five males killed a bull at 1040. When I returned to the kill at 1710, about 100 buffalo crowded around it, seemingly sniffing at the carcass while the lions waited 60 m away. Another time a herd of 25 buffalo stood around a bull that had died of disease while three gorged lionesses sat nearby.

I observed 7 attacks on buffalo, one of which was successful. Two of these are described in detail.

> (1) At 0815 I see a young bull buffalo on his side with a lioness biting at his thigh and rump as if trying to turn him on his back. I assume the buffalo is dead, but he suddenly raises his head, then lies still again. Two lionesses nearby make no effort to help kill him. At

0820 the buffalo suddenly struggles to his feet. The lioness rears up beside him, one paw on his neck, the other over his back, as she bites his shoulder in an effort to pull him back down (fig. 31), but he plunges ahead, whirls around, and charges. She flees into a gulley 15 m away and he follows. There they face each other at 2 m. He lunges, emitting a long-drawn grunt, and she dodges aside, growling harshly. Another lioness joins the attack. Both make feint rushes at the bull, seemingly trying to get behind him, but he twists around each time. He flees up the embankment and one grabs his rump, then falls off. For the next half hour both lionesses chase him and are chased in return inside of a thicket. After that, until 0925, all is quiet. Then one lioness returns to the attack. He bolts into the open and faces the lioness, and each lunges at the other as close as 3 m. At 0938 he walks off, a deep cut in his nose, a few claw marks on his body, and a tail that has been bitten through at the base, but otherwise not injured.

(2) At 0800 I find 14 lions of the Magadi pride at the edge of a marshy area about 20 m from a bull buffalo standing up to his belly in mud and water. Deep lacerations cover his muzzle and rump, his hocks are shredded, and his shoulders are full of bites, all the result of an earlier attack this morning. He faces the lions and grunts each time one moves. One lioness approaches to within 5 m of him but retreats when her paws get wet. At 0925 hours five nomadic males chase the pride away, then return to the buffalo at 1010 and lie 6 to 10 m from him. Fifteen minutes later he walks slowly toward the lions, a suicidal gesture. One male grabs his rump, another places a paw over his back and bites his shoulder. The buffalo sinks to his knees. A lion then clambers up on the lower back of the animal, bites him there and leans to one side as if attempting to turn him over. Meanwhile the other lion first licks blood off the old wounds on his shoulder, then bites there again. The buffalo bellows, yet makes no attempt to defend himself. The two males then pull him on his side, slowly, methodically without violent movements. One grabs a foreleg and turns him fully on his back. At this moment the third and fourth males join: one bites the buffalo in the throat, the other holds his nose and mouth shut with his teeth. The fifth male does nothing. One male eats the bull's testicles. After 10 minutes, at 1040, the buffalo dies.

Attacks by buffalo on lions were witnessed several times in Manyara Park. One lioness attacked a lone bull. She bit him in the lower neck but he continued to walk or trot until she released her hold, whereupon he charged her, chasing her into a tree. "On one occasion, we came on four lionesses and a male lion in an acacia tree with about 200 milling buffalo below it. The lions had killed a buffalo, but they were unable to descend and eat until evening when the herd finally departed. Another time two park rangers saw a herd of buffalo chase a lioness with three cubs. The mother and two cubs escaped into a tree but the third cub was trampled and killed by the buffalo. When we arrived on the scene, the buffalo were moving around the tree, paying no attention to the mangled cub" (Makacha and Schaller, 1969).

Beyers (1964) photographed a lioness as she first straddled a cow buffalo (fig. 31) but then was thrown off and gored while attempting to hold on to the animal's neck with her forepaws. At that moment several bull buffalo attacked and drove the lioness and two others away. In a more successful hunt, described by Leyhausen (1965b), two male lions circled a buffalo cow until one was able to rear up behind her and hold her with one forepaw along the back and the other on the side by her shoulder. The buffalo circled, but the lion retained his grip, finally biting her in the lower back and pulling her down, The second male then bit her shoulder and turned her over.

Relation of Prey Size to Sex of Lion Killing It

It might be hypothesized that male lions, because of their large size, can kill buffalo, eland, and giraffe more readily than can lionesses. Such selection should become apparent from the kills on which lions are found eating. Thomson's gazelle and other small kills cannot be used in such computations because males often appropriate a whole carcass from females, something they seldom do with other prey, a finding contrary to the supposition made by Kruuk and Turner (1967) on this point. Table 62 shows that females are on medium-sized prey over twice as often as on large prey; males show a reverse pattern, though less clearly. This suggests that females attack large prey less often or at least less successfully than do males. It may also indicate that the relatively slow-moving males are not as adept at capturing agile medium-sized prey as the lionesses. Both males and females were present at more than half of the kills, and in most instances I did not know which caught the animal. On eight occasions we observed wildebeest being captured when lions of both sexes were present; females accounted for seven (87%) of these captures. For eight zebra the figures were the same. I have no comparable data for buffalo, eland, and giraffe. Once A. Root arrived just after the Masai pride had killed a buffalo bull in the middle of a pool. All the lionesses were wet, but the two males were not, indicating that they had merely watched the proceedings.

LEARNING TO HUNT

Cubs have few opportunities to stalk and kill prey. Gazelle are not captured by cubs less than 15 months old, and large animals such as zebra are usually attacked by adults rather than the youngsters in a group. At the age of two years most cubs have probably never hunted large prey successfully on their own, although they may have participated in the stalking and killing. Some female cats promote the learning of skills connected with hunting in their young by creating situations where these can be practiced. Cheetah, for example, may bring a gazelle fawn to their young and release it, and a tigress may pull down a buffalo and then let her cubs kill it. Such behavior was extremely rare in lions; only one female was seen to carry a living gazelle fawn to her cubs. However, young lions have the opportunity to learn stalking techniques and killing methods by observing adults. They trail along on hunts when only a few months old, and on one occasion, when a lioness captured a zebra in a streambed, 13 cubs lined up along the bank and watched her strangle it. That house cats are able to learn a task rapidly by observing another animal perform it has been repeatedly shown in laboratory experiments (Adler, 1955; John et al., 1968). Thus cubs may have learned the hunting techniques before they have had much experience. Errors in the actual performance are not critical to the survival of cubs, for their association with adults continues until they are at least 2½ years old, longer than for any other cat. The role of the lioness is largely one of providing "the correct situation for evoking the developing repertoire of responses of the young who are thus enabled to educate themselves" (Ewer, 1969).

KILLING

There are many opinions about the way in which lions kill their prey. "Lions generally kill by seizing the animal by the nose with one paw, dragging the head down and biting through the back of the neck" (Brocklehurst, 1931). Or the lion seizes the nose of the quarry and "pulls the head sideways and downward with such force as almost invariably to break the neck at once, or else gives the beast a tremendous bite at the back of the head" (Tjader, 1911). Percival (1924) found that kills usually showed "a bite in the neck, usually at the back, but sometimes also in the throat." Guggisberg (1961) noted that "small animals are knocked over with a quick blow of the paw, and finished off with a bite in the neck or throat," and that large ones are often killed by pushing the nose down with a paw so that "in falling forward the animal breaks its neck." According to Eloff (1964), lions in the Kalahari Gemsbok Park kill a gemsbok by biting it in "the haunches and with a jerking motion upwards" breaking its back.

The killing process consists, first, of bringing the animal down and, second, of actually killing it. The methods used to accomplish these tasks vary with the size of the prey. Small animals, such as Thomson's gazelle and reedbuck, are either slapped on the thigh, causing them to fall, and then clutched with both paws, or they are simply grabbed with both paws, often while being knocked down by the violence of the contact. In either case, the lion immediately pulls the prey to its mouth and bites. An animal was in many instances killed so quickly that I was unable to observe the exact location of the bite, especially since dust or grass often obscured the scene. The killing bite was directed at the back of the neck on 16 of 26 occasions when I saw the details, at the throat six times, at the head twice, and at the back and chest once each. Not included in these figures are small gazelle fawns which were merely bitten through the body or whatever part the lion happened to grab. Three warthog were bitten in the nape and one in the chest. Frequently several lions simply tore prey to pieces without using a specific bite.

Medium-sized prey is pulled down in several ways. Generally a lion runs up behind or beside the fleeing animal, grabs it by the rump with the forepaws, and drags it down (plate 28). A galloping animal is precariously balanced, judging by the ease with which a lion can haul it off its feet, and hunts end abruptly when it crashes onto its side. The following incident is representative of this method.

> As several zebra dash from a river, one lags and veers aside, closely pursued by a male lion. The lion lunges and places both forepaws on the zebra's rump, which buckles with the impact. Both fall, the male with a paw on his quarry's shoulder as he bites the nape, seemingly in an attempt to hold the struggling animal on its side. Suddenly he spots my car and releases the zebra, which flees at a slow trot. But another male appears and after a 50 m chase pulls it down, again by grasping the rump.

Five other zebra, five adult wildebeest, and one topi were dragged down in a similar manner. On one occasion, a lioness leaped against a large zebra foal, knocking it down; once a lioness clutched the neck of a zebra and bit into its nape (plate 29). On yet another occasion a lioness ran up beside a wildebeest and placed one paw over its shoulder and another against its side, clinging to her quarry while she bit its throat; when it fell, she jumped aside. A pull with the paws is in most instances sufficient to bring prey down, as these examples illustrate, but a lion may obtain further leverage by biting the nape, shoulders, or back. A lion usually keeps its hind feet on the ground during an attack, a necessary requisite for pulling effectively, although it may leap during the initial assault or cling to the prey and use its body weight to drag the animal down. Prey is seldom attacked from the front because many species carry weapons there and because an animal is usually in flight with a lion in pursuit. Furthermore, lions may find it difficult to stop a fleeing animal from the front. One night a zebra herd bolted toward several crouched lionesses, which merely stood and looked: to grab a zebra would have been like jumping on a fast-moving train from a standstill. The points of attack are also evident from the marks on kills. Long scratches are characteristically on thigh and rump and sometimes on shoulders, side, neck, and face as well. The neck, shoulders, and back often have tooth punctures. Of several hundred kills I examined, none had a broken neck as far as I could discern.

As soon as the animal is on its side, a lion lunges for the neck or nose and bites one or the other. A hold there confers the following advantages: (1) horns cannot be used effectively in defense by prey; (2) thrashing hooves cannot reach a lion; (3) little effort is required to prevent an animal from regaining its feet; and (4) it is a highly vulnerable area of an animal's anatomy. Unlike small prey which is usually killed by a bite in the back of the neck, wildebeest, zebra, and others are usually grabbed by the throat. The teeth of lions are not built for crushing bones, nor are they particularly long, making it difficult for them to reach a vulnerable part by penetrating the layers of muscle, skin, and bone at the nape. On the other hand, the throat is unprotected, and a lion may grasp it for as long as 8, 10, and 13 minutes to mention the three longest periods noted, until prey ceases to move (plate 30).

The throat of zebra and other thin-skinned prey is often lacerated by the canines but damage is usually limited to deep punctures. However, the skin on the throat of wildebeest is up to 1 cm thick and in many instances the canines fail to penetrate it; four slight indentations, made by the tips of the canines, may be the only evidence of the throat hold. Death in this and other large species appears to be due to strangulation. Another method of killing is suffocation; a lion places its mouth over the muzzle of its prey and holds it shut until the animal dies (plate 31). Of 11 zebra I witnessed being killed, 10 were held by the throat and in one the upper canines were pressed in the throat while the lower ones were in the nape; of 12 wildebeest, including 3 calves, 7 were held by the throat, 3 by the muzzle, one by both nose and throat (involving two lionesses), and in one only the upper canines were at the throat. I examined several hundred kills and all had tooth marks either on the nose or throat and often the hair of the latter was also matted from the lion's saliva. While one lion grasps the throat, others usually begin to eat, and the animal may die from loss of its blood and viscera rather than from strangulation or suffocation.

Buffalo, eland, and giraffe are pulled down and killed in the same way as wildebeest and zebra judging by the carcasses I examined and the one killing I saw. One buffalo had frayed Achilles' tendons, the result of having been grabbed there by lions, my only such observation for any lion kill. G. Adamson told me that he once found a buffalo that had been hamstrung.

After prey has been pulled down, especially if this has been done quickly, it struggles surprisingly little, sometimes even failing to thrash its legs. An extreme instance of such behavior was an uninjured buffalo that lay on its side while a lioness chewed on his tail. Animals in such situations seem to be in a state of shock.

After observing several species of cats kill prey in captivity, Leyhausen (1965b) concluded:

> Genets and cats clearly aim their bites towards the contraction formed by the neck between the head and shoulders of a mammal. This "neck-shape taxis" is innate. By means of this taxis and the preference for pouncing down on the back of the prey from behind, a young cat succeeds in biting the nape of the neck in most of its first serious attempts at killing prey. It then quickly grasps the advantages of biting the nape rather than other parts of the neck, and soon it has learned to aim its bite purposely and exclusively at the nape. This learned orientation is an entirely new mechanism, not a modification or adaptation of the innate neck-shape taxis, which persists unaltered. The cat henceforth has the choice of either way of orientating its killing bite.

In contrast to house cats and other small felids which usually kill with a bite through the nape, lions may aim their bite at one of several places, the site depending largely on the size of prey. Large animals are usually held by the throat, the most effective killing site. Small animals, with weak bones, on the other hand, tend to be bitten through the nape which, during a pounce from above, is easier for a lion to grab than the throat. That convenience affects the location of the bite is evident among playing cubs. Some of their movements resemble those of prey-catching, among them a bite directed at the throat. Out of 109 such bites tallied, 84 were directed at the throat, mainly because the cub that was being attacked often rolled on its back.

BEHAVIOR AT THE KILL

> Six nomads, two of them males, rest at 1635 near Naabi Hill when a sick wildebeest stumbles by. One lioness pulls it down, and while she suffocates it a male begins to feed at the groin. The animal dies after 4 minutes and the male eats alone, the others waiting until he finishes at 1710. After that the lionesses feed, but 15 minutes later the male suddenly chases them from the carcass and guards it until 1840 when he permits the others to join him; all gorge themselves by about 2000. They then sleep until 0300 at which time a male briefly snacks on the remains. At 0550 ten hyenas circle the kill while two lionesses clean the last meat off the bones. Then, at 0615, they leave the area while hyenas carry the bones away. Fourteen hours after the killing all that remains at the site is a dark blood stain and a mound of rumen contents. Each lion has eaten some 16 kg (35 lbs) of meat.

This is an almost classic sequence of events. However, the behavior of lions at a kill is variable and I describe here the different ways of disposing of a carcass including the rate of food consumption.

Carrying and Dragging the Kill

Lions often move small prey after killing it, sometimes to eat it in the protection of a thicket, at others to carry it to their cubs or to escape from another lion intent on a share. The carcass is either held by the throat or nape and half dragged or lifted off the ground by the shoulder, rump, or back, the place being mainly one of convenience. Pieces of meat, weighing as much as 40 kg, from a large carcass are carried similarly and with apparent ease.

Wildebeest and zebra kills are seldom moved far, not only because of their weight but also because any attempt by one lion to drag a body away often may cause others to pull in the opposite direction, as if afraid of being deprived of a meal. Consequently a kill in a pool or other awkward spot may remain there until it can be dismembered. Large kills are transported in two ways, as shown by a male which had just killed a zebra. First he backed up, tugging violently at the throat of his kill, and moved the body 20 m in this fashion. Then he bit it in the nape, and, with the neck of the carcass between his forelegs, pulled it 23 m into a ravine, taking a succession of rapid steps before pausing briefly (plate 32). A lioness moved an adult female wildebeest 160 m in 27 minutes by the latter method, taking the kill 10 m at a time, then panting a minute from the effort. Another lioness killed an adult female zebra in a riverbed. She disemboweled it, and the drag mark showed that she had hauled the carcass up the embankment, 100 m along the edge of it, down into the river again and up the other side, where she tucked it under a bush, a prodigious feat of strength. Although two lions may turn a carcass over in unison, giving the impression of a cooperative effort, I observed no instance of several lions pulling a carcass together to a new location. On one occasion a lioness struggled for 5 minutes without being able to haul the remains of a wildebeest up a steep riverbank. The rest of the group stood around without helping. When she gave up the attempt, a male carried the piece up easily. Large kills may be moved to shade, to a more comfortable eating place, or to dense cover where they can be protected from scavengers and other predators more easily than if they were in the open. For instance, one male rested beside a wildebeest in the plains but when 16 hyenas gathered around him, he dragged the kill 100 m to a patch of reeds where they hesitated to enter.

Feeding

After having killed, a lion either begins to eat immediately or else moves the carcass to another location. On several occasions, prey was caught in high grass by a lion which then sat down and looked around casually for as long as 5 minutes, as if its hunt had been unsuccessful. It gave the impression of trying to conceal the presence of the carcass from the others that had taken part in the hunt, for as soon as these lay down or moved away it began to eat.

On a few occasions, lions plucked out long hair, such as the beard of a wildebeest, with their incisors before starting to feed, and they sometimes licked a portion of the hide, particularly a bloody portion, before biting into it. Cutting skin, fascia, and other tough parts with their carnassials, tearing chunks of muscle and viscera with canines, and pulling off small pieces with their incisors, lions bolt meat so rapidly that if many are present only the skeleton of a zebra may be left after 30 minutes.

When eating small prey, a lion usually eats the hindquarters first, followed by the forequarters and lastly the head. However, when several lions are present, the kill is torn to pieces and each animal retreats with its portion. All parts of the body are eaten except for rumen contents, horns, teeth, and a few bone splinters and scraps. Some 14% of the liveweight of Thomson's gazelle consists of digestive tract contents (Ledger et al., 1967); and bone and other waste comprises another 8 to 10%, indicating that on the average about three-quarters of the weight of such an animal is consumed.

Many authors (for example Stevenson-Hamilton, 1954) stated that lions characteristically disembowel large prey and drag the digestive tract away and bury it. "The paunch is buried in the sand, or at least covered with earth, grass or leaves" (Guggisberg, 1961). Such behavior was rare in the Serengeti. In fact, the intestines of wildebeest, zebra, and others are usually eaten before the skeleton muscles. Often a lion gorges itself on viscera alone, an interesting preference when the low caloric value of intestines is considered (plate 33). Most prey species deposit little fat, but the mesenteries around the viscera have a concentration of it. Possibly lions eat the viscera to satisfy their nutritional requirements for fat and also vitamins. Leopards, too, often eat viscera first, but cheetah reject them and tigers tend to consume the meat from the rump and thighs before the intestines. At times lions use a special technique to remove vegetable matter from intestines before eating them: one end of the intestine is placed on the rough tongue and drawn in slowly with a light lapping movement past the incisors which serve to squeeze the material out of the other end. Some of it may also be shaken out. Leyhausen (1960) observed similar behavior in the house cat. The thighs and rump are eaten after the digestive tract, followed by the forequarters. Crouched or standing, lions surround a kill and eat wherever they can in a writhing and growling mass until the carcass can be dismembered. Except to pull a desired portion closer, the forepaws are little used in such a situation but individual pieces are handled with dexterity (plate 34). Propping a bone between the pads of the forepaws, a lion rasps the last shreds of meat off it with the tongue, or holding a chunk of meat down with the large dewclaw, it cuts and tears at it with the teeth.

When lions have more meat than they can eat in one meal, they usually remain with the kill and feed again a few hours later. Feeding periods vary considerably in length. Two nomadic males killed a zebra at 2105, and one ate six times between then and 0820 the following morning, in periods lasting from 10 to 80 minutes, for a total of 4 hours 5 minutes; his companion ate four times, in periods lasting from 20 to 160 minutes, also for a total of 4 hours and 5 minutes. Another male ate three times during a night for periods of 65, 70, and 145 minutes.

A lion may ingest a prodigious amount of meat, feeding until its abdomen balloons grotesquely. Guggisberg (1961) noted that lions can eat 18 to 30 kg of meat in one meal. I attempted to measure the intake precisely by giving a lion small weighed pieces, but the animal merely grabbed a portion and fled: it obviously was unused to receiving food without having to fight for it. When I presented a lion with a large piece, other lions usually joined and shared. However, one male ate 33 kg (73 lbs) during a night and could probably have eaten more, for he was not lean when he started to feed. A male may thus be able to eat as much as 40 kg, or one-quarter of his body weight, in one meal. Some other figures tend to confirm this estimate. One night a male ate an adult male Thomson's gazelle between 0220 and 0450 and an adult female gazelle between 0545 and 0750. Taking an average adult weight of gazelles and subtracting from it 8.4 kg which the male did not consume, he ate about 27.4 kg (62.5 lbs) in 5½ hours on top of a zebra he fed on two days previously. Two males and a female consumed about 150 kg of zebra in 1⅔ days; four lionesses ate much of a male zebra in 5 hours, at least 25 kg of meat each; and two males cleaned all meat off an adult male topi in 20 hours, eating an estimated 43 kg of meat each.

Occasionally a group is unable to eat all the meat, and when the sun becomes warm most animals may retreat to shade. One or two remain with the kill and function as guards, preventing vultures and other scavengers from taking the meat. The males of a group usually do the guarding. When the Masai pride killed five wildebeest, one male and six females and cubs departed while two males remained throughout the day by the carcasses where they were rejoined by the others in the evening.

Lions usually begin to eat as soon as they arrive at a kill, regardless of the time of day, and continue until gorged or the meat is finished. On a few occasions, however, a lion merely rested for unknown reasons by the kill without eating. In one such instance, I gave a lean male the hindquarters of a wildebeest to measure how much he could eat. He stayed beside the meat from 1510 on October 22 through October 23 and 24 until around 0500 on October 25, two other males took the meat from him.

Lions consume large kills as completely as possible. Of the viscera only the contents of the digestive tract and occasionally the lungs are discarded. The vegetal matter in the digestive tract comprises about 10% to 16% of the liveweight of most Serengeti prey species with as high as 20% in the wildebeest (Ledger et al., 1967). The skin is eaten, often even that on the lower legs and on the head; the small soft bones on the nose, the edge of the scapulas, and the ends of the ribs are cracked and swallowed. But large bones and hooves are discarded. I weighed the bone and skin remains of 14 wildebeest; these comprised about 17% of the liveweight for adult females and 19% for males. Adding to this the digestive tract contents, about 35 to 40% of a wildebeest is not eaten, and the figures are similar for zebra and other large species.

Although lions usually consume all edible portions, especially if a group feeds together, they may abandon a kill after having gorged, thereby leaving a considerable amount of meat for scavengers. A total of 3 zebra and 12 wildebeest, or 1.5% and 4% of the total number of kills of these two species, respectively, remained uneaten or almost so. For example, one lioness captured a wildebeest calf, but then joined another lioness on her kill, leaving the calf to several hyenas. On another occasion, four lionesses had two wildebeest kills. They ate one almost completely but left the area that night, abandoning the second carcass.

Covering the Kill

A puma covers its kill beneath leaves, branches, and snow (Hornocker, 1970) and a tiger may conceal the remains of a carcass beneath grass and dirt (Schaller, 1967). I observed or saw evidence of such behavior at only 13 lion kills, all of them large except for one gazelle. The vegetation at the site consisted of tall grass on six occasions, of almost bare ground on six occasions, and of sand once. Attempts were made to cover the whole carcass three times, some viscera four times, and only rumen contents six times. In covering the remains, a lion rakes in earth and vegetation with its forepaw, actions which vary from a few cursory backward strokes to persistent scraping. Even a light layer of dirt covers rumen contents fairly well, and twice some viscera were quite hidden in grass except for the conspicuous claw marks around the site. One lioness disemboweled a zebra in a riverbed and covered the viscera completely with sand. Both males and females may paw and once a 7-month-old cub did so.

Such pawing is generally interpreted as being an attempt to conceal the carcass from avian scavengers, and indeed this may be so among tigers and pumas. It is a futile gesture for lions to attempt to cover large prey in open terrain; in woodlands the thickets provide suitable places for concealing a carcass. However, it may be questioned that lions paw to hide a kill. Pawing is usually directed at a waste product, such as rumen contents or viscera, rather than at the carcass itself. The behavior is most prevalent after the lion has finished eating and is abandoning the remains, not before a temporary absence. I think that pawing in lions functions mainly as a means of marking the kill site by visual and olfactory means, even though the pattern may have had other functions originally.

Interactions with Hyenas, Jackals, and Vultures

Several hyenas, a jackal or two, and vultures often complement a scene of lions on a kill, all waiting for a share of the spoils. For this and other reasons lions interact quite frequently with these species and usually not amicably.

Hyenas. As noted earlier, lions readily appropriate kills from hyenas. In most instances hyenas merely withdraw from approaching lions and either drift away or wait patiently 30 to 100 m away until the lions have finished and they can take the scraps. In one instance several hyenas waited 35 hours for the remains of an eland (Kruuk, 1972). Occasionally a hyena advances to within 10 m of a kill, but at that distance a lion may rush at it with a cough or growl and chase it as far as 60 m. But a hyena can easily elude its pursuer. At other times, hyenas become exceedingly bold (plate 35). On one occasion, six hyenas stood within a meter of a male lion which had taken a wildebeest from them; another time, after a subadult male had scavenged a wildebeest, eight hyenas tugged at one end of the carcass while he tried to eat at the other. Instead of attacking, he departed with a growl. Such hyena behavior is also evident when lions feed on their own kill. Once two hyenas ate together with a lioness on a zebra for several minutes until she left. Considering the irrascible nature of lions, it is noteworthy that so few hyenas are seriously mauled. Of four dead hyenas I found that had definitely been killed by lions, only one was beside a carcass, where a male had strangled it.

I was often surprised that a few lions were able to take a kill with impunity from many hyenas. For example, one subadult male took a carcass from 17 hyenas, and 2 lionesses appropriated one from 31 hyenas. A communal attack by the hyenas could easily have driven these lions off, and this indeed happens occasionally as the detailed descriptions by Kruuk (1972) show. Hyenas, like lions, are less bold in daytime than at night, and some of the interactions I witnessed were probably influenced by this. Early in the study I noted that gorged lions on the plains or at the edge of the woodlands usually failed to keep their kill through the night even though they could not have eaten it up. To find out what happened, 23 nights were spent beside such kills. Only wildebeest, zebra, and eland kills with little eaten from them and with no more than four lions by them were chosen. Seventeen percent of the kills were abandoned after the lions had eaten a portion of the carcass; 39% of the carcasses were consumed by the lions themselves, usually after others joined those already present; the remaining 44% were taken over by hyenas before the lions were finished with them. Usually hyenas gathered around the kill, circling it, whooping, drawing closer then retreating, playing, it seemed, a game of psychological warfare, until the lions became uneasy and departed. On three occasions they drove lions off forcefully. Two examples illustrate the behavior:

> (1) Four males lie by a partially eaten eland. At dusk hyenas begin to drift in quietly and encircle the kill until 25 of them are present. At 2030 they draw the circle closer as they approach with their bushy tails raised, emitting roaring sounds, but the males ignore them. Some whoop after that but 5 minutes later all is silent. At 0130 they try again. Whooping rapidly they surround the eland at 10 m and then stand there and call until three males depart. One remains. All hyenas lie down but 20 minutes later they call once more, 30 of them now. The male charges them once. At 0250 they repeat the performance and this time the male relinquishes the carcass and walks away.
>
> (2) Two lionesses have an adult wildebeest on which they eat off and on between 1950 and 2120, at first watched by only one hyena, then by several as more and more arrive silently. They sit or lie, waiting. At 2300 both lionesses suddenly begin to feed rapidly as if anticipating the loss of their kill. Within a minute 17 hyenas begin to whoop as if on signal and move in, circling the kill and drawing ever closer until at 2315 their tails whip up and they rush at the lions. These trot off, emitting harsh growls as several hyenas chase them for 30 m.

The antipathy between lions and hyenas is mutual. The two species may attack each other at times even when a kill is not involved. On several occasions a lion stalked a hyena and chased it 100 m or more for no obvious reason. Once Kruuk and I saw 12 hyenas approach a male lion and two lionessess. One lioness chased a hyena and swatted its leg, bowling it over, then bit it in the rump and abdomen. The other two lions rushed up and mauled the hyena in the face and hindquarters. However, they released it and as it loped away, bleeding, a lioness batted it once more. Kruuk (pers. comm.) observed some 20 hyenas mob 2 lionesses one night. One escaped into a tree but the other was briefly covered by the pack before she freed herself and ascended a small tree where she sat while several hyenas tried to jump up at her. On two occasions I observed hyenas mob a lion in daytime for no obvious reason by following it in a tight pack while whooping. On another occasion, I found male No. 159 crouched in the plains while 17 hyenas stood near him, some within 30 m. When he rose, the hyenas whooped immediately and trotted toward him. Emitting a growly miaow, he lay down with his head on his forepaws. An attack seemed imminent but at that moment five lions appeared 200 m away and the hyenas dispersed.

Jackals. Black-backed or golden jackals, or both species at the same time, are often at kills, trotting alertly around, jumping back as soon as a lion moves and darting in to snatch a morsel when an opportunity arises. Usually there are only one to three jackals near a carcass, although I once saw 19 black-backed ones together and another time 13 golden ones. Lions generally ignore jackals, even when these are only a few meters away, except for an occasional jerk of the head or growl and, rarely, a brief chase. Only once did I find a jackal that had been killed by a lion. A jackal may follow a walking lion, barking as it does so, possibly a means of proclaiming territorial rights; it may also investigate a resting group as if checking on a possible kill, but such behavior usually elicits no response from the cats.

Vultures. Whereas vultures may feed on a kill with hyenas and drive a cheetah off its prey, they are extremely cautious around lions. Hooded vultures are often the first to land, not necessarily because they have seen the kill first but because these small birds are so agile that they can pick up scraps near the carcass without harm, leaping lightly into the air to avoid the occasional lunge or rush of a lion. On the other hand, the large species of vultures can take flight only after a slow, cumbersome run—not always fast enough to escape an attacking lion, especially when the vultures are packed tightly around a carcass. On one occasion, seeing vultures on an eland, a lioness rushed them, grabbed a white-backed one, bit it, dropped it, then bit it once more and left it lying quietly but still alive. Another time, a cheetah left its kill when it saw a lioness approaching. Several white-backed vultures descended and tried to eat quickly, but the lioness pounced on one, slapped it, and then ignored the bird as it limped away. On a third occasion, several lions left their kill, but when vultures landed by it one lioness ran back and leaped nearly 2 m into the air in a futile effort to grab one. As a result of such attacks, large vultures either wait on the ground or in a nearby tree for a lion to finish, or they keep the carcass under aerial surveillance.

Lions generally ignore vultures unless the latter are within 20 m of them. A typical sight around a large kill is single birds and small clumps of them, joined at times by such other avian scavengers as marabou stork and tawny eagle, moving slowly closer as the lions approach the end of their meal until finally the carcass is abandoned and they scramble for the remains. White-backed and Rüppell's vultures may not land at a small kill, although they may circle high overhead, because lions eat the carcass so completely that for them to wait would be futile.

Amount of Prey Killed and Eaten

Some writers have attempted to estimate the average number of animals killed by a lion in the course of a year. Stevenson-Hamilton (quoted in Pienaar, 1969) put the figure at 10 to 12, Wells (1934) and Pienaar (1969) at 15, Guggisberg (1961) at 20, Talbot and Talbot (1963) at 35, Wright (1960) at 36.5, and Roosevelt and Heller (1914) at 52 to 73. Such figures have limited value unless the size of prey is specified, which it usually is not, for a lion obviously has to kill more Thomson's gazelle than zebra to survive. Nor is it noted if cubs are included in the computation, a factor which may influence the results considerably. For example, the Cub Valley group consisting of 2 males, 3 females, and 8 small cubs killed 4 adult wildebeest in 13 days. Excluding cubs, each adult killed at the rate of 21 wildebeest per lion per year, but including cubs the rate was 8 wildebeest per lion per year. In trying to find out the food intake of lions, three measures are relevant: the physiological requirements of the animal, the amount of meat needed to keep a lion healthy in captivity, and the amount of prey killed and consumed by free-living lions.

Lamprey (1964) discussed the relation between energy requirements and body weight in several East African species and concluded that the average metabolic rate in Calories per hour, although proportional to the weight$^{0.73}$, is roughly equal to the weight of the animal if it weighs between 45 and 450 kg. For a lion weighing 320 lbs (145 kg) he gave a figure of 389 Calories. An average lioness weighing 250 lbs (113 kg) would require 303 Calories per hour or 7,272 per day. A male weighing 350 lbs (159 kg) would require 428 Calories per hour or 10,274 per day. The Calories in meat vary with the species and the parts of the body, beef heart containing 850 Cal per kg and lamb chop 3,200 Cal per kg (Davis, 1954). Considering the low fat-content of lion prey and the propensity of the cats to eat viscera, skin, and other parts, the average caloric content of a kilogram of meat is estimated to be 1,500. A moderately active lioness would, according to these calculations, require about 5 kg (11 lbs) per day and a male 7 kg (15.5 lb).

Lionesses at the New York Zoological Park are fed 3.1 to 3.9 kg per day and males 4 to 4.7 kg (Crandall, 1964), usually with meat containing much fat. Captives are less active than free-living animals and even at this rate of food intake some become obese.

Using the figures of 5 kg for a lioness and 7 kg for a male as the average daily requirement, the former would need 1,825 kg (4,015 lbs) and the latter 2,555 kg (5,621 lbs) of meat per year. However, 25% of the weight of small prey and 40% of large prey, for an arbitrary average of 33%, is inedible. To obtain the necessary amount of meat, each female and male lion have to kill 2,724 kg (5,993 lbs) and 3,813 (8,389 lbs) of prey respectively. These figures presume that lions eat all edible portions of a carcass, but, as described earlier, kills are sometimes abandoned, portions of carcasses are wasted, and hyenas appropriate some remains, all factors which tend to raise the number of animals the lions have to kill. Furthermore, it is assumed that lions eat only as much as they need, which may not be true. How much, then, does a lion actually consume?

To estimate the food intake of lions is difficult. With a lion able to consume enough meat for at least five days in one meal and able to fast for more than a week without obvious discomfort, observational samples of only a few days may give misleading results, and to follow one individual continuously for many days presents other problems. A figure for killing frequency may have little relevance, particularly on the plains, where some lions obtain much meat by scavenging. In calculating food intake, I assumed that a male required 35 kg to gorge himself and a female 22 kg.

Male No. 134 was followed for 9 consecutive days on the plains and he ate on 7 of these (table 14) ingesting an estimated average of 9 to 10 kg of meat per day; male No. 159, another nomad, ate on 7 out of the 21 days we followed him (table 15), ingesting an estimated 6 to 7 kg per day. The 2 males and 3 lionesses of the Cub Valley group killed 4 adult male wildebeest in 13 days. The 5 adults were killing at the rate of 11.6 kg per day and they had an average of 6.8 kg (15 lbs) of meat available per day. Also present were 8 very small cubs, and if these are included in the computations the lions were killing at the rate of 4.4 kg per day. Several of the cubs ate some meat but the amount was negligible. One group of 2 males, 7 females, and 2 cubs captured 3 zebra and 5 wildebeest in 10 days near Naabi Hill, a killing rate of about 14 kg per lion per day, including cubs, and of this 8.4 kg was edible.

On a total of 59 days for which I have detailed information, nomads on the plains ate something on 27, or once every 2 to 2½ days on the average. An indication of how often lions eat can also be obtained in another manner. The size and tautness of the abdomen of those lions that have gorged themselves the previous night are obvious, as are those animals that are still at a kill. In 1967 and 1968, 1,815 nomads on the plains were classified. Of these 20% were on a kill, 13% were gorged, and 67% were neither on a kill nor gorged. One-third of the lions had eaten the previous night, an average of one meal every 3 days. Males, females, and subadults showed similar percentages in each category. Lions which ate only a little during the night are not distinguishable from those that had nothing, making the earlier figure of a meal about every 2½ days the more accurate one.

The Seronera and Masai prides led a life of feast or famine. Once 2 males, 7 females, and 13 small cubs killed 14 wildebeest and zebra and scavenged a further 3 wildebeest in 6 days, a provisioning rate of 11.7 kg per day including cubs and 28.6 kg per day excluding cubs. If for purposes of determining food consumption each cub is considered equivalent to ¼ adult, then 14¼ adults each ate 11 kg of meat per day. Between November 1 and 23, 1968, two males of the Masai pride participated in the eating of 6 zebra, 5 wildebeest, and 2 buffalo. They ate on 13 of the 23 days and their average daily food intake was about 10 kg. More typical was the consumption figure for 3 lionesses: they gorged themselves on a wildebeest one night, ate nothing the following 4 nights, then killed a zebra foal the sixth night, an average of about 5 kg of meat per lioness per day. Once, during the dry season, 7 lionesses and 13 cubs caught 3 gazelle in 4 days. The cubs obtained no meat and the lionesses each an average of 1.4 kg per day. A total of 30 whole days were spent with the Seronera lions at all seasons, and in this time they had 25 prey totalling 1,435 kg available. If the total number of lions present each day are added, then 198 of them, including 93 small cubs, killed at a rate of 7.2 kg per day; the adults alone killed at a rate of 14 kg. If 4 small cubs are considered equal to one adult, then each lion had 7.8 kg of meat available per day.

Of 866 lions I checked in the woodlands, mostly during the dry season, 21% were on a kill, 7% were gorged, and 72% were neither on a kill nor gorged, suggesting that the animals obtained something to eat every 3½ to 4 days. Since some lions feed without gorging during the night, the actual figure is probably about 3 to 3½ days, a somewhat longer interval than on the plains.

Most of my information on food intake is based on lions with ample available prey; I have no data from the woodlands during the rains, when prey is scarce, because I was unable to follow certain individuals continuously for many days. On the whole, lions on the plains ate slightly more than their estimated need. They readily abandoned carcasses with meat still on them, secure in the knowledge that they could kill or scavenge more within a few days. At certain times of the year, particularly when wildebeest migrated through an area, the Seronera and woodland lions killed and ate almost twice as much as they needed; at other times, they just fulfilled their daily requirements or were unable to meet them. The Serengeti lions were generally in good physical condition throughout the year, indicating that they had enough to eat, except for occasional brief periods when they probably drew on their fat reserves. Occasionally young subadults had obviously been starving during the rains, but this may be attributable more to their ineptness in hunting than to a shortage of food. Cubs sometimes starved to death, particularly at Seronera. This was not due to the unavailability of prey, but to the small size of it and the refusal of lionesses to share. Adults remained in excellent condition while cubs died. Periods when lions kill and eat more than they need probably are compensated for by times when they obtain less. I think that the estimated daily food intake of 5 kg for an adult lioness and 7 kg for a male corresponds reasonably well to the actual consumption. Subadults are smaller than adults, especially between the ages of 2 and 3 years, and their requirements are correspondingly less, perhaps 4 kg for a female and 5 kg for a male. Large cubs are arbitrarily assumed to require half an adult female portion and small cubs one-quarter of a portion. The population composition of lions was about 20% adult males, 32% adult females, and 12% each of subadult males, females, large cubs, and small cubs (see table 26). On this basis, the 1,450 to 1,700 lions in the park consumed 2,401,990 to 2,822,865 kg of meat per year, and the 2,000 to 2,400 lions in the ecological unit ate about 3,306,900 to 3,968,280 kg. With 33% of the average carcass inedible, the lions had to kill or scavenge 3,585,060 to 4,196,812 kg in the park and 4,935,671 to 5,922,807 kg (10,847,476 to 13,030,175 lbs) in the ecological unit to meet their requirements—considerably more than the 1,260,000 kg calculated by Kruuk and Turner (1967).

III Other Predators

9 The Leopard

"Secretive, silent, smooth and supple as a piece of silk, he is an animal of darkness, and even in the dark he travels alone" (Edey, 1968). Because of this, leopards remain the least known of the large African cats, their food habits the only aspect of their biology that have received more than passing mention (Mitchell et al., 1965; Kruuk and Turner 1967; Pienaar, 1969). Turnbull-Kemp (1967) summarized the existing information on leopards, and his book illustrates well this paucity of knowledge.

Most Serengeti leopards were shy, but those around Seronera were quite indifferent to vehicles. My intermittent contacts with these leopards totalling no more than 20 hours of observation, provide the information on which this short chapter is based.

Distribution and Numbers

The most adaptable and widespread of the large cats, leopards range from Manchuria and Korea through China, Malaya, Burma, and India, to Arabia and Syria as well as over much of Africa where they inhabit semidesert scrub, moist evergreen forest, moorlands on high mountains, and other vegetation types. The Serengeti leopards favor riverine forests, kopjes densely overgrown with shrubs, and the extensive thickets in the northern part of the park. They were sparse on the plains but there were usually one or two around Naabi Hill and Lemuta Hill, the Gol and Simba kopjes, and other scattered sites including several around Lake Lagaja. Early in 1967 seven leopards were resident around Seronera within an area of 200 sq km, and the same number was there in December, 1968, and July, 1969. Assuming an average density of 7 resident leopards per 200 sq km for the woodlands area of the park as a whole, the total population would be 389. Some leopards are nomadic, others frequent the plains, and in the north their density may be greater than around Seronera. My guess for the whole park is 500 to 600 leopards, or one per 22 to 26.5 sq km, and possibly as many as 800 to 1,000 in the ecological unit as a whole. Pienaar (1969) estimated a population of 650 leopard, one per 29 sq km in Kruger Park.

Movements

Many leopards were resident in the same locality for several years. The Seronera population consisted of the following animals in March, 1967: an adult male which spent much of his time along a 6 km stretch of the Seronera River, a female with a semi-independent male cub which used the watercourses penetrating the plains in the direction of Lake Magadi, a female with one small cub near the airfield and Downey's Dam, a third female which generally ranged toward the Masai kopjes, and a fourth female about which I knew little. The female and the large male cub disappeared and one female was killed by lions, but the male and two other females were still there at the end of this study. Other leopards also used the area. One female with a large male cub was seen several times near the Masai kopjes late in 1968, but their usual haunts appeared to be in the woodlands north of there. A female which I did not know appeared around Seronera in 1969, but her disregard of vehicles suggested that either she had already been around for several months or that she had returned after a period of absence.

Figure 43 shows the locations at which three residents were seen. The male was seldom encountered and his range undoubtedly was larger than indicated. One female had cubs in the Masai kopjes late in 1967 and remained around there until she moved toward Mukoma Hill in about March, 1969. The other female had one litter by the Seronera River early in 1967, and in 1968 she centered her activity near Seronera. Late in 1968 and during the first part of 1969 she roamed widely but then confined herself to the vicinity of a kopje where in May she had another litter. The minimum ranges of these females were 40 and 60 sq km respectively. Turnbull-Kemp (1967) stated that leopards usually roam within an area of 18 to 26 sq km with a maximum of 260 sq km. In the Wilpattu National Park, Ceylon, leopards range over only 8 to 10.5 sq km (Eisenberg, 1970).

The ranges of resident leopards overlapped considerably, although each animal tended to focus its activity in an area little used by others at the time. Two adults were occasionally within .5 km of each other, and during one period three females and a subadult male hunted along the same 5 km-long stretch of river, yet I only once saw two adults

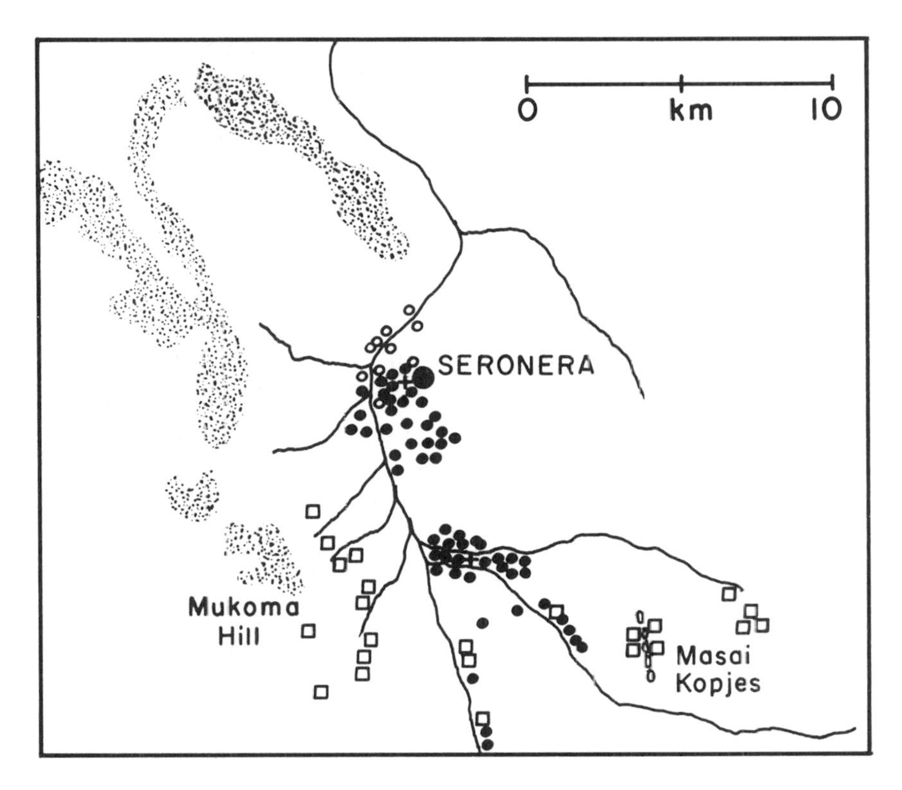

Fig. 43. The ranges occupied by a male leopard (open circle) and two females. Each symbol represents one sighting. The two crosses mark the sites where one female (closed circle) gave birth to her litters.

together when they were not courting. This indicates a strong mutual avoidance probably based both on direct visual contact and on such indirect methods as marking with scent. There was no evidence that these females actively defended a territory, but the fact that only one adult male used the area suggests that he possibly did so. Similarly, Eisenberg (1970) noted that in the Wilpattu National Park "there was little or no overlap between the home ranges of adult males, though a given male's home range overlaps an adult female's." The land tenure system of the leopard closely resembles that of the puma, or mountain lion, as reported by Hornocker (1969): "Resident male lions occupied distinct winter territories without overlap, but resident females shared some common areas. Male territories overlapped those of females. Lions exhibited a high degree of tolerant but unsocial behavior. No evidence of territorial defense was noted. Transient lions of both sexes moved freely through occupied territories. A mutual avoidance behavioral mechanism acted to distribute lions in both time and space. Visual and olfactory marks serve to facilitate avoidance between lions."

Population Dynamics

Males can readily be distinguished from most females by their huskier build, broader head, and often by their larger size. Meinertzhagen (1938) found that 6 males averaged 63 kg (60–65 kg) in weight and 3 females ranged from 44.5 to 58 kg; 3 males reported by Robinette (1963) varied from 37 to 56 kg; 5 males weighed by Wilson (1968) averaged 49 kg (37–59.5 kg) and 6 females 34 kg (28–36 kg) in weight. One male collected by Sachs (pers. comm.) in the Corridor weighed 35.8 kg, and his total length was 204.5 cm of which 81.5 cm consisted of tail. The weight of a hand-reared male cub increased from .6 kg at birth to 1.2 kg at 4 weeks, 2.7 kg at 12 weeks, 32 kg at 52 weeks, to 45 kg—his adult weight—at 96 weeks (Crandall, 1964). Two cubs whose development I observed intermittently in the wild followed a roughly similar growth schedule, reaching almost adult size by the age of 2 years.

Population Composition

Around Seronera, the only area for which I have a sample, subadult and adult females outnumbered males by a ratio of about 2:1. In March, 1967, two of the four females had cubs with them, and in July, 1969, both adult residents had cubs but a 2½-year-old subadult did not.

Reproduction

Sadleir (1966) observed 34 estrous periods in three captive leopards and found that these averaged 6.7 days in duration. The average length of the inter-estrous periods was 45.8 days with a variation of 20 to 50 days. The size of 27 litters born in the Zoological Garden of London varied from 1 to 3 (Zuckerman, 1953) and that of 10 litters born in San Francisco Zoo from 2 to 3 (Reuther and Doherty, 1968). Two Seronera females were each accompanied by a large male cub when I first saw them; a third had one female cub and her next litter consisted of two cubs which I was unable to sex before the study ended; a fourth female also had two cubs, both males. In no instance was it possible to find out how many cubs were actually born. In the Wilpattu Park, Ceylon, "only one juvenile in three different mother-young associations" was noted (Eisenberg, 1970).

One female gave birth to at least one cub in January, 1967. She raised it and in late May or early June, 1969, had another litter. Assuming a gestation period of 98 to 105 days (Crandall, 1964), the female conceived in late February, some 25 months after the birth of her previous litter. Another female had young in September, 1967, and she had not given birth again in late August, 1969, some 24 months later.

Mortality

Captive leopards may reach an age of 21 years (Crandall, 1964) but none of those I saw in the wild seemed old. Keymer (1964, quoted in Krampitz et al., 1968), found *Hepatozoon* sp. in a Rhodesian leopard, but one examined from the Serengeti for this blood parasite was negative. One leopard had pentastomids (*Neolinguatula nuttali*) in its nasal passages (v. Haffner et al., 1969) and tapeworm of the family *Diphyllobothriidae* in the intestines (Sachs, 1969); tests for brucellosis were negative (Sachs et al., 1968). Murray (1967) autopsied a leopard from Nairobi National Park and noted severe pulmonary hemorrhage and necrosis, symptoms similar to the so-called Nairobi bleeding disease found typically in the local dogs. Pienaar (1969) listed anthrax as a cause of death in Kruger Park.

On April 3, 1967, visitors came upon eleven lions of the Masai pride as they milled around and pawed a freshly killed female leopard: she had been bitten through the lower back and throat. The same month a leopard was found dead in the plains, possibly hit by a vehicle. A sub-adult male had a large flap of skin torn from his paw but the wound healed. One cub from a litter of two apparently starved to death, judging by the emaciated condition of both cubs, during the prolonged rains of 1968 when little prey remained around Seronera. Turner gave me the following excerpt from his field notes: "On 17th September, 1960, at 8 A.M., the local pride of lions was noted lying under a tree near the Seronera River with a female leopard high in the tree above them obviously very nervous. . . . The leopard attempted to descend but was promptly chased up again. Suddenly the lions converged on a grass clump and pulled out two small leopard cubs about 6 weeks(?) old. They were immediately torn to pieces and consumed."

Social Structure

Leopards lead a remarkably self-contained existence. I encountered adults 155 times and on only 3 of these were two together, the same courting pair twice, and a pair on a kill. Adult females never associated. Two females or a male and a female were together on several other occasions but in each instance it was a mother accompanied by her semi-independent offspring. Leopards of all ages were seen on 201 occasions; 62% of these animals were alone, 29% were in groups of 2, and 9% in groups of 3.

Communication

Postures, tactile gestures, and facial expressions of leopards are so similar to those of lions that a further description of them is not necessary. The leopard's tail provides a more striking visual signal than the lion's. Usually the tail is carried limply, but when, for example, a female leads her cubs she loops her tail up, thereby exposing the white terminal third of it.

Leopards were seldom heard to vocalize. In many nights spent along the Seronera River, I never heard them rasp, the coughing call which has been likened to the sawing of wood and which is analogous to the roar of lions. Once a leopard rasped by our house, once by our camp in the Salai plains, and another time by our camp along the Duma River. Captives may rasp up to 20 times in succession (Ulmer, 1966). One leopard growled at an approaching lion and a mother snarled at her playful cub. Leopards are also said to hiss when angry, grunt when alarmed, and caterwaul when treed by dogs (Turnbull-Kemp, 1967). A. Root told me that small cubs emit a soft *urr-urr* when calling their mother, and I heard a big cub moan and then miaow once when greeting its mother.

A scent-marking leopard typically approaches a bush or tree, wipes its face against the leaves or bark, then turns and squirts; sometimes it ascends a tree and marks the branches. Female leopards seem to squirt more frequently than lionesses do. One female scraped the ground with her hind paws and a male raked his claws on a tree. Feces are not used in any special manner. Eisenberg (1970) described marking by leopards in Ceylon:

> It scratches trees, scrapes ground and sprays urine. Scratching trees are repeatedly visited by the same leopard and one tree in particular appeared to be very near the boundary of two adjacent home ranges. A scratching tree may show a lean or have a very large limb approximately six to eight feet off the ground. When a leopard approaches such a tree, it sniffs at the base and then springs rapidly onto the branch or climbs up the sloping trunk. It pauses to sniff at previous scratch

> marks, extending itself along the branch. Then suddenly tensing its shoulder muscles, it begins scratching with its forepaws, or it may crouch and scratch backwards with its hind feet. It may spray urine around the base of the tree and impregnate the branch with exudates from the paws or from the animal's body as it scratches or reclines on its ventrum. While reclining at full length the leopard may rub its chest or perineal region on the limb or rub its cheeks on points which apparently bear traces of odor.

In general, the marking system of the leopard resembles that of the lion.

Interactions between Leopards

On one occasion a female walked along followed at 10 m by a male. When he scent-marked a tree, she approached him and rubbed cheeks, then crouched while he mounted, biting her nape. But she turned with a snarl and he jumped aside. The following day this courting pair was still together. One morning at 0655 a male rushed several drinking gazelle and caught one which he stored in a tree. At about 0830 a female arrived, took the gazelle into a patch of brush, and was joined there by the male. The next day only the male was near the kill, the remains of which once again were up in the tree. These were my only observations of interactions between adults.

Females keep their litters secluded, using, for example, kopjes and hollow *Kigelia* trees around Seronera. Even after the cubs are mobile they tend to remain around thickets. One female cub, born in January, 1967, began to travel independently of its mother when 13 months old, although the two continued to associate. On August 18, 1968, they were 100 m apart, each with a gazelle kill in a tree. Between August, 1968, and May, 1969, I encountered the mother 13 times but the cub was never with her after November 23, when it was 22 months old. In mid-1969 the female had another litter.

A male cub, born in September, 1967, remained closely attached to his mother until December, 1968, when he was two-thirds her size. In the morning of December 15 they played together: he stalked her, rushed and straddled her, then bit her lightly in the nape; she in turn pounced on him and they wrestled lightly, clasped in each other's arms, until she growled and they broke apart. Early in 1969 the male was sometimes alone, sometimes with his mother, but in May, at the age of 20 months, he had become essentially independent. Yet he still hunted in the same area as his mother and the two met occasionally, as, for example, on July 24 when A. Root saw him take a kill from her with impunity. In this instance, too, the association between mother and offspring lasted at least 22 months even though the male was fully independent by then.

Another male cub remained with his mother for a similar length of time, judging by his size. Once, when he was about 17–18 months old, I watched his mother approach a group of trees in which he was lying When he saw her, he bounded up, rubbed heads, then sinuously moved his body against hers several times while draping his tail over her shoulder and back. They ascended a tree and reclined on a branch, where they greeted again and she licked his head and neck for several minutes (plate 36). Both disappeared from the area four months later.

Predation

Only occasional anecdotes about hunting behavior have been published (Denis, 1964; Turnbull-Kemp, 1967), and little can be added here. My main efforts were devoted to finding kills as a basis for evaluating the role of leopard as a predator in the area.

Food Habits

The leopard's diet is more varied than that of lion and cheetah, and food items in the Serengeti include python, several kinds of birds,[1] hare, hyrax, various small and medium-sized antelopes, and a surprising number of other carnivores, a total of 24 species (table 63). In addition, Bell told me of finding a leopard with an eland young prior to my study, and a tourist reported one with a buffalo calf, a record I was unable to verify. A leopard once entered a chicken house at Seronera and killed 11 of its occupants (Turner, pers. comm.). Estes (1967) mentioned a leopard in Ngorongoro Crater which killed two Grant's gazelle. "But this leopard's main prey appeared to be jackals, of which she brought back eleven in a matter of three weeks." Mitchell et al. (1965) reported 22 kinds of prey among 96 kills in the Kafue Park—including porcupine, cane rat, genet, civet, vervet monkey, and catfish. Some of the 31 food items from Kruger Park include klipspringer, steinbuck, bushpig, leopard, aardvark, and ostrich (Pienaar, 1969).

Any casual list of kills is strongly biased in favor of large animals,

1. The large number of European storks in the sample point to the fact that these winter visitors are rather naive about large feline predators, a danger not present in Eastern Europe where they nest.

which leopards tend to store in trees. I collected most kills around Seronera and the edge of the woodlands during dry times when Thomson's gazelle were present, which was, on the average, for two-thirds of the year; almost no data were obtained during the rains. Kruuk and Turner's (1967) kill record cannot be compared directly with mine for it was largely collected by Turner over a period of years throughout the park. In the north, where gazelle do not occur, duiker and oribi probably have a prominent place in the leopard's diet.

Among the large prey items were Thomson's gazelle, which weigh at most 23 kg, reedbuck 66 kg, impala 64 kg, and Grant's gazelle 70 kg (Sachs, 1967; Wilson, 1968). In addition, a female topi and hartebeest, weighing about 109 and 126 kg respectively, were killed, as was a yearling male wildebeest weighing an estimated 130 kg. Kruuk and Turner (1967) reported a yearling topi, a yearling female wildebeest, and an adult female wildebeest among their large kills. Other prey consisted of small species or the young of large ones. Leopards seem to prefer prey in the 20 to 70 kg size category with an upper limit at about 150 kg, two to three times the weight of the cat itself. Seventy-eight percent of prey in Kruger Park consisted of impala, and Pienaar (1969) noted that most of it does not exceed the weight of a leopard. Similarly, in the Kafue Park, duiker, puku, and reedbuck were the principal prey species among the medium-sized antelopes and "five full grown hartebeest and a past prime kudu are the largest recorded kills" (Mitchell et al., 1965).

Since leopards either hid kills in thickets or rested beside them in trees, it was seldom possible to collect the teeth for aging purposes. The abandoned remains usually fell to the ground at night and scavengers took them. Judging by the size of horns and body, 59% of the Thomson's gazelle kills were adult and 25% were subadult; 16% were not classified. Most age classes were represented in the small sample of jaws (table 64). Among 19 reedbuck kills were 15 adults, 1 young and 3 unidentified ones; the 11 wildebeest kills consisted of 10 calves four months or less old and 1 yearling; and the 11 Grant's gazelle comprised 7 adults, 2 young, and an unidentified animal. Of 791 impala kills aged by Pienaar (1969), 76% were adult.

Among the 70 sexed adult (class VI–X) Thomson's gazelle in the kill sample, 51 (73%) were males and 19 were females, proportions considerably different from the 31 males and 39 females that would be expected if leopards preyed unselectively on the population. As noted earlier, males without territories often frequent the vicinity of watercourses which makes them particularly vulnerable to predation. Five of the seven adult Grant's gazelle kills were males. All 15 of the adult reedbuck kills were females. A sample of 319 reedbuck around Seronera and Lake Magadi revealed a ratio of 53 males to 100 females. I am unable to explain this high vulnerability of females. In contrast, of 13 reedbuck captured by leopards in Kafue Park, 6 were males and 7 females (Mitchell et al., 1965).

One Thomson's gazelle suffered from sarcoptic mange, the only animal obviously in poor condition among the kills.

Killing Rate and Food Consumption

Leopards in zoos are fed 1 to 1.2 kg per day (Crandall, 1964) or 365 to 438 kg per year. Assuming that, on the average, 25% of a kill consists of inedible portions, then leopards have to kill 487 to 584 kg per year to survive. One leopard entered a small game park in Zambia and ate two reedbuck, one bushpig, and a duiker before it was shot 20 days later. This leopard was killing at the rate of about 7 kg per day and he had "approximately 160 lb of edible meat" (Mitchell et al., 1965), or an average of 8 lbs (3.7 kg) per day. This leopard killed at the rate of 2,555 kg per year. The actual figure lies between the two extremes, especially when given the propensity of leopards around Seronera to kill more than they need at times. One leopard had a reedbuck and two gazelle stored in a tree and another had three gazelle. An average killing rate of 1,000 to 1,200 kg per year is probably of the correct order of magnitude, which, if true, would indicate that the leopards around Seronera whose food consists of about two-thirds Thomson's gazelle for 8 months of the year would each capture in that period about 40 to 48 gazelle weighing an average of 11 kg, or slightly more than one animal per week. Estimating 800 to 1,000 leopards in the ecological unit, 20% of them less than a year old and therefore requiring only half the amount of food of an adult, the total kill per year is then between about 720,000 kg and 1,080,000 kg.

Hunting Behavior

I observed nine leopard stalks in daytime. One of these attempts, at a waterhole, culminated in a successful rush. Other observers told me of six rushes, three of them successful. On one occasion a female stalked a herd in the plains and was within 25 m of it when the gazelle sensed her, but instead of fleeing they merely snorted and stamped their forelegs. On another occasion, a young male crouched in a patch of grass as several gazelle passed him at 10 to 12 m. He followed them in a crouching walk for 100 m, but when he was only 6 m from them and visibly tensing for a rush they suddenly scattered. Turner (pers. comm.) saw a leopard stalk and rush a reclining reedbuck which escaped by dodging under a fallen tree. A visitor observed a leopard leap on a warthog but fall off when it lunged ahead and fled. One leopard found a Thomson's gazelle fawn crouched in the grass. It "beat the fawn's hindlegs with a paw till it fell down. Then the leopard waited till it arose, beat it again and again. Finally the leopard grasped the fawn around the neck . . . and carried it off" (Walther, 1969). One cat attacked a female impala, "grabbed it around the neck and shoulders and bit its throat" (Kruuk and Turner, 1967). S. Trevor (pers. comm.) watched three adult male baboons, in a reversal of roles, chase a leopard away from the vicinity of their group along the Seronera River.

I was unable to observe an actual killing. One goat in India was grasped by the throat and an axis deer fawn had been bitten through the nape (Schaller, 1967). Kruuk and Turner (1967) also mention the throat and nape bite. Turnbull-Kemp (1967) stated "that the leopard anchors itself with the forepaws and bites through the neck from above." Many freshly killed reedbuck and gazelle had bloody throats, with the hair matted from saliva, but the leopard may have grasped the animal there while hauling it into a tree.

A leopard drags large prey by grasping the animal's neck and straddling the body between forelegs, a position which is also used to scramble with a carcass up the trunk of a tree. To wedge a kill into some branches may require several minutes of effort, as the limp body often slips and has to be lifted to another and more secure spot. Such storing of carcasses in trees helps keep them out of reach of jackals and hyenas and usually of lions too. Vultures seldom bother kills in trees although I saw several feed on an unattended one in an acacia. On one occasion, a leopard immediately hauled its kill 12 m into a tree when a hyena approached. Lions climbed trees and took a kill on three occasions, but they failed to find it on several others. Once three lions circled a tree for 15 minutes, sniffing the trunk and seemingly staring directly at the carcass clearly visible in a fork 5 m above them, but they finally walked off. Another time, an injured and obviously hungry lioness behaved similarly when a reedbuck hung a mere 4 m away. Once an elderly male rested at least 12 hours beneath an acacia tree in which a leopard reclined with two kills. He behaved as if waiting for something to fall down, looking up intermittently, but he made no attempt to climb; another time a lioness behaved similarly. The lions seemed aware that there was a kill near the tree, but in most instances they seemed unable to recognize the motionless, distorted carcass by sight.

Leopards begin to feed either on the viscera or on the meat of the thighs or chest, the site probably depending partly on the position of the carcass in the tree. Occasionally they pluck the hair from part of the body before eating. Unless disturbed, they remain by a kill until all edible portions have been devoured. An adult Thomson's gazelle may be consumed in a day or two, but more often it lasts several. On two occasions, a leopard fed on a female gazelle for 4 days; another female gazelle lasted 3 days and on the fourth the leopard killed a reedbuck. One leopard remained with a reedbuck for 5 days. According to Turnbull-Kemp (1967) a leopard can eat from 8.1 to 17.6 kg of meat in 12 hours. One Seronera leopard killed an adult female gazelle at 0830. By late afternoon she had eaten only a little off one thigh, but she consumed nearly all edible parts during the night, probably around 10 kg of meat.

10 The Cheetah

With its small round head, deep chest, trim waist, and long slender legs the cheetah is the most atypical of the cats, an animal built for speed rather than for power. Man has tamed cheetah for centuries and hunted antelope with them for sport. For example, Akbar the Great, Moghul emperor of India in the sixteenth century, kept 1,000 cheetah with which to hunt blackbuck antelope, a sport which survived in that country until the middle of this century. Considering the long intimate contact which man has had with cheetah, remarkably little became known of its habits. When I began work in the Serengeti, the literature consisted mainly of cursory popular accounts (Denis, 1964) and lists of prey animals killed by these cats (Mitchell et al, 1965). Graham (1966) attempted to analyze the social structure of cheetah, but "as practically all the information recorded by the survey was drawn from people's memories many mistakes could have been made." Recently Eaton (1969, 1970a, b, c, d) studied cheetah for 4½ months in Nairobi Park and McLaughlin (1970) did so for 1½ years in the same area. Valuable information on tame but freeliving animals was obtained by J. Adamson (1969).

With published information scanty, I attempted to learn as much as possible about cheetah in the Serengeti. Most cheetah there are exceedingly shy, sometimes fleeing from a car at distances of up to 1 km, but several have become used to vehicles and permit themselves to be approached to within 10 to 15 m. Two sisters became so indifferent to persons and vehicles that they commonly used a car's hood as a vantage point from which to spot prey and on one occasion permitted me to approach on foot to within 3 m. Although my contact with cheetah was limited to about 150 hours of direct observation, other persons, particularly my wife, Kay, D. Baldwin, A. Braun, A. and J. Root, S. and L. Trevor, and A. Laurie spent many hours with them and generously reported their findings to me.

Distribution and Numbers

Within historic times cheetah were found in arid regions from India westward through Iran to Arabia and over most of Africa, except in the driest deserts and dense forests, a distribution similar to that of the lion. The cheetah became extinct in India in 1952 but remnant populations of the Asiatic subspecies may survive in parts of Turkmenistan, Afghanistan, Iran, and possibly in northern Saudi Arabia (Simon, 1966). Cheetah in North Africa are rare (Hoogstraal et al., 1966–67) but south of the Sahara they are found sparsely from Chad and Sudan through parts of East Africa to Rhodesia, Angola, Southwest Africa, and South Africa where the Vaal River is at about the southern limit of their distribution. They are most widespread in wooded steppe, grass steppe, and undifferentiated woodlands, using the terms as defined by Keay (1959).

Cheetah occur throughout the Serengeti ecological unit but they prefer the plains and the woodlands-plains border, judging by how seldom I saw them in the Corridor and Northern Extension. In contrast, the cheetah in Kafue Park "do not venture on to extensive open plains" (Mitchell et al., 1965). Their distribution in the Serengeti seems to be less influenced by the vegetation than by the movements of Thomson's gazelle. With the onset of the heavy rains in 1967, cheetah disappeared from Seronera, together with the gazelle, and none were there from March 28 to July 4. After that date and until December they were again common. With the renewed rains both species moved to the plains, and I saw no cheetah around Seronera between December 8, 1967, and June 7, 1968. Their appearance and disappearance correlates with the gazelle migration (see fig. 37).

The extensive movements of cheetah in and out of my study area made it difficult to census the animals, especially since some undoubtedly traveled out of the park to the Salai plains and other areas. The cats were uncommon in the woodlands, and even on the plains I often spent many days without finding any. Occasionally cheetah concentrated temporarily in a locality, but such aggregations bore no relation to their density in the park as a whole. On May 14, 1967, for instance, I found 14 large cheetah around the Gol kopjes, and on December 19, 1967, I counted 10 there in contrast to most other days when none were seen. I guess that there are about 200 to 250 cheetah in the ecological unit, or about one animal per 102 to 127 sq km, as compared to one animal per 72 sq km in Kruger Park (Pienaar, 1969). The densest known cheetah population existed in Nairobi Park whose

115 sq. km were used at one point in 1969 by 15 adults and 11 cubs (McLaughlin, 1970). At the other extreme, Lamprey (1964) made only 11 cheetah sightings in four years in the 20,800 sq km Masai steppe of Tanzania, and he estimated a population of "less than 20."

MOVEMENTS

As mentioned earlier, many cheetah follow the movements of Thomson's gazelle but I have no information on the total extent of their wanderings. One female had a litter in the Masai kopjes in July, 1967. As soon as the young were mobile, she moved toward Seronera and remained for about 1½ months within an area of 10 sq km. The family disappeared on December 8, having spent the season within an area of about 60 sq km. On June 26, 1968, the mother and two cubs reappeared in their old haunts where they roamed widely between Mukoma Hill and Seronera, using at least 65 sq km of terrain. The cubs separated from their mother in mid-October but all remained in the area until early December. On February 14, 1969, I found one of the grown cubs near the Barafu kopjes, 40 km east of Seronera. One of the cubs was seen back at Seronera on about April 20, the second cub in mid-May, and the mother in June. The mother was accompanied by a new litter, and one of the cubs had a litter about 10 km from the place at which she was born. Both remained within the area they had occupied the previous two seasons, at least until late September. These observations indicate not only that some cheetah tend to remain localized after migrating off the plains, but also that they may return to the same area each year. In Nairobi Park, cheetah occupied permanent and overlapping ranges. One female with cubs used 76 sq km of terrain, another 82 sq km, and two males roamed together over 102 sq km (McLaughlin, 1970).

Other cheetah remained for a few days or weeks within a small area but then moved away without returning. This was particularly true of subadults which had just become independent of their mother. For example, one group of 3 such animals stayed at Seronera from late January to early March, 1967, another group of 3 was there from mid-July to early August, 1967, and a third group, again of 3, from mid-June to mid-July, 1969. One female with two cubs, 3 months old, was first observed on August 24, 1968. She remained in the same locality until December, then disappeared until April 18, 1969. After her return she was seen almost daily, but suddenly vanished on about May 12.

Several cheetah occupied the same general area around Seronera during the dry season, some as temporary residents others only as transients. In mid-July 1969, for instance, a total of 2 males, 5 females, and 8 cubs, singly and in groups, frequented at one time or other the area lying between the Seronera River and Mukoma Hill; of these only 2 females with cubs were resident for the season. Cheetah merely avoided contact when they saw each other, and there was no evidence for any form of territorial defense. The animals spaced themselves out by centering their activity in a locality not much used by others at the time and by avoiding meetings both by visual and olfactory means (see below). In Nairobi Park, where cheetah are resident, Eaton (1969) saw no evidence of territoriality either. He wrote: "It appears that there are loosely-defined boundaries to a home range. In a two-day period, 13 cheetah of three groups used the same area, indicating how the home ranges overlap."

Population Dynamics

For convenience I divided cheetah into several age classes. The pelage of newborn cubs is black with spots only faintly visible; in addition, there is a long blue-gray mantle of hair on the head, neck, and back, a distinct natal coat unique to cheetah among cats.[1] The mantle begins to disappear and the black coat gives way to the tawny black-spotted one of adults when the young are about 3 months old. Cubs grow rapidly; by the age of 6 months they are half the size of their mother and by 12 months over two-thirds her size. The deciduous canines are replaced by permanent ones at about 240 days of age (J. Adamson, 1969). Young aged 0 to 3 months were designated as black cubs, those aged 3 to 6 months as small cubs, and those 6 to 12 months as medium-sized cubs. Cubs older than 12 months were termed large cubs until they separated from their mother. These cubs could be distinguished from an adult mainly by their more slender build and a small ruff on the nape.

Males tend to be larger and stockier than females. Three males measured by Shortridge (1934) were 236, 221, and 220 cm long, and one female was 190.5 cm long; the length of 3 males reported by Stevenson-Hamilton (1954) varied from 193 to 211 cm. The weights of 4 males ranged from 58 to 65 kg and that of one female was 63 kg (Meinertzhagen, 1938); one male reported by Wilson (1968) weighed

1. I am unable to explain the adaptive significance of this coat.

54 kg, another by Pienaar (1969) 45.5 kg. None of the adults I was able to observe closely were old, judging by tooth wear, even though the potential longevity of cheetah is at least 15½ years (Crandall, 1964).

Population Composition

I was able to sex five litters of black cubs but only after some young had already disappeared. The survivors consisted of 4 males and 7 females. Nine out of 11 cubs sexed by McLaughlin (1970) in Nairobi Park were males. There were 11 males and 7 females in 7 captive litters (van de Werken, 1968; Manton, 1970). I further sexed 14 different litters of small to large cubs and these comprised 10 males and 18 females. An equal sex ratio at birth seems probable, but more data are needed to confirm this.

It was difficult to obtain an unbiased sex ratio of adult cheetah because the animals around Seronera would be represented too often and because females with cubs were easier to identify at a distance than solitary individuals. Consequently I excluded all Seronera cheetah from the computations and considered females with young separately. Among the cheetah tallied were 58 males, 92 females, and 94 unidentified ones. Assuming that the ratio of males to females in the unidentified sample is the same as in the identified one, the total adds up to 94 males and 150 females. In addition, 68 females with litters were encountered. Females thus outnumber males 2:1, but whether this is due to a differential death rate or is merely the result of emigration of males into areas surrounding the park could not be determined. In contrast, 58% of 471 cheetah sexed in Kruger Park by Pienaar (1969) were males. My sample also included 68 litters totalling 146 young, all less than 16 months old. Thirty-one percent of the females, including a small number which were not yet reproductively active, were accompanied by cubs. The population as a whole consisted of 21% males, 47% females, and 32% young. Some 44% of the young were large ones, 12 to 16 months old.

Reproduction

One of Akbar the Great's numerous captive cheetah is said to have had a litter of 3 cubs. No other captive births were reported until 1956, when a litter was born in the Philadelphia zoo. Since then a dozen others have been born in various zoos (for summary up to 1967 see van de Werken, 1968).

One tame but free-living female conceived for the first time at the age of 22 months (J. Adamson, 1966). Of two Serengeti females born in July, 1967, one had her first litter on about July 10, 1969; it was conceived in April when she was 21 months old, assuming a gestation period of 90 to 95 days (Asdell, 1964; Spinelli and Spinelli, 1968). Her sister courted on May 19, 1969, at the age of 22 months.

The size of 10 captive litters varied from 1 to 4 with an average of 2.7 (van de Werken, 1968; Manton, 1970; San Diego zoo, pers. comm.). The tame Adamson cheetah had litters of 3, 4, 4, and 4. Foster and McLaughlin (1968) mentioned litters of 4 and 5 in Nairobi Park, Eaton (1970a) noted 5 litters of 5 cubs each in the same area, and McLaughlin (1970) reported on 6 litters which ranged in size from 3 to 6 cubs with an average of 4.3. Graham (1966) cited several litters of 8 large cubs but these observations need to be confirmed. One newborn Serengeti litter consisted of 4 cubs. The average number of cubs in 14 other litters of black cubs was 3.0 with a range of 1 to 5, but some young probably had already died. The true average probably lies between 3 and 4. Fourteen litters of small cubs averaged 1.9 animals in size, 12 litters of medium-sized cubs averaged 2.0, and 23 litters of large cubs averaged 2.1, indicating that some one-third to one-half of the cubs died between the ages of about 5 to 6 weeks and about 3 to 4 months; after that, their mortality was low. These observations were confirmed by direct observation. For example, one female had 3 black cubs when seen for the first time on September 16, 1967. By September 25 one cub had disappeared, but she raised the remaining two. She appeared with another litter of three on June 6, 1969, but within a month one weak cub that lagged behind the group vanished. The daughter of this female had 4 cubs on about July 10, 1969. One of her cubs disappeared between August 15 and 19, a second one between September 4 and 16. Pienaar (1969) noted that cub mortality in Kruger National Park is about 50%, a figure close to the 43% given by McLaughlin (1970) for Nairobi Park. One tame but free-living cheetah lost 2 of her 4 litters within the first weeks of life, one possibly to hyenas (J. Adamson, 1969).

A female may come into estrus again soon after losing her litter. The Adamson cheetah once mated within 3 weeks after her 6-week-old litter disappeared and once within about a week after her 13-day-old young vanished. Estrus is held in abeyance while cubs are small, although a captive female came into heat 4 months after the birth of a litter (Spinelli and Spinelli, 1968). One female raised her cubs to an age of 16 months before mating again (J. Adamson, 1969). Foster and McLaughlin (1968) wrote: "Two male cheetah, survivors of a family of four born to a female in May, 1966, separated from her in July, 1967. In November, 1967, she gave birth to another five cubs," an interval of 15 months between birth and the next conception. Other birth intervals in the same area included 17, 18, and 19 months (McLaughlin, 1970). One Serengeti female conceived again 18 months after giving birth.

My sample of 14 litters of black cubs is too small to show a breeding season. Birth months were evenly distributed between January and August, but no litters were known to have been born between September and December. According to Eaton (1970d), births in East Africa "mainly occur from March to the end of June."

Mortality

As noted earlier, about half of the cubs die within the first few months of life, but the causes of death remain largely unknown. Turner told me that one litter died in a grass fire. Several cubs were sick, judging by their slow movements and unsteady gait. Four cheetah young from Nairobi Park showed postmortem findings similar to those found in domestic cats suffering from infectious feline enteritis, a fatal viral disease (Murray, 1967). Pienaar (1969) noted that "malnourished or vitamin-deficient cheetah cubs" are frequently found in Kruger Park. Small young are also vulnerable to a variety of avian and mammalian predators, and in fact Eaton (1970a) postulated that most deaths in Nairobi Park were caused by lion, leopard, and hyena.

The cheetah population in the park is surprisingly low considering the available prey and the success the cats have in catching it (see below). While the number of cheetah around Seronera may have increased in recent years (Turner, pers. comm.), there is no evidence that the park population as a whole has done so. The fact that almost 15% of the population consists of large cubs 12 to 16 months old indicates not only that reproduction is adequate but also that mortality of adults is high, assuming a stable population. Possibly cheetah move out of the park and are shot and snared. While this may account for a few animals, other factors no doubt help to decrease the size of the population and finally keep it depressed at a level below the one which the park can support. The same is true in Kruger Park. That an area can tolerate a fairly high cheetah density is shown by Nairobi Park. I have no data that would help to account for high adult mortality. Turner (pers. comm.) found a cheetah that had been killed and stored in a tree by a leopard. One young adult female in emaciated condition, weighing only 23 kg, was killed by lions; there were many *Rhipicephalus carnivoralis* ticks on her. Pienaar (1969) found that cheetah in Kruger Park may die of anthrax. Murray (1967) noted bacillary particles, identical to *Eperythrozoon felis* which causes hemolytic anemia in domestic cats, in a cheetah from northern Kenya. The same animal was infested with a spiruroid nematode, *Spirocerca lupi*, which may cause damage to and rupture of blood vessels.

Social Structure

In this section I describe, first, cheetah group structure, a point about which published information is confusing, and then those aspects of communication and other interactions between individuals which are of particular interest for comparison with what is known about lions.

Group Size and Composition

Taking the sample of 244 individuals not accompanied by cubs which I used in the computations of sex ratio, 52% were solitary, 31% were in groups of 2, 14% in groups of 3, and 3% in groups of 4. I once saw a female with 4 large cubs and once one with 5, and such individuals would all have been labeled as adults by the casual observer. A female with young was never accompanied by other adults or by her cubs of the previous litter. I never saw adult females together except once after two had met inadvertently. One or more adult males were seen with a female three times when courting and twice for a cursory visit only. Adult males, on the other hand, may form companionships in the manner of lions. Most occur in pairs, but in Nairobi Park four adult males hunted together in 1965 and 1966 (Foster and Kearney, 1967). All other groups consisted of litters which had broken away from their mother but had not yet split up. Such groups were either of the same or of mixed sex. Females probably leave such groups and become solitary before their first estrus, but males may remain together. One such group consisting of two males and one female were together most of July but late that month the female left. Two female cubs left their mother on October 18, 1968, and were still together on November 17. On February 14 one was seen alone and in April it conceived. The sisters never again associated, to my knowledge, nor did they have contact with their mother even though all three used the same area. These observations show that adult females are unsociable, except when in estrus and when they have cubs, but that adult males may form social bonds with others of their sex.

My data differ strikingly from some published ones. According to Pienaar (1969), "cheetah are more sociable creatures than leopards, and although single cheetahs are sometimes seen, they are more often encountered in pairs or in family groups up to 8 in number." Of 1,794 adult cheetah unaccompanied by young, tallied by Graham (1966) from questionnaires, 27% were solitary, 34% were in groups of 2, 19% in groups of 3, and 20% in groups of 4 to 12; he also found that 63% of litters were accompanied by one adult, 21% by 2 adults, and the rest by 3 and 4 adults. In Nairobi Park, one group of 3 adult males and 2 adult females hunted together, as did one of 2 males and 1 female, according to Eaton (1970c); however, McLaughlin (1970) reported no such associations. If cheetah in these areas have a different social system from those in the Serengeti, or if the discrepancies are the result of careless observation, cannot be decided without further research.

Communication

Many gestures and facial expressions of cheetah are so similar to those of lions that further description would be redundant, but several patterns are worthy of note.

Gestures and postures. During agonistic interactions with other cheetah or with other species of predator they use the head-low posture in the manner of lions. Slapping, which in lions is usually with a sideways motion, consists in cheetah of such a rapid and forceful downward thrust either with one paw or with both in unison that on several occasions the animal hit the ground with a thump.

Head rubbing, a method of greeting prominent in *Felis* and *Panthera*, was not observed in free-living cheetah although one may touch the face of another with its nose or cheek. However, Leyhausen wrote me that the gesture occurs in captive ones.

Around Seronera the cheetah were harassed by tourists to such an extent that even tame individuals began to avoid vehicles by hiding. A lion in such a situation crouches and a cheetah usually does too, but on two occasions one lay flat on its side in high grass; A. Root observed similar behavior. On another occasion, a female had just killed a gazelle when she saw a lion 150 m away. She reclined immediately on her side in the open, only her head raised.

Vocalizations. Cheetah use two distinct vocalizations with which they communicate at a distance. One is a chirp, reminiscent of a bird call or the distant yip of a small dog, given repeatedly and with varying intensity. It consists of a single modulated note with a frequency range between 1 and 3 kHz and a duration of about .2 seconds (fig. 22[I]). The sound can be heard for several hundred meters in spite of its ephemeral quality. Females chirp when separated from their cubs after a hunt, and cubs use the vocalization when looking for their siblings or their mother. Once, for example, a cheetah with two cubs left a third one asleep in the shade of a tree. It woke up when the others were 150 m away and chirped. The female returned and led it to the others. Cheetah also chirp when excited, as when, for instance, two adults meet, when they court, and when cubs are around a kill. The sound resembles the miaow of a lion cub spectrographically, and indeed it possibly should be classified as such.

The chirr is emitted once or several times in succession. It is a staccato sound, spectrographically reminiscent of the lion's growl, and it may last from .3 to 1.0 seconds at a frequency of up to 4 kHz (fig. 22[H]). This sound is often given in conjunction with the chirp. One female called her cubs thus: 5 chirps, 1 chirr, 2 chirps, 1 chirr, 18 chirps, 1 chirr, and so forth. At other times chirrs predominate. The function of the two calls seems to be similar, denoting "Here I am," except that the chirp can be heard farther than the chirr. In this respect the two calls are analogous to the soft and loud roars of the lion except that the calls are discrete rather than graded.

On one occasion small cubs emitted nasal bleats interspersed with chirps when separated from their mother, and two cubs bleated and hissed when fighting. A female gave several growly bleats as she circled a male lion after he had taken a kill from her. It is probable that a bleat denotes distress, just as it does in lions.

Cheetah purr loudly in the manner of house cats when they lick each other or when cubs rest contentedly by their mother.

In agonistic situations, cheetah growl, snarl, hiss, and cough but they do so infrequently. Growling at kills, for example, is rare, and hisses are directed mainly at lions and hyenas when they approach a kill on which a cheetah is feeding. Cheetah also moan, a loud *uuuu*, which they emit when approached by a lion or leopard. A female also moaned at a cub from another litter. The vocalization, sometimes given in the head-low posture, functions mainly as a threat. According to Eaton (1970b), cheetah mothers emit an *ughh*, which deters cubs from following, as she sets out on a hunt. Although I was often close to cheetah in such situations, I never heard this sound.

Marking. Male cheetah, and also females on occasion, lift their tail and squirt fluid against tree trunks and tufts of grass. Once a male rubbed his face in the moist area, and animals frequently sniffed the scent after depositing it, something seldom done by lions. According to Eaton (1969), "groups of cheetah on the move mark the same trees that they have on prior days . . . when cheetah come across a marking, they spend several minutes smelling the marked area before they mark it themselves." In fact, Eaton (1970a) felt that groups tend to orient their movement according to the location of marking sites. Cheetah may scrape soil alternately with their hind paws, and Eaton (1970a) noted that two males then "defecated just a small amount and/or urinated onto the scrape." I have several times seen cheetah defecate on termite mounds but it was not clear to me whether the behavior was deliberate or fortuitous. But once I watched a cheetah detour 20 m to a termite mound and defecate on it. Pienaar (1969) noted that cheetah defecate on isolated boulders, and Hanström (1949) observed two animals climb 4 m into a tree and deposit their feces on a branch. Cheetah also claw tree trunks on occasion.

Interactions between Adults

Adults watched each other on several occasions at a distance of 200 to 600 m without joining or retreating. A mother and her daughter, both with cubs, spent several weeks in the same area and sometimes saw each other but to my knowledge never associated. On September 2, 1968, a male approached a female with two large cubs and briefly sat 3 m from them without eliciting a response. On September 20, at 0835 the same male was 300 m from the family. By 1730 he had moved to within 8 m of it, according to Kay, who watched the event. He sniffed the ground then advanced a step at a time, his lowered forequarters supported on his elbows, a posture resembling one used by hyenas in submission (Kruuk, 1972). One cub lunged at him. He then stood broadside 2 m from the group, and when a cub moved closer he charged it. He sniffed the ground, circled the others, and reclined 3 m from them. By the following morning he was alone again. McLaughlin (1970) observed that males fought with family groups on 4 out of 6 encounters.

At 0840 on July 21, 1969, a female with 2 cubs about 3 months old killed a gazelle. She called until her cubs came. Suddenly she crouched and sneaked away. Attracted by the chirps, another female, also with two 3-month-old cubs, arrived. One of her cubs ran to the others and mingled with them freely. However, the new female advanced in the head-low posture to within 2 m of the first female which sat and chirped loudly. The new female then left at a trot followed by her two cubs. A cub from the other litter chased after them and attacked a cub. The new female rushed to the defense of her young which in turn caused the other female to trot closer. At this the new female fled with her cubs—as well as with a cub from the other litter. The first female returned to her kill, ate, then rested, seemingly oblivious to the fact that one of her cubs was missing.

The new female noted that she had an extra cub but was unable to distinguish her own from the stranger. Each time a cub came near her, she slapped it with one or both paws, hissed at it, or lunged at it with bared teeth. After being hit several times, all three cubs cringed and crept around the female just out of reach of her attacks. By 1140, she was 130 m from the cubs. She chirped occasionally but crouched and faced any cub that came near, then fled again. The cubs recognized the stranger, for they hissed at it and sometimes exchanged blows with it. At 1715, the female called the cubs once more but hit one hard when it ran to her. Somehow during the night the families sorted themselves out, for each female had two cubs on subsequent days.

One or more males may court an estrous female. Akbar the Great is said to have trapped six males in pursuit of one female (Ali, 1927). Once I saw two males follow a female closely, and I presumed that she was in heat. On another occasion I saw two males briefly rear up and slap each other while a female stood nearby, but all fled when they perceived the car; a fourth cheetah, possibly another male, stood in the distance. Males have been known to kill each other in such fights (Stevenson-Hamilton, 1954). Eaton (1969) wrote that "there was no aggression between three males which all copulated with the same female," but no further information about this incident is given. A 14-month-old cub tried to mount his mother but was deterred by a slap (Eaton, 1970a).

I observed courtship only once, at 1000 on May 19, 1969, but since the animals chose the Seronera airfield for this activity they were constantly interrupted. The male sat 13 m from the female and both chirred and chirped constantly. When she walked, he followed closely; when she reclined beneath the wing of an aeroplane, he lay 2.5 m from her. When several persons rolled a barrel past them at 10 m, the male retreated 50 m but she remained. When the male returned, the corners of his lips far retracted, she slapped him gently and rolled on her back. Disturbed again by people working in his vicinity, the male stayed 70 m from her until 1755. At that time both entered the hangar. She rolled on her back in front of him, dashed away only to approach again and paw him. But when he advanced, she raked both forepaws past his face while he chirred intensively. She left the hangar with him trailing 1.5 m behind and climbed on the hood of my car and sat there while he waited at 5 m. They left at darkness and I could not find them the next day.

Interactions between Mother and Young

Females hide their newborn young well. For example, one female had a litter in a kopje densely overgrown with shrubs and another gave birth in tall grass but moved her young to a thicket two days later. Births were observed on January 13, 1966, and December 14, 1966, in a private zoo (Florio and Spinelli, 1967; Spinelli and Spinelli, 1968). In the first litter, consisting of a single cub, the birth lasted 2 hours, from the first contraction to the expulsion of the young. The placenta was discharged immediately and was eaten by the female. The cub weighed 300 gms. "It crawled soon after birth, stood up unsteadily at the age of one week. It started to walk 12–13 days after birth, but often fell down. Its eyes did not open until four days after the birth." The three cubs of the second litter were born at 0935, 1000, and 1020, and the female broke the fetal membranes of each one with her teeth. The weight of the young ranged from 250 to 280 gms. Their umbilical cords fell off at the age of 4 to 5 days, on the tenth day their eyes opened and they also stood up, on the sixteenth day they walked, and on the twentieth day the first teeth appeared. Similarly, J. Adamson (1969) found that cubs can stand at 9 days of age, that their eyes open at 11 days, and that they can walk well at 21 days.

I saw a heavily pregnant female on July 9, 1969, and three days later her four young were found in a patch of grass. The eyes of the cubs were closed. They tried to crawl away and emitted soft chirring sounds when I handled them. The female had just moved one cub to a thicket 300 m away and she returned at a trot, picked up a second cub by the back, and carried it off too. The third was handled in a similar manner, but the fourth and last cub was carried by the upper arm, a rather haphazard method of transport reminiscent of the one used by wild dogs rather than the precise grip used by lions. After moving her cubs, the female returned once more, sniffed silently around the site, walked 30 m toward her cubs, then checked the former resting place a final time. The cubs remained in that thicket until August 15, although S. Trevor told me that he saw the cubs following their mother unsteadily on August 7. On August 19, at the age of $5\frac{1}{2}$ weeks, the female took the cubs to their first kill. The gazelle was not dead and the cubs were frightened of it, according to S. Trevor who witnessed the event.

"At the age of 18 days the cubs started to eat donkey meat regurgitated by the female" (Florio and Spinelli, 1967). Similarly, J. Adamson told me that she observed a cheetah feed her cubs with regurgitated meat. I never saw such behavior. It seems likely that regurgitation is practiced by a female only when her cubs are less than 6 weeks old, too small to follow her to a kill. Regurgitation, a canid trait, has not been reliably reported for other cats. Spinelli and Spinelli (1968) noted that their captive litter was weaned at 5 months of age. I never saw cubs older than 3 months suckle, indicating that cheetah young are fully weaned at a much younger age than lions. However, J. Adamson (1969) noted that a cub almost 14 months old attempted to suckle on its mother.

Once they are fully mobile, cubs follow their mother at all times, becoming separated from her only when she chases prey. Their contacts are remarkably free from friction, compared to those of lions, and aggressive interactions are rare even when a group feeds on small prey. One cub grabbed a meaty bone from its sibling and 5 minutes later the former casually took it back. After a meal the mother often licks the face of her cubs and sometimes two animals lick each other mutually, a utilitarian as well as perhaps a social gesture. Yet in spite of the fact that a mother and cubs constitute a closely knit social unit, the animals give the impression of remaining aloof, of lacking the intense social orientation of lions.

Cheetah cubs play occasionally in the morning and evening and while traveling from one rest site to another. Solitary play consists of dashing along in a zigzag run, tail raised, and of climbing a meter or two up the trunks of trees and bushes. Although some social play involves crouching, stalking, and pouncing, most of it consists of chases with one swatting at the flanks and rump of the other in the typical manner of an adult bringing down prey. Such play was first seen at the age of 11–12 weeks in a captive litter (Encke, 1960). Occasionally two cubs face each other and hit with rapid downward thrusts of a paw. Wrestling, a common form of play in lion cubs, is only occasional and brief in cheetah. A female may paw her cubs gently, and once I watched a mother and two large cubs dash around for several minutes as each chased first one and then another.

> Three cubs, about 15 months old, were seen playing with a young Thomson's gazelle. When the cubs found the fawn hidden in the grass, they crouched around it, their faces 15 cm from that of the gazelle. It fled after a few seconds. One at a time, a total of 10 times, the cubs swatted the fawn and bowled it over yet it continued its attempts to escape. The mother cheetah suddenly rushed up, bit the fawn in the neck, but then released it. Again the fawn tried to run away but, after being knocked over twice more, it merely crouched. The three cubs surrounded it and one grabbed its throat (Schaller, 1968).

Two female cubs, born in July, 1967, were still closely associated with their mother on October 17, 1968, at 1815. The next day at 1825, when Kay found the cheetah again, the cubs had separated from their mother. They never came together again to my knowledge. They were 15 months old at the time. The abrupt severing of social bonds was dramatic, especially since I was unable to anticipate the break by such behavioral means as an increased level of aggression or temporary separations prior to the final one. Other families behaved similarly. One female was with her two large cubs on June 19 but not on June 21; another family split between August 17 and September 23. McLaughlin (1970) noted the abrupt separation of two litters both at the ages of 16 to 17 months. The sudden transition from dependence to independence in the cheetah is strikingly different from the gradual one in lion and leopard.

The litter of the tame but free-living cheetah observed by J. Adamson (1969) behaved differently than the Serengeti litters. At the age of 11 months the young wandered away from the female for as long as 2 to 3 hours, and by the age of 16½ months they remained away for longer periods than that. The final break with the mother occurred when the cubs were 17½ months old.

During their 15 to 17 months of association with their mother, cubs must learn to hunt well. In this they pass through three broad stages. At first they ignore the hunting postures of the female while they play around her and trot ahead, thereby alerting prey at times. She in turn may reduce such interference by her behavior, as one hunt illustrates.

> The female stalks slowly toward some gazelle 300 m away. Her three cubs, 9 weeks old, run playfully ahead of her. She sits. When the

> cubs return and cluster by her, she immediately advances, but the youngsters once again range in front. She sits 5 minutes and the cubs settle by her side. The gazelle have drifted behind some bushes. She walks rapidly, trailed by the cubs, to within 100 m of the gazelle, suddenly sprints, and advances to within 50 m before the gazelle become aware of her and flee. She pursues an adult female, almost loses her when she dodges, and then both paws flash out and miss. After another 35 m she slaps the flank of her quarry and it crashes on its side. She overshoots 5 m, dashes back and grabs the throat. She drops the gazelle after 2 minutes and pants after her 270 m run. Then she pulls the gazelle by the throat or flank about 50 m toward her cubs. But a lioness trots up and takes the carcass.

Cubs 3 months old or older usually remain behind their hunting mother either walking slowly or waiting until the kill has been made. Small cubs may be introduced to live prey by their mother: "D. Baldwin and my wife watched a cheetah carry a small Thomson's gazelle to her two cubs, four months old. When she dropped the fawn, it jumped up and fled. While the female watched, the cubs pursued it and once knocked it down, but they were unable to catch it. The female then killed it. Kruuk and Turner (1967) also related an instance of a cheetah providing her cubs with the opportunity to chase a gazelle fawn" (Schaller, 1968). Similarly, Laurie (pers. comm.) watched a cheetah bring a fawn to her cubs which mauled it for 10 minutes before it died.

By the age of about 8 to 12 months cubs may initiate stalks and capture prey by themselves. A female may even provide cubs with the opportunity to capture prey by refraining to take part in the chase herself (Eaton, 1970b). I watched two 1-year-old cubs catch a gazelle fawn, but most such hunts ended in failure because of ineptness. For example, on August 21, 1968, a 13-month-old cub knocked a gazelle fawn down several times until finally the mother cheetah ran up and killed it. In three subsequent hunts observed, the female did the killing. On October 1, the same cub, now over 14 months old, again failed in its attempt to kill and the female finally dispatched the fawn. In spite of these failures, this cub and her sister separated from their mother on October 18. On October 28, one of these cubs caught a gazelle fawn, but it escaped and another 100 m of pursuit were required to bring it down. Suddenly a hyena ran up and appropriated the kill. The two cubs became very lean in the month following the separation from their mother. Obviously cubs are far from experienced hunters when they become independent, yet their learning proceeds rapidly for I found no evidence of death from starvation in that age class.

PREDATION

Most information on predation was collected around Seronera, an area which provided me with 70% of the kill records as compared to 3% for the woodlands and 27% for the plains. There were usually many Thomson's gazelle around Seronera when cheetah were there, and this together with the open terrain had undoubtedly a marked influence on food habits and hunting behavior. In a preliminary paper (Schaller, 1968) some aspects of cheetah predation were described. This account includes an additional two years of data.

Food Habits

A total of 261 kills were found, 91% of them Thomson's gazelle and the rest Grant's gazelle, wildebeest, impala, hare, and several others (table 63). Thomson's gazelle obviously are the most important prey of cheetah in the area. Similarly, of 23 kills reported by Kruuk and Turner (1967), 13 were Thomson's gazelle and 6 were wildebeest. The preponderance of gazelle kills reflects the seasonal movements of cheetah in pursuit of that species. Of 1,092 kills reported by Pienaar (1969) from Kruger Park, 68% consisted of impala, and among the 23 other kinds of food items were two young giraffe, two young buffalo, aardvark, porcupine, jackal, and other cheetah. Impala, Grant's gazelle, Thomson's gazelle, and hartebeest, in that order of importance, constituted some 85% of the kills of cheetah in Nairobi Park (McLaughlin, 1970).

Size of prey was an obvious factor in its selection. Excluding Thomson's gazelle, which seldom weigh more than 20 kg, all wildebeest, the hartebeest, one Grant's gazelle, and one topi kill were less than 2 months old; adult prey included four female Grant's gazelle, two female impala, and a male and female reedbuck, all weighing less than 60 kg each. However, large prey may be attacked, especially if two or more cheetah hunt together. Two male cheetah killed a yearling topi weighing about 90 kg. N. Tinbergen (pers. comm.) observed three cheetah wait beside a sick adult zebra at least four hours and one once darted in and bit its anus. Kruuk and Turner (1967) mentioned an adult female wildebeest, a male hartebeest, and yearling zebra as having been killed by cheetah. Foster and Kearney (1967) noted the following kills in Nairobi Park: "1 zebra, 7 Grant's gazelle, 1 Thomson's gazelle, 4 impala, 1 waterbuck, 5 kongoni and 1 ostrich. All of the large prey were killed by 4 male cheetah which hunted together."

The preference for small species or young of large ones weighing 60 kg or less is related in part to the difficulty cheetah may have in subduing large prey. Tracks in the sand showed that a cheetah and reedbuck scuffled for some 25 m before the latter was vanquished. Kruuk and Turner (1967) observed a cheetah attack a male Grant's gazelle: "it grabbed a large male and tried to drag it to the ground whilst biting it around the muzzle, the Grant meanwhile butting at the cheetah with his large horns. The fight lasted several minutes and ended by the cheetah suddenly leaving its victim, maybe frightened away by our presence." Adult prey may also defend itself or its young. Kay watched a cheetah knock down a topi calf but an adult attacked and drove the cat off; when it made another attempt, two topi drove it away. Trevor (pers. comm.) saw wildebeest behave similarly.

Table 64 shows the ages of 228 Thomson's gazelle kills. These consist of 55% fawns (classes I–IV), as compared to about 35% of them in the living population. However, small young are usually not represented fully in a kill sample. One female we watched intermittently throughout the day for 26 days killed 24 gazelle in that time of which 62% were fawns. The actual hunt was observed in 66 of the 228 kills (excluding 10 unaged and unsexed ones) and the results were as follows:

Age class	*Killing seen*	*Killing not seen*
I–IV	80%	45%
V–X	20%	55%

The number of small young in a kill sample will vary with the season, but judging from these figures it is likely that at least two-thirds of the gazelle a cheetah captures are less than a year old, a higher proportion than expected. McLaughlin (1970) noted that 52.4% of all kills in Nairobi Park consisted of juveniles. Table 78 compares yearling and adult kills with a sample from the living population. The figures show that classes IV and V are relatively invulnerable to predation, that classes VI to VIII are taken about as often as expected, and that old animals (IX and X) are killed slightly more often than expected.

If cheetah prey unselectively on the sexes, and 22% of the population consists of adult males and 28% of adult females, then about 44 males and 56 females should occur in a sample of 100 kills (classes VI-X). The actual figures were 35 males and 65 females, about 25% fewer males than expected. As table 64 shows, predation was more or less equal on the sexes in all age classes except in the two oldest ones. In those classes more females than males were taken ($P = <.05$), but if this was because they were more vulnerable or because there is a disproportionate sex ratio of old animals in the population I do not know.

Two gazelle and a wildebeest kill were heavily infected with sarcoptic mange. The other prey appeared to be healthy, and during the hunts I observed none of the pursued animals showed evidence of being in poor condition, which, of course, may merely indicate that the cheetah could detect slight weaknesses which I could not. The marrow of five adult gazelle was checked in the dry season of 1969 and three were depleted of fat.

In contrast to other predators, cheetah were not seen to scavenge, although Pienaar (1969) reported several instances of it in Kruger Park. Being low in the interspecific predator hierarchy, cheetah are probably too timid to investigate possible sources of meat in most instances.

Killing Frequency and Food Consumption

From October 13 to November 5 and from November 14 to 19, 1967, a female cheetah with two cubs, 3–4 months old, was kept under intermittent observation throughout the day for 26 days to obtain some idea of how often cheetah kill. She was not observed at night, but, judging by the fact that she was usually located in the morning in the same place where she was the previous evening, she did little or no traveling at night. During the period she killed 24 Thomson's gazelle and one hare. She failed to capture prey on three days but on each of two days she caught two gazelle (Schaller, 1968). A. Laurie watched the same female from July 10 to 15, 1969, when she had another litter of two cubs, about 3 months old. In that period she failed to kill on two days, but captured a total of two adult female gazelle and two fawns on the others. Between August 13 and 15 she caught three adult female gazelle, one each day. Thus in 35 hunting days, this female captured 31 gazelle and a hare. Her grown daughter with three small cubs was observed for 6 days in August and in that period she killed a gazelle on each of five days and on the sixth one she was so persistently bothered by tourists that she was unable to hunt (Trevor, pers. comm.). I observed a female with two large cubs for five days between August 21 and 28, 1968. Six gazelle were caught in that period. In the Seronera area, female cheetah accompanied by cubs kill almost at a rate of one gazelle per day, or about 341 per year. I have no precise information on how often solitary cheetah kill but McLaughlin (1970) found out that they do so every 2 to 3 days or about 150 kills a year.

An estimated 200 to 250 cheetah use the ecological unit. Of these, 28%, or 57 to 70 of them, are solitary, each killing at a rate of 150 animals per year. The rest are in about 54 to 72 groups, some consisting of females and cubs and others of independent young or adults, each group killing at the rate of about 341 prey per year. Thus a total of 26,964 to 35,052 animals a year are taken by cheetah. Around Seronera and on the plains most of their food consists of Thomson's gazelle, but impala, dik-dik, topi young, and others may contribute much to their diet in the woodlands. Possibly 60% of all kills are Thomson's gazelle, which would indicate an annual kill of 16,178 to 21,031.

The average amount of food actually consumed by a cheetah over a period of days could be calculated from the kill record of the female that was observed for 26 days (Schaller, 1968):

> By using the average weights of adult male and female gazelle presented by Sachs (1967) and by estimating the weights of fawns visually, it was determined that the total weight of prey killed by the cheetah was 261.3 kg, an average of 10.0 kg/day or 3,650 kg/year. However, only about 8.8 kg/day were available to the cheetah because a lion scavenged her kill twice and a hyaena once. Part of each kill is not eaten, primarily the digestive tract and its contents, most bones, and most of the skin. The weight of an adult female Thomson's gazelle averages 16.2 kg. The remains of two females weighed 6.4 kg and 6.8 kg, respectively, after the cheetah finished eating. About 60 per cent. of the weight of an adult female Thomson's gazelle is therefore consumed, and a similar amount seemed to be taken from adult males and from large fawns; perhaps a somewhat higher percentage is eaten from small fawns. Although the cheetah had 8.8 kg of prey per day available, they ate only 5.3 kg. Each cub took probably at least 0.5 kg, leaving 4.0 kg/day for the female. This rate of food consumption is over twice the 1.3 to 1.8 kg/day needed to keep a cheetah in healthy condition in a zoological garden (Crandall, 1964).

Assuming that an adult has on the average 4 kg of edible food available each day and a cub 2 kg, including meat of large kills that is wasted, and that a third of the 200 to 250 cheetah are cubs, then these animals used a total of 243,090 to 292,950 kg per year. About 40% of a large prey and 30% of a small one consists of inedible portions and the amount actually killed is therefore about 35% higher than that which they have available as food, or about 373,985 to 450,692 kg.

Since these figures are based on several assumptions, they are at best of the correct order of magnitude.

Hunting Behavior

Cheetah spend most of the day resting. Around Seronera, where prey is usually plentiful, they seemingly wait for it to wander into their vicinity rather than roam in search of it. They seldom travel more than 3 to 5 km in a day, but in Nairobi Park they are said to move almost 8 km on the average according to Eaton (1969) and 4.3 km according to McLaughlin (1970). Their various forms of activity, including stalking, feeding, and walking at their normal speed of 3 to 4 km per hour, usually require less than 4 hours of the day. Although cheetah are primarily diurnal, they may also travel and hunt on moonlit nights. The 147 exact and estimated times of killing prey as well as the times of 32 unsuccessful attempts are combined in figure 44 to show that the animals may hunt at any daytime hour between 0600 and 1900 with peak activity periods between 0700 and 1000 and 1600 1800. In addition, once cheetah were on a kill that was made in the dark at about 1930, and on another occasion a gazelle was captured in moonlight at 0200.

Around Seronera, gazelle are typically scattered over the plains in small groups of 2 to 20 individuals, with cover during the dry season consisting mainly of occasional shrubs and patches of grass. Cheetah typically hunt for prey either by walking alertly over the plains, sometimes climbing into the low branch of a solitary tree or on a termite hill to scan, or by waiting until a gazelle walks by. When a cheetah tried to approach prey undetected or ran after it, I considered it a hunting attempt. I observed 48 such attempts of which 16 failed before a rush could be made, eight runs were unsuccessful in spite of the fact that the cheetah pursued at great speed, and 24 (50%) ended in a kill. Other observers told me the outcome of 55 actual chases they witnessed. The 16 hunts that were terminated prior to the rush failed for several reasons: in 8 instances the cheetah bounded or trotted several meters toward gazelle then halted, presumably because it did not see a vulnerable individual; in 4 instances it stalked then ceased to do so for no obvious reason; in 3 instances it stalked but the prey sensed danger and looked up; and once prey moved inadvertently out of the cheetah's reach.

A complete hunt can be divided into an approach, chase and capture, and finally the kill. A cheetah may approach prey in several

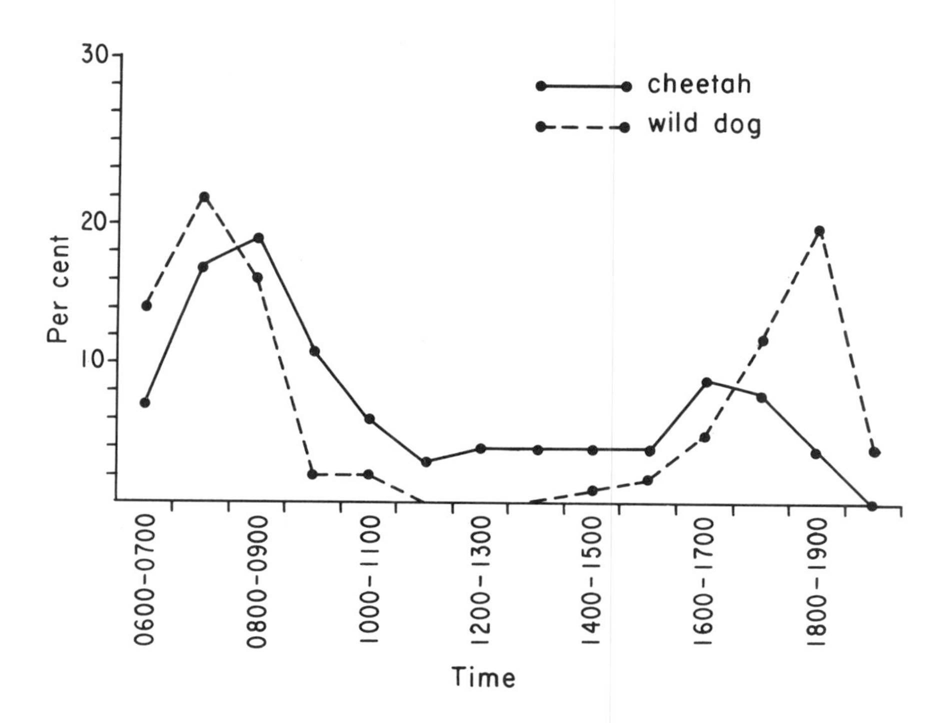

Fig. 44. The time of day when cheetah and wild dog killed their prey.

ways. Occasionally it walks toward gazelle without concealing itself. The gazelle may stand and watch and even trot closer, to within 60 m to 70 m. The cheetah may then run at one or another group, giving the impression of searching for a suitable quarry. Another kind of approach is to bound toward an unsuspecting herd from at least 100 m, and by the time the prey becomes aware of danger the cheetah is close enough to select an individual. If a cheetah sees a small fawn it may begin its pursuit from as far away as 500 m to 600 m, according to Root who saw two such hunts. Many hunts involve a variety of techniques from walking closer, waiting and stalking, to actual pursuit at various speeds. Cover is used for concealment but no attention is paid to wind direction. Eaton (1970a) stated that "the cheetah does not crouch like most cats," but it does so readily in the Serengeti. If a gazelle looks up from its feeding, the cheetah may halt, standing motionless until the animal lowers its head again.

In 17 successful hunts which I observed from beginning to end, the cheetah began its approach from an average distance of 180 m (50–300 m). On 7 of these it stalked closer for an average distance of 125 m before rushing, on 3 of these it walked and trotted some 60 m to 130 m closer, and on the remaining 7 hunts it ran closer without preliminaries. Usually a gazelle spotted the cheetah immediately and fled, but on three occasions the animals did not see the cat until it was 20 m, 30 m, and 50 m away, respectively. Six unsuccessful hunts began from an average distance of 125 m.

Cheetah can attain a speed of about 104 to 112 km per hour (65–70 mph) according to Howell (1944), and even 114.6 km per hour on a race course (Demmer, 1966). The fact that these cats can readily capture gazelle, which I have clocked as they ran beside the car at 70 to 80 km per hour, suggests that they may certainly reach a speed of 95 km per hour. However, a cheetah can use its full speed only to draw close to a gazelle, and after that it has to follow at a reduced rate in order to follow each twist in the unpredictable route of its quarry. With danger imminent, the gazelle may zigzag but seldom does so more than 3 to 4 times. When close enough, the cheetah slaps the hindleg, thigh, or rump of its prey with a sideways or downward sweep of its forepaw; occasionally a cheetah attempts to hit a gazelle with both paws in unison. Eaton (1970b) noted that large prey such as hartebeest may also be slapped on the foreleg. The slap may leave a cut in the skin, presumably caused by the large dewclaw. When hit, the gazelle crashes to the ground or it may even flip over in mid-air. One Grant's gazelle apparently broke a foreleg in such a fall and another Grant's gazelle a hindleg. "Racing up alongside its quarry the chita springs with unerring aim at the throat or on the back" according to Stevenson-Hamilton (1954) but such behavior was not observed.

Many runs begin at a moderate speed while a quarry is being selected, and after that the cheetah accelerates to a final burst which never exceeded a distance of 300 to 400 m. The total distance of the chase varied from 35 to 460 m. Small fawns (classes I and II) were caught after an average run of 190 m in 12 hunts observed, but 11 runs after large fawns and adults required an average distance of 290 m. Ten unsuccessful runs after adults were 270 m long.

Once a cheetah slipped and fell when making a sharp turn after a gazelle, on another occasion a hunt failed because the cheetah was unable to select an individual from the compact herd, and on a third occasion the cheetah lost sight of its prey after it mingled with others. Twenty-three other unsuccessful runs terminated when a cheetah gave up the chase after 200 to 300 m, particularly after it failed to follow a sharp turn by a gazelle. It seemed exhausted, panting at a rate of some

150 breaths a minute, and seldom attempted another rush for at least half an hour unless it spotted a small fawn.

Hunting success varies with the age of the gazelle selected. Of 31 runs after small fawns (class I and II), 100% were successful. This figure does not include three crouched fawns that were snapped up nor does it include one instance, related to me by a visitor, in which a cheetah caught a fawn but then released it alive when a lion approached. Of 56 runs by cheetah 14 months old and older after large fawns, yearlings, and adults, 30 (53.5%) were successful. The success rate for gazelle of all ages was 70%. These percentages do not include stalks and other hunting attempts which failed to culminate in fast pursuit. McLaughlin (1970) saw 236 chases of 25 m or more after a variety of prey with a success rate of 37%.

Several factors play a role in the selection of a particular gazelle. (1) Small fawns are selected undoubtedly because a cheetah is assured of a meal, even if a small one. (2) Cheetah prefer prey which is separated from others of its kind. When, for instance, a gazelle veered from a fleeing herd, the cheetah pursued it. It was my impression that groups of two and three gazelle were favored by cheetah over large aggregations. (3) Gazelle which are near cover, behave inattentively, or are situated in such a way that the cat can approach undetected are often chosen; final selection depends on size and on distance from the cheetah, but not primarily on the animal that flees first as has been suggested by Kruuk and Turner (1967) and Walther (1969). (4) Fleeing animals seem to be chosen more than stationary ones, so that territorial males, which tend to stand and watch the cheetah rather than escape immediately, may be passed by. In many instances a quarry is not selected until both predator and prey have run a distance and the physical condition of the latter may, by then, have influenced the cheetah's final choice.

On several occasions a female assisted her cub when it was unable to subdue a fawn. Trevor (pers. comm.) once watched a large cub bound at several gazelle; while their attention was drawn in one direction the female sprinted in from another and caught one. But such cooperation seemed fortuitous, and communal efforts in securing prey as practiced by lions and wild dogs were not observed.

Three examples of hunts provide a more complete picture of the behavior than presented so far.

(1) A female approaches several gazelle through high grass for 130 m in a stalking walk, but they spot her when she is 45 m away and she

gives up her attempt. Twenty-seven minutes later she sees about 15 gazelle at 250 m, approaches slowly to within 150 m and suddenly runs toward them, the first 130 m moderately fast as she chooses a victim, and then 160 m at full speed after a yearling. She slaps but misses as the gazelle dodges, and the hunt is a failure.

(2) A female climbs 3 m into a tree and looks around. She spots 10 gazelle 230 m away and walks 140 m toward them with head held low and then lies watching for 5 minutes. One female grazes somewhat separated from the herd; the cheetah rushes her and is within 35 m before the gazelle turns and flees. After a chase of 160 m including a 180° turn, the gazelle suddenly flips forward, hind feet in the air, apparently tripped by the cheetah which lunges in and grabs her throat.

(3) A female sees about 12 gazelle grazing on burned stubble 175 m away. She first walks toward them 100 m then bounds, chasing the herd at moderate speed for 175 m before selecting the smallest individual. She sprints after it, follows three zigzags closely, and after 130 m catches it in a cloud of dust. She holds its throat for 5 minutes before it dies.

One or both forelegs and the chest may be used to hold the struggling animal down. The cheetah tries to grab the throat of its quarry, usually remaining behind the animal, away from the sharp hooves, while doing so. In 21 detailed killings observed, 20 were with a throat hold, the cheetah either standing, sitting, or lying; one small fawn was bitten in the nape (plate 37). Although the skin was often punctured by the canines and occasional chewing movements were made by the cheetah, death in most instances seemed to be due to strangulation. An average of 4.5 minutes (2 to 11) elapsed before the gazelle ceased to move. Occasionally the cheetah dropped the body, but when the animal continued to kick it was picked up again and held by the throat several additional minutes. If a bush or a tree is near, the body may be dragged or carried as far as 250 m into the shade with the cheetah holding the carcass by the throat, nape, back, rump, or thigh. The hunt and the moving of the kill seem to exhaust a cheetah so much that it may rest half an hour before eating. Cubs less than 3 months old are often unable to cut through the skin of prey and the mother may do so first and then rest while her offspring feed.

Cheetah usually eat the meat off of one thigh and then off of the abdomen and ribcage before starting on the other thigh, the forelegs, and on the liver and heart. Blood in the body cavity may be lapped up, useful behavior in areas where water is scarce. (Cheetah around Seronera drank at the river irregularly in daytime as well as at night.) All that typically remains of a gazelle is the articulated skeleton with most of the skin attached as well as the digestive tract. Most bones and skin of a small fawn may also be eaten. Meat from a Thomson's gazelle is seldom wasted, but a considerable amount may be abandoned when an adult Grant's gazelle or other large prey is killed. Cheetah were never observed to take more than one meal from a prey. When one or more large cheetah feed on a gazelle fawn, the remains are abandoned within 15 to 35 minutes, but a female with small cubs may eat on such a kill for as long as 70 minutes. A female with two small cubs required a total of 50, 65, 75, and 120 minutes, on four occasions, to consume adult gazelle. According to J. Adamson (1969) cheetah may scrape dirt over the carcass after eating.

Cheetah feed rapidly, stopping occasionally to look around, as if nervous, probably because other predators frequently appropriate the kill. Of 238 kills, 20 were known to have been taken by lions before the cheetah was finished with the meal (plate 38), 11 by hyenas, and one by leopard, a loss of 12%. On at least 11 of these occasions lion and hyena were attracted by vultures that descended to the kill. In fact, on two occasions a solid phalanx of white-backed vultures advanced to within 1 to 1.5 meters of a cheetah, which abandoned the remains before finishing. Usually, however, the cheetah ignores vultures or rushes at them, occasionally leaping into the air and swatting at an escaping one. Jackals, too, may be chased from the vicinity of a kill, and in one instance a female pursued one 100 m and slapped it. Hyenas meet little resistance when taking kills from cheetah although in one instance two hyenas were attacked and driven off and on another occasion a cheetah hit a hyena in the face as it took the kill. More typical, however, was the following sequence of events.

> A female cheetah sees several gazelle 200 m away, among them a small fawn. She trots, then bounds, some 175 m toward them, then chases the fawn 100 m, her tail held almost vertically, and catches it. As she holds it by the throat, a hyena lopes up. She drops the fawn, jumps back 3 m and hisses while the hyena carries the carcass 50 m and eats it in 10 minutes. A few minutes later she spots another fawn 170 m away and catches it, thereby providing the same hyena with another meal.

Cheetah have no defense against lions, and whenever one approaches a kill, they either retreat immediately or circle the lion at distances of 10 to 20 m while growling, moaning, hissing, and occasionally lunging.

11 The Wild Dog

Small in size and with a splotchy white, black, and yellow coat, the African wild dog or hunting dog does not seem to be the kind of animal that could arouse the passion of man to such an extent that it has been relentlessly and irrationally persecuted throughout its range. "I hate the hunting dog . . ." wrote Percival (1924); "I maintain these pests should be outlawed wherever found," wrote Hunter (1960) in a conservation magazine. Wild dogs occasionally attack livestock and are said to annihilate wildlife populations or at least disperse them. "When hunting dog move into an area, the game move away" (Brooks, 1961). In the past few years studies by Kühme (1965), Kruuk and Turner (1967), and Estes and Goddard (1967) have elucidated the social system and some aspects of predation of wild dogs, yet in many respects the animals remain enigmatic. I attempted to find out as much as possible about them.

Wild dogs are scarce in the Serengeti, with months sometimes elapsing between sightings of a pack, except when it is localized at a den. The animals usually disregard the presence of a vehicle, permitting its approach to within 10 m or less, and sometimes they even bite the tires or otherwise investigate it. On one occasion, when my car fell into a warthog burrow, a pack stood 12 m away and watched me dig it out.

I observed dogs for about 190 hours, a time which does not include eight nights of intermittent observation at a den; in addition A. Root and several others generously reported incidents of interest to me. Most observations were made around Seronera and in the plains. The data in this chapter apply to that area rather than to the park as a whole.

Distribution and Numbers

Wild dogs are widely though sparsely distributed in Africa south of the Sahara, except in deserts and moist evergreen forests, ranging upward into the mountains as high as the summit of 5,894 m Mt. Kilimanjaro

(Thesiger, 1970). Packs are found throughout the Serengeti ecological unit, but are rare judging by the infrequency with which they are seen. Kruuk and Turner (1967) estimated that "possibly 150" occur in the park. My guess for the ecological unit is 250 to 300 dogs (excluding pups in dens), or one dog per 85 to 102 sq km, as compared to about one dog per 57 sq km in Kruger Park (Pienaar, 1969).

Movements

Packs with small young remain in the vicinity of a den. The animals leave the den to hunt and afterwards return to it, a pattern which restricts them to a certain radius of activity. One such pack hunted for 2½ months within an area of about 160 sq km except for occasional excursions farther afield; a second pack used about 210 sq km of terrain around the den during the 3 weeks we observed it; a third pack ranged over at least 110 sq km during my eight encounters with it. One pack watched by Kühme (1965) hunted within an area of 156 to 208 sq km.

After the young are mobile, packs travel so widely that I was unable to delineate their range. Those that use the plains usually retreat with the prey to the woodlands during the dry season. For example, one pack denned near Naabi Hill in June and July, 1966, and it was still in that area in August. On December 5, it was at the Moru kopjes; on February 11, 1967, it was back at Naabi Hill but late that month and in March it roamed the plains southeast of the Moru kopjes only to move to Lake Lagaja in April. The area encompassed by the 15 sightings of this pack was about 620 sq km (230 sq m). Another pack denned from March to at least May, 1968, near Naabi Hill, and between August and November of that year it was 50 km west of there near Seronera. In March, 1969, it was on the plains once more only to return to the edge of the woodlands during the dry season, having used an area of at least 710 sq km (fig. 45). One pack seen near Seronera in December was at the Simba kopjes in January and at Olduvai Gorge, 72 km from Seronera, in April.

Ranges overlapped extensively and in a few instances probably completely. Unfortunately I never saw packs meet, even though five of them, including two with dens, used the same 250 sq km of plains in April, 1967. Kühme (1965) observed one pack with pups bark at and retreat from another pack. While it is possible that packs defend the area immediately around their den, there is no evidence to suggest that they are territorial at other times. The scarcity of packs alone tends to space animals out, and olfactory cues, such as the powerful

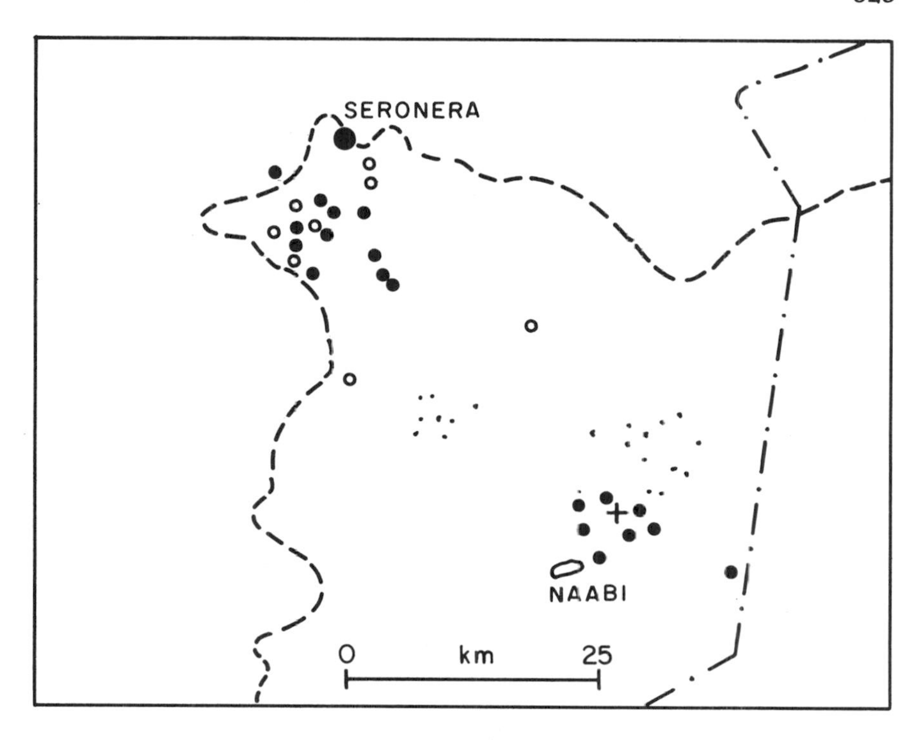

Fig. 45. Sightings of a wild dog pack in 1968 (closed circle) and 1969 (open circle) showing extent of movement. The cross marks the 1968 den site.

body odor of the animals, may also help in this respect. For comparison, Murie (1944) presented evidence that wolves are territorial and Mech (1966) noted that "Isle Royale's packs also seem to be territorial, at least in winter." However, the ranges of the packs on Isle Royale overlap so much that the behavior when packs meet may be hierarchical rather than territorial in nature. In Algonquin Park, "the spatial distribution of wolf packs was fairly uniform throughout the study area. Specific packs were repeatedly observed in certain areas, but ranges were not entirely exclusive as temporary union of two separate packs was occasionally observed" (Pimlott, et al., 1969). No such spatial pattern was discernible in wild dogs.

POPULATION DYNAMICS

Wild dog pups begin to actively clamber around when about 3 weeks old (Dekker, 1968), and it is at this age that they probably venture regularly to the den entrance. The den is abandoned and the pups accompany the adults at the age of about $2\frac{1}{2}$ months. Until they are about 9 months old the young can readily be distinguished from adults by their smaller size and, after that, by their lanky build. Wilson (1968) found that in Rhodesia three males ranged from 18 to 28 kg in weight and two females were 17 and 23.5 kg, but, on the average, East African animals seem to be smaller (Estes and Goddard, 1967).

Population Sex Ratio

Estes and Goddard (1967) commented on a possible preponderance of males in the adult segment of the dog population in Ngorongoro Crater where one pack they studied consisted of six males and one female. "Of a total of 237 wild dogs destroyed in carnivora control operations [in Kruger Park] during the period 1954–1960, 60.34 percent were males and 39.66 percent females" (Pienaar, 1969). On the other hand, de Leyn (1962) found fewer males than females, at a ratio of 32:68, in the Kagera National Park. One pack in the Kafue National Park consisted of 5 males and 7 females, another of 2 males and 6 females (J. Hanks, pers. comm.). A total of 33 packs were classified in the Serengeti. Some animals may have been tallied several times because packs changed their size and composition, making it difficult for me to recognize some after a lapse of months. Of the 327 dogs over 6 months old involved, 58% were males and 42% females ($p = < .05$). Nineteen packs contained more males than females, in 6 packs the sex ratio was equal, and in 8 packs there were more females than males. An unequal sex ratio is already evident among pups at the den: 4 litters included 22 males, 15 females, and 7 unidentified pups that died before I could sex them—a ratio of 59:41 among the survivors. One 5-month-old litter not at a den consisted of 7 male pups. One litter in Ngorongoro Crater was composed of 8 males and 1 female (Estes and Goddard, 1967). The large pups, 6 to 12 months old, in table 65 had a ratio of 57:43 in favor of males. In a sample of 16 litters from the Giza, Krefeld, Nairobi, Peking, and Okahandja zoos there were 64 males, 52 females, and 9 unidentified young, a ratio of 55:45, not significantly different from an expected 1:1 ratio. A preponderance of males in a carnivore population is unusual and the significance of it remains obscure when the biology of the species as a whole is considered (see below).

Pack Structure

Packs ranged in size from 2 to 32 with an average of 9.9, excluding pups in the den, a figure similar to the average of 9.2 given by Kruuk and Turner (1967). Dangerfield told me of once seeing a pack of about 43 animals in the Serengeti. Average pack size in Kagera Park was 11 (de Leyn, 1962), and it varied in Kruger Park from 8 to 11, depending on the area (Pienaar, 1969). Excluding 2 solitary males, 9 packs in the Kafue Park ranged from 3 to 14 animals in size with an average of 8.8 (Hanks, pers. comm.). These averages show a remarkable uniformity, suggesting that they represent the optimum for a pack.

Composition varies from small packs which sometimes contain members of only one sex to large ones with several males and females (table 65). Small packs never had large young with them, whereas those containing 11 or more individuals usually had some. Of the 153 dogs in the sample, 35% were large young. Since young tend to remain with the adults even after the birth of a new litter, it is likely that most packs consist of closely related individuals comprising at least three generations.

Some packs retain their composition for long periods, except for births and deaths. The members of one pack remained the same from March 27, 1968, to September 10, 1969, those of another from at least December 1968, to June, 1969, and those of a third from July to December, 1966. Changes, however, may occur. One pack of 5 adult males and 1 lactating female, as well as 5 male and 3 female young of the previous season, were at a den for a week. Suddenly the 3 young females disappeared while the others remained there for 2½ months and raised a litter. At another den, there were 9 males and 5 females yet none of the latter had the pendulous breasts typical of lactating mothers even though 4 small young were present. But perhaps the mother had died.

Reproduction

In the course of the study, 8 heavily pregnant females and 6 litters less than 3 months old were seen. The probable birth months of these were: October (2), November (1), December (1), January (1), February (1), March (4), and April (4). Root told me of seeing a lactating female in late August. The litters studied by Kühme (1956) were born in January, the one by Estes and Goddard (1967) in March. Thus, young may be born throughout the year, with most of them conceived between August and February, assuming a gestation period of 69 to 73 days (Cade, 1967; Dekker, 1968). The birth peak in March and April coincides with the time of year when prey is abundant on the plains, and the dogs there have few difficulties in killing near the den. Packs occupy a den only once a year.

No yearling females had pups. Many packs contained more than one adult female yet in most instances only one of these gave birth; of 9 adult females observed at these dens only 3 had young. However, Kühme (1965) reported 2 litters, born a week apart, in the same den area, and one pack seen during this study had 2 pregnant females in it. The evidence indicates that at least half of the adult females are not reproductively active, but if this is due to a failure to come into estrus, failure to conceive, or a high death rate of fetuses is not known. Packs usually have pups in successive years but I do not know if the same female does. Wolves, too, may fail to reach their reproductive potential. Pimlott et al. (1969) in a study of wolves in Algonquin Park noted that "wolf populations which are not exploited, stabilize and maintain a much lower percentage of young than they are capable of producing." Woolpy (1968) found that, among captive wolves, the dominant female in the pack attacks any other female which attempts to mate with the result that only one litter was born in each of five years.

Litter size at four dens was 4, 10, 14, and 16 after the pups appeared at the entrance. Kühme (1965) reported litters with 4 and 11 pups, and Estes and Goddard (1967) a litter with 9. The average of these 7 litters is 9.7. Thirty-eight litters born in various zoos (Cade, 1967; Dekker, 1968; Reuther and Doherty, 1968; S. Holz, pers. comm.) varied in size from 1 to 13 with an average of 7.5; seventeen litters born in a South African zoo contained an average of 7.8 pups with a variation of 2 to 12 (Brand, 1963). Given the propensity of captive mothers to eat their young, these figures may be somewhat low.

Mortality

Of 18 packs seen between September and February, half had failed to raise a litter the previous season. Between April and August, 1967, I knew five packs on the plains. Two had young in dens, one contained only yearling females, one lacked young for unknown reasons, and one included two pregnant females both of which somehow lost their litters. The fact that Sachs et al. (1968) found a titre of 1:160 in one out of three wild dogs sera tested for *Brucella abortus* might be significant with respect to these two females. Some females probably lost their whole litter after birth. One female in a small pack had obviously been lactating but there were no young. A heavy rain flooded a den I was watching one midnight. The female carried each of her 14 pups from the den and they then huddled near its entrance for the rest of the night. The exposure harmed them, for 5 died by the following evening. The mother ate parts of 4 of these pups. One pack had 4 pups when it left the den but two weeks later one was missing.

Yet 35% of the packs consist of large young, a fact which not only points to a reasonably high reproductive success but also to a high death rate of adults if, as seems to be the case, the dog population is stable. Turner told me that the park's population has not increased noticeably since 1956. Some animals undoubtedly are shot when they leave the park—as happened to a pack of 22 in 1957 (Turner, pers. comm.). One dog was found dead and eaten by vultures only two hours after it was with a pack and seemingly healthy. However, such deaths are incidental.

On March 27, 1968, we found a pack of 10 adults and 16 pups at a den. All pups were raised and the animals were last seen in healthy condition on May 12. On August 16 the pack appeared near Seronera with 14 pups, some of them obviously ill, very thin and with weak hindquarters. Two more pups disappeared by August 23. Between August 25 and 31, one adult and 7 pups were lost and one adult and 2 pups staggered as they walked. There was a grayish mucus in the corners of their eyes and mouths. I caught one sick pup at 1730 but at 1930 he whimpered, collapsed, and died within 3 minutes. By September 8 only 12 dogs were left and by November 4 only 10—5 adults and 5 young—out of the original 26. The following year 7 pups were raised and the 17 animals were last seen on September 10, 1969. The external symptoms, including a violent muscular twitch in the hind legs of one animal, suggested that the disease was canine distemper. Schiemann, who autopsied the dead pup noted in his report to me: "My thorough post mortem revealed a haemorrhagic gastritis and enteritis with heavy blood loss, a typical picture of the gastro-intestinal form of distemper." Two other packs also declined suddenly in number and the same viral disease may have been the cause. One pack declined from 29 to 22 animals between February 11 and 28, and one individual had a spastic leg. A year later the same pack consisted of nine dogs, but, of course, factors other than disease may have been responsible for the decrease. In November, 1967, I found a pack with a sick young which had symptoms similar to those later diagnosed as distemper.

With neither food availability nor the land tenure system being responsible for the low number of dogs in the park, disease appears to be the main population depressant. While a disease such as distemper can cause a decrease in numbers, the high reproductive potential—a litter of 10 to 15 pups by every adult female every year—would rapidly compensate for such losses especially since those adults which have had the disease mildly may be immune to further attacks. Some other factor must be operating on the population to keep it stable at the low numbers found in the park. The breeding potential has been reduced in several ways, by an unbalanced sex ratio favoring males, by a large percentage of reproductively inactive females, by a high death rate of pups; but, in addition, an unknown mechanism prevents the population from fluctuating. Something may, however, tip the balance of this mechanism to cause a slow increase. According to Pienaar (1969) a rickettsial disease in Kruger Park "caused very serious mortality amongst the wild dog population in the southern district of the Park, with the result that these predators were all but wiped out in this part of the Park during the period 1927–1933. Since 1936 there has been a slow but progressive recovery in the population and large packs are again found in all their old favourite haunts." The wolf population on Isle Royale fluctuated only between 20 and 28 individuals between 1959 and 1966. Few young entered the population and turnover was low. "Survival of pups and, more generally, growth of the population is limited by food available to the family group during the rearing season" (Jordan et al., 1967), a situation quite different from that found among dogs.

Social Structure

The social structure of a wild dog pack provides a useful comparison with that of a lion pride, for the two species so similar in certain respects regulate their existence by strikingly different means.

Communication

With packs being nonterritorial cohesive social units, gestures and vocalizations are used primarily for communication within the group. Methods of signaling over a long distance are limited to rather inadvertent marking with urine and body odor and to hooting when pack members have become separated.

Postures and gestures. When two dogs approach each other or the animals become active prior to a hunt, they greet with a distinctive posture (plate 39). Walking parallel, bodies touching, their necks lowered, ears retracted, and with lips drawn back horizontally to expose the teeth, they nibble at and lick each other's mouth. One dog may lower the forepart of its body, wag its tail, and push its muzzle into the corner of the mouth of the other; at times it may place the forepart of its body beneath the abdomen of the other dog. Partners are frequently changed as each animal runs twittering and whining from pack member to member. If a dog continues to rest while others greet, several may nudge it, nip the side of its neck, maul its throat, roll on it, burrow beneath it, tug its tail, nibble at its belly and groin, and jab it with one or both forelegs, as if trying to induce it to join by this show of friendly aggressiveness. Similar actions were observed toward a sick pup unable to keep up with the pack, and it was a common form of play which alternated with chasing and rearing up. The throat biting, nipping, and jabbing with the feet are typical components of aggressive interactions in some canids (Fox, 1969).

As Kühme (1965) pointed out, the greeting represents ritualized food-begging, for pups solicit food by poking their nose along the lips of adults. Adults sometimes get their signals mixed, so to speak. One gorged animal received chunks of regurgitated meat from two hungry ones after a greeting. Begging pups do not expose their teeth whereas adults do so, and in this aspect the greeting among adults conveys passive submission as it does in coyotes (Fox, 1969). The lowered body, together with the fact that an animal sometimes urinates and then rolls on its back, further indicates that the display is in part an appeasement gesture.

Dogs often display another posture, the stalk, as they approach each other prior to greeting. They stand or walk slowly with neck lowered and muzzle pointing forward and ears retracted, but suddenly their whitetipped tails flip up and they bound toward each other. Kühme (1965) interpreted this posture as expressing inhibited aggression. Dogs behave similarly when approaching a herd of zebra or wildebeest (but usually not when approaching Thomson's gazelle), the pack walking tightly bunched, an action which functions to lower the flight distance of prey. Pups also chase adults from a kill, and adults compete for food, by adopting the stalking posture, except that the necks may be held even lower than prior to greeting and the lips are retracted to expose the teeth in an expression which resembles a low-intensity threat gape, as described by Fox (1969) for several canids.

Male dogs have a conspicuously large and hairy penis sheath, suggesting that it has some signal function, but it does not feature prominently in any particular display.

Vocalizations. The vocalizations of dogs are largely discrete. The most distinctive call is a bell-like hoot—*hoo-hoo-hoo*—which begins with a rapid slur and continues for about .3 seconds at a frequency range of .5 to 1.0 kHz (fig. 22[F]). Dogs hoot mainly when they have become separated from the pack. As Estes and Goddard (1967) rightly point out, the hoot functions as a contact call, not a hunting call (Kühme, 1965). One adult hooted for no obvious reason at a den and several pups, 3 weeks old, answered squeakily. Spontaneous communal hooting, analogous to, for instance, howling by wolf packs, was not heard in wild dogs. The bark (fig. 22[K]), resembling that of the domestic dog, is used in potentially dangerous situations, such as the approach of a lion or of a person on foot. Other pack members jerk to attention when hearing the sound and pups dash into the den. Given either in conjunction with the bark or by itself is a brief growl which elicits the same response as the bark. The twitter, a chittering sound made with lips drawn back and quivering jaw, is elicited by several situations—particularly during greetings, when chasing and killing an animal, and when competing for food. It apparently expresses excitement. Dogs in pursuit of prey may also yip, as do pups while roaming near the den and occasionally adults when calling pups to them prior to feeding. Three situations elicit the whine, a graded signal: a soft plaintive whine is given by pups that have become separated or are otherwise distressed; an abrupt, loud whine is emitted by adults at the den entrance to induce pups to come out; and a prolonged rather nasal whine is given in conjunction with the twitter during greetings and also by pups begging for food. This nasal whine shows a rising and falling modulation with definite harmonics between 1 and 6 kHz (fig. 22[J]).

Scent-marking. The strong body odor of dogs automatically marks any site at which they rest, particularly the mud wallows in which they sometimes lie. They also roll on certain objects such as bird dung, meat regurgitated by a hyena, a dead pup, and stomach juice thrown up by the dog itself. As far as I could discern, urine and feces were not deposited at any particular location. Urinating dogs merely tend to spread their hind legs, although one may lift a leg slightly; on rare occasions an animal stands briefly on its forelegs, both hind legs off the ground, and urinates, a posture also described in golden jackals by Golani and Mendelssohn (1970).

Interactions Between Dogs

The most striking aspect of dog society is the amity that exists between members. Even when the whole pack is crowded around a kill there is little overt strife; growling and snapping usually do not occur, as it does characteristically among lions. I never saw two dogs fight. In competition for a bone or in other situations in which a fight could erupt, both animals tend to assume the appeasement posture, thereby terminating the interaction. Packs seem to lack a rigid hierarchy, a conclusion also reached by Kühme (1965) and Estes and Goddard (1967). Dominance interactions between two individuals may occur as when, for example, one faces away from another in a competitive situation or an animal hesitates to crowd in at a kill, but I spent too little time with packs to find out if these interactions were restricted to certain individuals or if an actual hierarchical system was involved. Mounting between male dogs, particularly between yearlings, was common at kills, and such behavior was possibly related to rank. Possibly males compete over estrous females, but I have no observations on this point. One dog mounted a pregnant female repeatedly without response from other pack members.

Males and females share most tasks equally, including hunting and feeding the young, although certain individuals assume special functions. For example, some adults lead in the hunt, whereas others, especially yearlings, trail behind. However, the spoils are divided among all members. When pups are at the den, one or more members remain with them and function as guards. The returning hunters regurgitate meat for the pups as well as for the guards. On one occasion a lame dog dropped so far behind the pack that no meat remained when he reached the carcass. He begged from several members and received some.

All pack members share in the care of the pups, but the mother retains certain prerogatives. She sometimes drives dogs from the den entrance when they try to feed the pups, an antipathy shown particularly toward yearlings. She, of course, suckles the young, either standing or lying while doing so, but she often withdraws from them or snaps at them by the time they are 3 weeks old, and by the age of 5 weeks they are essentially weaned. Dogs occasionally shift their den—which usually consists of an abandoned hyena warren that may be modified with some cursory digging—and the task of moving the pups is primarily that of the mother. One pack used its den from mid-March to April 12, when it was flooded. The mother moved all nine pups herself to another den 60 m away, carrying them haphazardly by the head, a leg, the lower back, and one by the tail. On April 22 she moved them again, this time 300 m, and the next day once more. A final move was made on May 12, this time with the pups walking. Another pack, observed from March 27 to May 10, changed dens once. "Since all pack members contribute to feeding and protection of the young, the mother is not essential to their survival after the first few weeks" (Estes and Goddard, 1967). Indeed, these observers watched one pack in which the sole female died 5 weeks after the birth of the 9 pups, and 5 males raised them successfully.

As mentioned earlier, dogs have a division of labor in that some guard pups whereas others hunt. In one pack of 11 adults, the mother remained behind alone at the den in 23 out of 32 hunts observed. On seven occasions one to four other dogs stayed with her; twice, after the pups were at least 5 weeks old, no adult guarded. When not hunting, the adults rested singly and in small clusters within 100 m of the den, or, in the case of a hot day, inside of it. The adults at another den behaved somewhat differently. I did not observe the young until they were about 6 weeks old. At that time they were not guarded in 4 out of 9 hunts and on several occasions most adults rested as far as 1.5 km from the den.

Around the age of 7 weeks the pups undergo a striking physical change. Their muzzle lengthens and their legs do too; their blackish and white natal coat is replaced by the adult pattern. Now they roam up to 30 m from the den entrance, and when the adults leave on a hunt they may follow as far as 200 m. An adult then leads them back to the den, but they may run after the pack once more. The den is abandoned when the pups are about 10 weeks old. One pack observed by N. Myers (pers. comm.) left the den at 1615 and, without stopping, had moved 14.5 km by 1815. Another pack, however, hunted near its den at least ten days after leaving it.

Once pups accompany the adults on the hunt, they are seldom fed on regurgitated meat. Instead, they take almost complete precedence at a kill. Adults may grab a bite at the time the prey has been caught, but as soon as the pups arrive they step back and permit them to monopolize the carcass. Any adult that tries to eat is chased away by twittering young, aggressive behavior already evident at the den when large pups bite at the lips and legs of adults while begging for food. The adults stand around the feeding young thereby forming a protective circle that prevents hyenas and jackals from grabbing the meat (plate 40). Pups retain their priority to food until they are about 8 months old. One litter born in mid-March still took precedence on November 5, but on November 15 all ate together. Considering the high death rate of adults, any factors which contribute to a high survival rate of pups, such as having priority at the kill, would have selective advantage.

PREDATION

In some respects the wild dog lends itself well to a study of predation for it hunts predictably each day and permits a vehicle to follow it without altering its behavior. I observed the hunting and killing of wildebeest 29 times, of Thomson's gazelle 37 times, and of other prey 21 times; less complete information was gathered on a further 111 kills. Almost all this information was obtained in the plains and along the edge of the woodlands, as was that reported by Kühme (1965) and Kruuk and Turner (1967), with the result that the food habits of the packs in the Corridor and Northern Extension remain essentially unknown.

Food Habits

Of 198 kills tallied, 42% were Thomson's gazelle, 38% were wildebeest and the rest zebra (6%), Grant's gazelle (4.5%), warthog (3%), and a few other items for a total of 12 species (table 66). The kills reported by Kruuk and Turner (1967) and Estes and Goddard (1967) also show a preponderance of Thomson's gazelle and wildebeest. The dogs in the woodlands probably prey extensively on impala and topi when wildebeest calves and gazelle are not available. R. Bell (pers. comm.) saw 5 kills in the Corridor; these consisted of 2 impala, 2 topi, and one waterbuck. Pienaar (1969) reported on 2,745 wild dog kills from Kruger National Park of which 87% were impala, 5% kudu, and 3% waterbuck, to mention only the three most important of the 20 species. Of 96 kills found by Mitchell et al. (1965) in the Kafue National Park, 26% consisted of duiker, 25% of reedbuck, 15% of hartebeest, and 2% of impala, in addition to such items as two lions and a porcupine among the 15 species killed.

The food habits of the Serengeti dogs show a marked seasonal difference. From January to June, 57% of the kills were of wildebeest, but between July and December, after the wildebeest had migrated off the plains, they comprised only 1.5%, the dogs' diet then consisting primarily of Thomson's gazelle (table 66). From the point of view of the number of animals killed as well as availability throughout the year, the Thomson's gazelle was the most important prey of the wild dog in the area studied.

Wild dogs readily scavenge meat. One pack found a still-born wildebeest calf and ate it, and on seven occasions dogs took meat from hyenas, including 3 zebra, 2 wildebeest, and 2 gazelle. Several dogs ate grass and two ate the feces of wildebeest.

Given the available prey, the size and vulnerability of it were the main factors influencing the selection of a particular individual. Adult Thomson's gazelle weigh 12 to 23 kg (Sachs, 1967). Wildebeest calves, the dogs' second most common food, range in weight from about 16 kg at birth to 60 to 65 kg by the age of 4 to 5 months, the period when most are killed. Of the 198 kills only 22 (11%) weighed more than 65 kg, among them adult male warthog (87 kg), adult female topi (109 kg), adult female wildebeest (144 kg), and adult female and male zebra (219 and 248 kg). The preference for prey weighing 65 kg or less is no doubt related to the ease with which the animal can be killed. Dogs have fairly small and weak jaws. It took a pack 8 minutes to drag a large wildebeest calf down, and another pack required 5 minutes just to chew through the abdominal wall of an adult female wildebeest after she had been pulled off her feet. One pack observed by A. Root harassed a zebra mare for 1¾ hours without being able to kill her. Turner told me that a pack required 1½ hours to kill an adult male wildebeest.

In contrast to wildebeest calves, which provide a substantial portion of the dog's food for several months, zebra foals are seldom killed even though they are abundant at the same time as the wildebeest. Zebra foals are members of a cohesive family unit and a fleeing herd characteristically remains tightly bunched, making it difficult for the dogs to attack the foals. In addition, stallions tend to lag behind the herd and bite and kick at the dogs. Wildebeest herds, on the other hand, often scatter or break into small units when pursued, and the dogs can then isolate a calf easily. On the three occasions when a pack attacked a zebra foal, adults defended it twice, once unsuccessfully. Wildebeest mothers actively defended their calf with varying intensity on 5 out of 26 occasions. On three of these the female butted at the dogs once or twice before leaving, but another held the pack at bay for a minute. On one occasion, dogs bit at a calf until it was unable to stand. Its mother remained by it for 40 minutes and lunged at any approaching dog until the pack retreated. Then two hyenas killed the calf.

Predation on zebra requires some comment. Eight of the 11 kills were made by one pack between April 7 and May 12, 1967, in spite of the fact that wildebeest or gazelle were also available. The pack seemed to have a predilection for zebra. Once it left the den at 0600 and traveled 25.5 km in a large arc, passing through thousands of gazelle until they found zebra, one of which they killed (plate 41). At 1155 they returned to the den, having covered 40 km (25 miles).

Animals of all ages are killed. Table 67 shows that among Thomson's gazelle 45% were young (classes I–IV), most of them small fawns which the dogs almost invariably pursued as soon as they saw them. Yearlings and young adults comprise a somewhat smaller proportion of the kills than expected and prime adults (class VII) a somewhat higher one (table 78). Eighty-two percent of the wildebeest kills were less than 6 months old and the rest were yearlings and young adults; however, one female appeared to be middle-aged. Zebra kills included one adult male, 6 adult females, an unsexed adult, and 3 young less than 6 months old. All 9 Grant's gazelle were small young, and the warthog consisted of 2 adult males, 1 adult female, and 3 yearlings. There were also 2 adult male impala, an adult topi and hartebeest, and a yearling bushbuck. The 4 ostriches were small chicks from one brood. Taken together, 58% of the kills consisted of animals less than 1 year old. The comparable figure for Kafue Park was 36%. Pienaar (1969) found that of 739 impala killed by wild dogs, 69% were adult, 22% were subadult, and 9% were unidentified.

Of 42 adult (classes VI–X) Thomson's gazelle sexed, 20 were males and 22 were females. Since females also outnumber the males slightly in the population, the dogs showed little preferential predation on the sexes. On the other hand, Kruuk and Turner (1967) reported 10 females and 14 males in a sample of 24 adult kills. Estes and Goddard (1967) found that 18 of 24 adult gazelle kills in the Ngorongoro Crater were males, suggesting selective predation there unless the ratio in the living population was also greatly biased in favor of males in the area in which the dogs were hunting. The authors stated that the selection for males "is evidently the result of territorial behavior" but this conclusion was questioned by Walther (1969). Of 739 impala killed by wild dogs in Kruger Park, 39% were male, 51% were female, and 10% were unsexed (Pienaar, 1969), showing probably no selection if females there outnumbered males in the population as they did in the Serengeti.

Two adult male Thomson's gazelle were encrusted with sarcoptic mange, but the rest of those killed seemed to be in good condition. However, dogs were undoubtedly far more adept than I in spotting weakness, and some of the adults they finally selected after a brief chase may have been in ill health. All four of the yearling and adult wildebeest I observed being killed were sick: one was covered with sarcoptic mange, another stumbled as it walked, and the last two looked healthy but lagged behind the fleeing herd. One zebra was giving birth when killed (Root, pers. comm.). In 3 other kills seen by Root and myself, an adult zebra dropped behind the herd after a short chase, a sign of some disability, and the dogs then attacked it. The evidence indicates that when killing large prey the dogs select primarily sick animals.

Killing Frequency and Food Consumption

A. and J. Root, various members of the Serengeti Research Institute, and I spent 20 days with a pack of 11 adults and 14 pups (5 of which died) between April 6 and May 19, 1967. Thomson's gazelle and zebra were abundant in the vicinity of the den for the whole period and wildebeest for about half of it. The pack killed 14 animals in the morning and 17 in the evening for a total of 31, or 1.5 per day (table 68). A pack of 21 dogs followed by Estes and Goddard (1967) in Ngorongoro Crater averaged two kills per day, mainly Thomson's gazelle. The approximate weight of the kills was determined by using the average figures for adults as presented by Sachs (1967) and by estimating the weights of subadults visually. The dogs killed a total of 903 kg in the morning and 1,088 kg in the evening, a total of 1,991 kg, or 100 kg per day. This averaged to 9 kg per dog per day, excluding the pups. Assuming that only 60% of the kill was usable by the dogs, the rest being mainly rumen contents, bones, and skin, each animal had about 5.4 kg available. Some of the meat was, of course, fed to the pups, but these probably required each only about 0.5 kg per day, still leaving nearly 5 kg to each adult. The daily food requirement of a dog in a zoo is about 1½ kg (Dekker, 1968), indicating that the animals had several times as much available as they needed even if their greater energy expenditure in the wild is taken into account. Mech (1970) noted that wolves consume considerably more meat than they require. Only one meal was eaten from a carcass, except once when the dogs returned to a kill near their den for a second meal an hour later. Meat was never cached, as is done by wolves (Murie, 1944), and on only two occasions was meat carried to the den by mouth, a gazelle fetus and a live fawn.

To obtain some idea of food consumption when prey was relatively scarce, I observed two packs for five days in November, 1967, as they hunted along the edge of the woodlands among scattered herds of Thomson's gazelle, topi, and a few impala (table 69). The dogs caught 5 Thomson's gazelle, 1 impala, and 4 ostrich chicks. One chick was grabbed by a lion and another by a hyena before the dogs could eat them. The total amount of prey killed was about 140 kg, or 2 kg per dog per day. With about 70% of the carcass of such small prey edible, each dog had 1.4 kg of meat available, a figure probably close to its daily requirement. Estes and Goddard (1967) estimated that each dog in Ngorongoro Crater had 2.7 kg of meat and that "two to three times as much food per day is available to wild dogs as is given to domesticated dogs of the same size." Wright (1960) calculated an average food intake of about 1.6 kg per dog per day.

It is difficult to estimate a killing rate for dogs because it varies, obviously, from area to area and season to season. Some kills are appropriated by other predators before the dogs finish eating—4.5% of the kills were taken by lions and 4.5% by hyenas—causing them to kill more than they otherwise would have. The dogs that we followed for 20 days were killing fairly large prey at the rate of 51 animals per dog per year, excluding pups in the den. Those that we watched for 5 days killed small prey at the rate of 58 animals per dog per year. The pack studied by Estes and Goddard (1967) killed at a similar rate. Assuming that each of the 250–300 dogs in the ecological unit kills 55 animals a year, then a total of 9,125 to 10,950 are taken. With most prey consisting of gazelle, wildebeest calves, impala, and other animals of similar size, their average weight probably does not exceed 20 to 25 kg, which would indicate a yearly kill of about 182,500 to 273,750 kg.

Hunting Behavior

Dogs hunt mainly in the morning and evening. After lying huddled in small groups in the vicinity of the den all night, they set off with the first light at about 0600, although at times they fail to leave until 0700 and later. The evening hunt, if any, rarely starts before 1700. Peak killing times are between 0700 and 0800 and between 1800 and 1900 (fig. 44). Occasionally packs hunt on moonlit nights: one pack left the den at 2300, another at 0030.

Setting out on a hunt, the dogs often move in single file but soon travel loosely spread over the terrain, a useful formation for finding crouched gazelle. Occasionally one or more dogs drop behind to sniff around a tuft of grass, lap some water, or harry a hyena. In one pack observed by Estes and Goddard (1967) "the same adult male was consistently the leader; he usually led the pack on a hunt, selected the prey, and ran it down." Kühme (1965) noted no consistent leader in the pack he watched. Those packs I observed had a leadership core consisting of some two to four adults, any one of which was likely to initiate pack activity. During the actual hunt, one or another led in the chase. Estes and Goddard (1967) found that "discipline during the chase was so remarkable among all pack members that even gazelle which bounded right between them and the quarry were generally ignored." I did not observe such adherence to a leader. One or more dogs sometimes veered aside in pursuit of prey so that at the end of the hunt the pack could be widely scattered. However, packs undoubtedly vary in their behavior, depending in part on their composition, the prey they select, and other factors.

The time a pack spends hunting depends, of course, on the availability of prey. On the whole, the animals rarely are active more than 4 to 5 hours a day—traveling, killing, and feeding. Of 48 hunts observed, from the time a pack left its den or rest area to the first kill, an average of 30 minutes elapsed. Some hunts were successful within 10 minutes, and all except 5 were successful within 60 minutes; after that, second and third kills may be made. One unsuccessful hunt covered 20 km in 130 minutes, when the dogs stopped because of darkness. The edible portions of a Thomson's gazelle are consumed within 5 to 10 minutes, although bones are often gnawed clean for several more. The dogs gorge themselves at a large carcass within 15 to 30 minutes.

Dogs typically trot at a speed of 8 km per hour, and the distances they cover in a short time can be large. Actual distances which a pack traveled from a den until its return to it varied from 2.2 to 40.3 km with an average of 9.7 km (6.1 miles) in 16 hunts measured. On days when two hunts are made the total distance is on the order of 15 to 20 km in areas with a reasonably large amount of prey. Packs without a den do not need to make a return trip, but they often travel without hunting actively and the total daily distance they move is probably similar to those with a den.

Hunting dogs may be trailed by one or more hyenas which later attempt to snatch the kill or at least a morsel of it. The dogs in turn may rush at the hyenas and, in the event that these have managed to appropriate some meat, may nip them in the rump until the food is dropped which the dogs then retrieve. When attacked, subadult hyenas often crouch instead of fleeing, and such individuals are at times permitted to feed on the kill (Kruuk, 1972). Several hyenas may drive dogs from a carcass, yet the reverse happens too. The interactions between these two predators are highly variable, depending probably more on the number of animals involved and the extent of their hunger than on any rigid hierarchical relationship. In fact, when not competing for food, dogs and hyenas tolerate each other more than any other two predators, even to the extent of resting side by side in the shade of the same tree.

The dogs' methods of hunting vary with the species of prey.

Thomson's gazelle. Gazelle sometimes graze within 100 m of a resting pack, but when the dogs trot or bound gazelle may flee at 300 m and more (Walther, 1969) and continue running for as far as 1.5 km even though not pursued, a flight reaction of an intensity shown to no other predator. In the plains, where it is difficult to approach a herd undetected, I observed two methods by which a pack increased its chance of success. In rolling terrain, the dogs sometimes ran over a rise and on seeing prey in the valley approached at full speed, both momentum and surprise being used to narrow the gap between predator and prey to 100 m or so. Estes and Goddard (1967) described similar behavior. On several evenings a pack loitered near a herd until it was dusky but still possible to distinguish individual gazelle. Suddenly the dogs raced at the prey and sometimes came quite close to it before being detected.

At the beginning of a chase dogs usually run in a broad front with a few of them trailing behind. Two or three dogs may then each pick a quarry and pursue it; at the end of the hunt the pack may be scattered over 1.5 km of terrain. A dog frequently chases a gazelle a few meters as if to test its reaction and stamina, then switches to another animal. Small fawns and obviously sick gazelle are almost invariably pursued. Sometimes the closest of several animals is chased, especially if the hunt is in tall grass which hampers the flight of gazelle. Since dogs must keep a quarry in sight to catch it, large compact herds are first scattered and animals that break aside or lag are then picked out. In some instances, however, I was unable to detect an obvious difference in behavior between one that was selected and one that was not. Once, for example, an adult male and female stood side by side and watched, as the pack approached to within 60 m, before they fled. The dogs ran after the female immediately.

Only one or two dogs usually chase a gazelle while others trail. The speed of the chase reaches 67 km per hour but more often it is about 55 km per hour. Often spronking at the beginning of the chase, the gazelle proceeds in a flat gallop as a dog draws closer and finally, with capture imminent, it zigzags and reverses direction. The latter tactic causes it to lose speed, and finally the dog grabs it by the side, rump, or thigh, or along the back, in the case of a fawn, and pulls it down. A bite into the lower abdomen disembowels the quarry, and other dogs usually arrive within seconds and tear it to pieces.

A pursued gazelle may run in a wide semicircle, and at such times the dogs hunt cooperatively. While one or two dogs follow the gazelle, most others cut across the arc and the gazelle suddenly finds itself surrounded. Brief relay hunts, with another dog taking the place of one that was chasing, were seen on three occasions; Kruuk and Turner (1967) described similar behavior. Most pursuits end after a short chase. Ten Thomson's and three Grant's gazelle fawns were simply snapped up as they lay crouched, and several others were caught after a chase of 300 m or less; however, one fawn eluded several dogs for nearly 5 km. Fourteen successful hunts for large young and adults, measured on the odometer from about the place where a dog selected its quarry to the kill site, varied from .6 to 2.7 km with an average of 1.7 km. The longest unsuccessful run was 3.4 km. On one occasion at least 3 female gazelle crouched in the manner of small fawns when pursued by dogs at dusk, thereby eluding their pursuers, behavior also described by Walther (1969).

Three condensed excerpts from my field notes describe hunts:

> (1) A pack of 13 dogs leaves the rest area at 1640. After traveling .8 km they see many gazelle about .7 km away. They first stand and watch for 5 minutes then walk closer until, .4 km from the gazelle, they break into a run. The lead dog dashes at three small herds in succession without selecting an individual, but after a chase of 1.4 km he chooses a female and pursues alone for 2.4 km before he can grab her thigh. Both tumble over and the dog grasps her throat and holds it until the others arrive and tear her to pieces. The time is 1715.
>
> (2) Sixteen dogs leave their rest site at 1745. Ten minutes later they chase a warthog 65 m until it escapes into a burrow. Coming over a rise, they see several hundred gazelle about 300 m away. These flee immediately except for an adult male. Three dogs chase him 1.6 km in a large arc before he suddenly seems to tire; he dodges, circles. Two more dogs arrive. He tries to jump over one of these, barely running now, but one bites him in the thigh and he is pulled down bleating. The time is 1757.
>
> (3) A pack of 12 dogs becomes active at 1640; the animals greet and chase each other for 15 minutes before they trot off. About .8 km away is a large gazelle herd. The dogs race toward it at 45 km per hour. Several dogs run after one fleeing group, the rest toward another, but when a female gazelle breaks to one side, one dog pursues her immediately. She flees in a wide arc, the dog 10 m behind. The rest of the

> pack veers toward them and two dogs cut in front of the gazelle. She swerves to avoid them only to face two other dogs; she slows to a walk, unable to find an escape route, and a dog grabs her rump. The time is 1706.

Wildebeest. Packs seem to search specifically for small wildebeest calves when they hunt this species. On one occasion several hundred adult wildebeest, yearlings, and large calves galloped along in single file while the dogs merely stood and watched them pass. But when a small calf, born late in the season, ran past, it was immediately pursued and killed. Usually wildebeest permit a pack to approach to within 50 m before retreating, but when calves are present they may flee at 200 m. Dogs often bunch up and walk toward a herd in a stalk before suddenly dashing closer when less than 100 m from it. Wildebeest draw together when pursued, behavior which makes it difficult for dogs to locate calves and single one out. To scatter a herd, the dogs run beside the herd, mingle with it, and, if it circles, some charge toward the advancing animals. If no calf is present, another herd may be attacked. If, however, one is seen, the dogs pursue it at 40 to 45 km per hour and attempt to separate it from the herd. The female wildebeest in turn often tries to place herself between it and the dogs, but once outside of the herd she rarely can hold off the whole pack. Once a calf butted a dog with its horns, the only instance in which one tried to defend itself. The average distance of 20 chases after calves was 1.1 km with a variation of .3 to 2.4 km.

To pull a large calf or yearling off its feet sometimes required several minutes of effort with one or more dogs hanging onto the nose and neck of the quarry, several tearing at the legs and the rest at the lower abdomen and anus, where the skin can readily be ripped and the intestines pulled out. On two occasions, a wildebeest—a yearling female and a young—sought protection from dogs by backing against the car from which I observed the hunt. One adult female backed against a fence and defended herself so successfully that the dogs left after harassing her for 25 minutes.

One example gives a more detailed impression of a wildebeest hunt:

> At 0655, I see 22 dogs trot across the plains. After traveling 1.2 km they approach a wildebeest herd in a stalking walk. The herd flees when the dogs are 80 m from it but when the latter stop, the wildebeest approach to within 50 m of them. Having spotted a single calf, three dogs rush at the herd, which stampedes. After a chase of .6 km, the dogs

halt, unable to see the calf in the dust cloud that envelops the area. The pack moves 1.4 km and sees 4 calves among 15 adults some 200 m away. At a distance of 130 m, three dogs begin their pursuit. One calf lags and at 0800, after a chase of .6 km, a dog grabs its thigh. Within seconds the others surround it and disembowel it. At 0815 only skin and bones are left.

Zebra. Zebra herds sometimes permit dogs to approach to within 40 to 50 m of them without fleeing, and, in contrast to other species, readily defend themselves. Once two zebra lowered their heads as if to bite after several dogs had walked to within 10 m of them, and on another occasion a dog leaped back when a stallion came to within 2 m of it. However, defense, if any, is sporadic and unorganized. Herds usually flee as a compact unit while the dogs pursue .3 to 1.5 km and then desist, if no zebra lag. Adults may place themselves between the dogs and their foals at such times. Occasionally the herd halts and faces the dogs, which then abandon their attempt and chase another herd instead until a vulnerable animal is found, a good example of rigorous selection against an expendable segment of a prey population. Two examples illustrate such selection.

(1) At 1745 a pack of 14 dogs leaves its den and, after traveling 1.4 km, chases a zebra herd for 1.4 km without attacking. The dogs move another .6 km and spot a herd which they pursue for 1.8 km before they are able to pull down a foal. It is 1818. The rest of the herd watches at a distance of 40 m. One mare, presumably the mother, returns to the foal. The injured foal struggles to its feet as the dogs retreat but, after following its mother 12 m, is pulled down once more. The mare then stands by the young, almost over it, but without attacking the dogs, which first pull their quarry by the nose and finally tear out the viscera. The mare drifts 6 m to one side of the foal, visibly startles when she sees the car, and walks off. At 1829 the first dog heads back toward the den, at 1850 the last one.

(2) At 0620 a pack of 14 leaves the den, passing many zebra .3 km away without chasing them, but some 2.3 km farther on is a herd of about 100 zebra. These they pursue at a speed of about 30 km per hour. The zebra remain in a compact herd circling again and again, never running at full speed in an attempt to elude the dogs; after running 2.1 km, a pregnant mare suddenly lags, obviously sick, her head held low. The dogs surround her at 0647, tugging at her legs and groin, and one or two hang on her nose. She makes no attempt to defend herself and at 0649 she falls on her side. Eight minutes later the first gorged dog heads toward the den, and at 0728 the final one.

Others. The warthog is the only large prey in the plains which could escape predation by going into a burrow. However, in two of the three kills I witnessed, the animal had reached the safety of a burrow but failed to remain in it. For instance, a pack chased a large boar which ran into a hole. The dogs sniffed around the entrance, then left. They were barely 20 m away when the boar popped from the hole and fled. Pursued, he raced back toward the burrow, passed the entrance, and continued on. The dogs surrounded him, biting at his thighs and rump, and he died 3 minutes later without having made a vigorous attempt to defend himself with his tusks. Once, when a pack chased a mother with young, the latter dashed into a hole while the female faced the dogs at the entrance. These sought other prey.

One pack pursued a pair of ostriches with 6 chicks. As the birds ran through a patch of tall grass, 4 of the chicks crouched there. Several dogs leaped repeatedly high into the air and landed with all four feet together, thereby flushing the chicks and killing them one at a time.

Once a dog flushed a mouse (*Arvicanthus?*) and then pawed it, snapped at it, and bounded in a tight circle around it. Three other dogs joined. The mouse squatted, paws raised, mouth agape; when a dog nudged it with a foot, it bit. Finally, after 5 minutes, longer than it usually took to kill a wildebeest, one dog grabbed it.

Hunting Success

I tabulated the success or failure of each pursuit by dogs after a specific individual, using both my data and those supplied by A. Root, S. Trevor, and a few other reliable observers. Because dogs frequently chased prey a few meters and then changed their minds, only rapid chases exceeding 100 m were arbitrarily considered to be pursuits, except in the case of small gazelle fawns which may be snapped up before they have the opportunity to flee. Mere chases after whole herds are not included in the computations.

Gazelle fawns are almost invariably captured (table 70). Three-fourths of the selected wildebeest calves fall prey. On 7 out of 9 unsuccessful chases after calves the dogs lost sight of their quarry in the herd or in the dust, and twice interference by the mother caused the pack to desist in its attack. In general, the success rate of dogs was considerably lower when they pursued yearlings and adults than when they selected young. For example, about half of the hunts after adult gazelle ended in failure, sometimes because the dogs lost sight of the prey but more often because they simply gave up the chase after .5 to

3.5 km. Zebra are not included in table 70 for, in most instances, a pack chased herds rather than individuals at the beginning of the hunt. Zebra herds were pursued vigorously on thirty occasions and five zebra were killed, a success rate of 17%. Attacks on two further individuals were unsuccessful. Given the high success rate that dogs have in capturing wildebeest calves, together with the short chase needed to capture one and the large reward of meat afterwards, it is not surprising that this prey is preferred when available.

Success rate is even higher than the average of 70% in table 70 when hunts as a whole are considered. Of 65 morning hunts, 95% were successful in that at least one prey was caught; of 47 evening hunts 81% were successful, a somewhat lower figure than the morning one because dogs sometimes hunt less persistently at that time, having gorged themselves earlier. The combined success of all hunts was 89%, a figure which is similar to the one of 85% reported by Estes and Goddard for Ngorongoro Crater.

12 Other Predators

The most abundant large predator in the Serengeti ecological unit is the spotted hyena, and an understanding of its ecology and behavior is essential when considering the role of predation in the regulation of prey populations. Kruuk (1966, 1970, 1972) studied hyenas intensively between 1964 and 1968, and his observations form the basis for this brief overview of their habits; in addition most other chapters mention various aspects of their behavior. Several Serengeti predators have not been studied so far even though one of them—man—affects the wildlife outside of the park considerably, not only by killing wildlife but also by competing for habitat by cultivating and grazing livestock. Most other predators are small and often rare and their impact on the large ungulates is minimal. Possible exceptions are jackals, which may in some areas prey heavily on gazelle fawns.

Spotted Hyena

The Serengeti hyenas reach their greatest abundance on plains. The woodlands population is sparse and the animals that are seen there are often visitors from the plains in search of a prey concentration. An estimated 3,000 hyenas occur in the ecological unit, a figure which does not include young less than 8 months old.

The basic social unit is the clan, which may consist of 10 to 60 or more hyenas of both sexes and all ages. Clans in Ngorongoro Crater tend to remain stable in composition except that individuals occasionally shift from one clan to another. Clan members are typically scattered singly and in small packs, much as is the case in lion prides. Pack size is smaller and solitary individuals are more common in the Serengeti than in the Crater. In general, pack size is adapted to availability and vulnerability of food: large packs are needed to hunt zebra but not to scavenge from lion kills.

The 260 sq km of the Crater were divided into eight distinct clan territories each with a centrally located communal den. "The animals fed as a rule on their own range; if a prey was killed outside it, the neighboring clan was on several occasions observed to chase the hunters off their kill. Any strange individual intruding in the range of a clan was met with much aggression" (Kruuk, 1966). Territorial boundaries are demarcated in several ways. The whoop (fig. 22[G]) acts as an auditory signal, analagous to the lion's roar. Packs patrol the borders and visit specific latrine areas at which animals defecate communally. A white secretion is also pasted on grass by dragging stalks between the hindlegs and past the anal opening.

Most hyenas are unable to maintain permanent clan territories in the Serengeti because their preferred prey, wildebeest and zebra, may be many kilometers from the plains. As the migratory herds retreat to the woodlands, clans on the plains tend to break up. Some animals become nomadic, following the prey. Hyenas from several clans may associate without animosity around a prey concentration and may even establish a temporary clan which dissolves as food again becomes scarce. These nomads return to the plains with the migratory prey, usually to the same area they occupied the previous season, a trait also noted among nomadic lions. Other hyenas, particularly females with cubs, move only to the edge of the woodlands with the advent of the dry season and from there may commute 30 km or more to some prey concentration, feed, and then return to their den.

Hyena cubs seldom obtain meat until they are at least 6 months old, in contrast to the young of cats which may be led to a kill at the age of 1½ months. Hyenas are fully grown at 1½ to 2 years of age, and they reach sexual maturity at 3 years. The expected annual recruitment to the Serengeti population was 37%, but the actual one was 5% due to a high death rate of cubs during the dry season when females had to travel far in search of food while leaving the young unattended at the den for several days. The population seemed to be stable in number.

Hyenas kill 68% of their prey and scavenge the rest (table 71). Because wildebeest are vulnerable and rewarding in the amount of meat they provide—an adult feeds an average of 12.6 hyena—this species is the preferred prey of hyena. Fecal analysis showed that the diet of hyenas consisted of 35.4% wildebeest, 32.5% gazelle, 19.9% zebra, 9.4% topi, hartebeest, and impala, and 2.8% miscellaneous items. When feeding hyenas were tallied, 65.1% ate on wildebeest, 16.5% on zebra, 12.6% on gazelle, and 5.8% on other items.

Hyenas spend about 20 hours of the day resting. They usually hunt at night, particularly in the early evening and early morning, although gazelle fawns and other vulnerable prey may be captured in daytime too. Quarry is merely chased; stalking and lying in ambush do not occur. A wildebeest herd may permit a hyena to approach to within 20 m or less before wheeling around and fleeing tightly bunched. The hyena in turn dashes at the herd as if to scatter it; finally one animal may be selected, often a calf if available, and pursued at speeds up to 40 to 50 km per hour. After a chase of 1.5 to 5 km, the hyena either gives up the attempt or is able to bite the legs or belly of the animal until it halts. I have seen a hyena run up beside a yearling and pull it off its feet with one bite in the side.

The decision to hunt zebra is apparently made by a clan before the hunt begins. Animals gather and synchronize their activity by, for example, visiting latrine areas. After that a pack may pass through wildebeest herds until it finds zebra. A herd occasionally permits hyenas to approach to within 5 m before it flees or a stallion attacks. The chase is at speeds of 15 to 30 km per hour, much slower than zebra can run. As with other prey, hyenas tend to select the physically least able animals from a herd.

Thomson's gazelle are usually hunted by one hyena. If a fawn is pursued, its mother may perform a distraction display by dashing back and forth in front of the hyena, behavior which, however, seems to have little effect on the success rate of hunts.

After pulling an animal down, hyenas rip open the abdomen and bolt first the viscera and then the rest of the meat. Fights seldom erupt in competition over food, although an animal that has a bone may be vigorously chased by others. Cubs have no priority at kills except when their mother protects them. On one occasion Kruuk observed 35 hyenas kill a female zebra and clear the site of all remains in 36 minutes.

MAN

Illegal killing of animals with wire snares, poisoned arrows, and rifles is particularly prevalent near the heavily settled areas north and south of the Corridor and west of the Northern Extension. Poachers operate both in the park and in the woodlands surrounding it, with the activity heaviest during the dry season when the migratory species are available and movement of meat to market is easiest. Resident prey is noticeably sparse and shy in some areas such as the Maswa one, indicating persistent heavy predation. The magnitude of the kill can be gauged by a few typical quotes from the monthly reports of the Chief Park Warden:

July, 1968. On 13th July in the western Duma section, the D.P.W. and a Patrol tracked down a poaching gang in long grass and made 5 arrests in a large camp. In this camp by careful count were the skins, bones and meat of the following freshly killed animals: 21 wildebeest, 7 kongoni, 2 impala, 11 zebra, 2 ostrich, 1 vulture, 1 warthog and 5 topi. Also a large pile of bones and skulls about 4 ft high of animals killed earlier were assembled and, at a conservative estimate, formed the remains of a further 36 animals. This camp had been operating for 2 weeks. All game had been killed by snaring.

In mid-July, a large party of Isenye took advantage of the immense wildebeest concentrations along the Grumeti river between Masabi and Kira Wira. 12 freshly killed wildebeest were discovered near the main corridor road, all killed by poisoned arrows, shot over the water-holes. Several arrests were made in this area and the river thoroughly patrolled.

On the lower Ruana, at least 5 vehicles are operating from Mugeta and Ikizu shooting game at night. These poachers are extremely difficult to apprehend, due to the many crossing places across the Ruana river. Efforts will be intensified to apprehend these people who run a lucrative fresh meat trade in the settlement.

August, 1968. On the lower Ruana, on the 14th August, our patrol intercepted a man on a bicycle heavily laden with game meat. When asked where he had obtained it he showed the patrol a full scale game meat butchery in action near the settlement. In the butchery, being sold for cash were the meat of 2 eland, 3 wildebeest, 1 topi and a cheetah skin. 5 arrests were made.

April, 1969. The increase in motorised poaching in the Western areas of the Park gives cause for grave concern. On 13th April at about 7 P.M. our Field Force patrol intercepted and captured a jeep vehicle containing two senior Government officials, an Italian contractor and two local people all hunting in the Park. When captured they had already shot: 1 cow eland, 2 wildebeest, 1 topi, 3 warthog, 2 Thomson's gazelle, 1 kori bustard, 1 zebra, 1 hyaena.

I accompanied M. Turner on an antipoaching patrol near the Duma River on August 31 and September 1, 1969. On the first day we found a line of 27 snares along a dry watercourse, with a zebra and a male lion as victims. One camp contained the butchered portions of 2 impala, 2 hartebeest, 1 reedbuck, 1 waterbuck, 2 giraffe, 1 eland, 1 buffalo, and 1 ostrich. On the second day, first a small camp with 1 eland and 1 Thomson's gazelle was found, followed by a second one with 2 eland, 2 Thomson's gazelle, and 1 bushbuck. Finally, a third camp, a large one, was found hidden under a *Grewia* bush; it contained 12 Thomson's gazelle, 5 impala, 6 warthog, 2 buffalo, 5 giraffe, 6 zebra, 2 hartebeest, 3 wildebeest, 1 roan, 1 eland, 1 topi, 1 reedbuck, 1 lion, and 1 white-backed vulture, in addition to three piles of burned bones which indicated that the camp had been operating successfully for several weeks. Thus a brief search in a limited area revealed the remains of 67 freshly killed animals. In spite of intensive efforts by Turner and his staff, many poachers manage to operate around the periphery of the park for several months of the year. Their depredations must account for at least 15,000 to 20,000 animals annually. Such predation is fairly unselective with respect to age and sex of adult animals taken, in contrast to that of other predators.

JACKALS

Jackals are major predators on gazelle fawns and possibly on impala young too, and since there are several thousand of them in the ecological unit they may have an appreciable influence on these species. The golden jackal inhabits primarily the short-grass plains although some individuals move to the woodlands edge during the dry season. The black-backed jackal is essentially a woodlands animal which penetrates, however, far into the plains where its range overlaps with that of the golden jackal; both species may congregate at the same kill. Pairs are territorial, defending an area of about 3 km in diameter against members of the same species (Wyman, 1967). Judging by concentrations of over ten jackals at some kills, territorial boundaries are not always maintained when a large food source is available. Golden jackals give birth between December and April, when migratory prey is on the plains, black-backed ones between July and October.

Jackals eat whatever they can catch or scavenge (see Grafton, 1965). Wyman (1967) estimated that jackals kill about 80% of their food themselves. He found that the remains of Thomson's gazelle and insects were particularly common in the feces of jackals around Seronera and in the plains; other food items included mice, rats, hares, birds, fruit, and carrion. On one occasion I saw blackbacked jackals kill an adult female Thomson's gazelle, possibly a sick one, and I have also observed them catch Thomson's gazelle fawns, mice, a hare, and a young bat-eared fox. The stomachs of two blackbacked jackals which had been killed by cars on the plains were both full of carrion. Gazelle fawns are only seasonally available in some areas. A female Thomson's gazelle may defend her young by placing herself between it and the jackal and by butting. In such a situation an attack by one jackal is less likely to be successful (16%) than by two of them (67%) according to Wyman (1967). A female may also walk slowly in front of a jackal thereby decoying it away from a hidden young.

A third species of jackal, the side-striped jackal, is seldom encountered. I saw it feed only once—on a lion kill in company with the other two jackal species.

Other Mammals

Most other mammalian predators are rare to uncommon in the park. Their largest prey consists of an occasional small antelope, particularly of Thomson's gazelle and impala young. I saw striped hyena only twice and caracal four times. The latter species is known to prey on hyrax and impala fawns in Kruger Park (Pienaar, 1969), and one stomach examined by Bothma (1965) in South Africa contained six mice, some duiker meat, and grapes. Sparsely distributed, shy and essentially nocturnal, the serval and wild cat are difficult to observe and I have no information on their food habits in the Serengeti. Bat-eared fox seem to prey mostly on invertebrates and small vertebrates: one stomach I examined contained many dung beetles, another one a lizard, 1 beetle, 1 grasshopper, 1 spider, and 6 ants; one fox was also seen to scavenge on a hare.

Baboons, the most abundant monkeys in the park, eat meat as an incidental part of their diet. My records include seven Thomson's gazelle fawns and one hare, and Turner provided me with four additional kills consisting of 2 Thomson's gazelle, 1 Grant's gazelle, and 1 impala young.

Reptiles and Birds

Turner (pers. comm.) twice found a python which had swallowed an adult female Thomson's gazelle. Pienaar (1969) noted impala, warthog, reedbuck, bushbuck, duiker, steenbuck, waterbuck calf, wildebeest calf, baboon, and jackal among the prey of this snake in Kruger Park. Judging by the fact that I never saw a python, the species is rare in the Serengeti. Crocodiles, potentially important predators, are generally scarce and limited in distribution to the few rivers which during the dry season retain pools. Being up to 4 m long, crocodiles may capture prey ranging in size from steenbuck and impala to wildebeest and buffalo as well as such miscellaneous items as vervet monkey and lion (Pienaar, 1969). My kill record for the Serengeti consists of a Thomson's gazelle, three zebra, and one wildebeest, and Turner told me of a poacher who tried to escape by swimming a river and was apprehended by a crocodile instead.

About 25 species of birds of prey occur in the Serengeti but of these I have only observed the martial eagle prey on an ungulate, a Thomson's gazelle fawn; other eagles, such as the tawny eagle, possibly take the newborn young of small antelopes too. Lappet-faced vultures kill Thomson's gazelle fawns on occasion, judging by my encounters with these birds on freshly dead carcasses.

IV SUMMARY AND CONCLUSIONS

13 Dynamics of Predator Social Systems

Each predator has a different social system, and on theoretical ground all features of these systems are adaptive or they would have been eliminated by selection. At present no generally accepted unitary description exists of the selective forces that shape a society, but it is clear that adaptiveness can only be assessed in relation to all factors operating on and within a species and that these can only be understood within an ecological framework. As a first step toward comprehending the dynamics of a society, a description of the species' way of life is necessary, particularly the way in which the animal differs ecologically and behaviorally from related species or from those with a similar kind of existence. Each aspect of behavior raises numerous questions. Why are lions more sociable than other cats? Why do male lions have a mane whereas males of other species do not? Why are lions territorial but not cheetah? Why are lions more aggressive at kills than wild dogs? Every behavior pattern can no doubt be explained by the selective forces that shape and maintain it, but these are not always apparent. Advantages may have concurrent disadvantages which prevent a pattern from being perfected and which obscure its survival value. Furthermore, when selection pressures affect one part of a social system, others may automatically adjust to the change. As recent primate studies have shown (Rowell, 1967; Gartlan and Brain, 1968), species may be so adaptable that they modify their social organization to meet local conditions, and this in turn may influence the expression of certain kinds of behavior such as the frequency of aggression, making it difficult to generalize about a species.

Social System

The big cats present an interesting array of systems ranging from solitary to social species. The lion is the most sociable cat with prides consisting of one or more males, several females, and cubs, numbering anywhere from 4 to 30 or more individuals. Such prides may remain constant in composition for several years except as they are affected by births and deaths and the emigration of some subadults. Members of a pride are not always together, being scattered singly and in groups within the confines of a circumscribed pride area. Tigers, though leading an essentially solitary existence, are social in that individuals may meet and share a kill and then part again. In contrast, adult leopards are solitary in the Serengeti and cheetah are too, except that several males may associate as companions. The gradient from leopard to lion can be considered steps in development of social behavior. From an essentially solitary species through a solitary one with intermittent friendly contacts to a social one seems to be a logical progression which required relatively little reorganization in the animals' basic mode of life. Lion society has probably been shaped and is maintained by several selective forces.

(1) The social lion, as well as wild dog and hyena, reach their greatest abundance in open woodlands and grasslands, whereas the solitary cats frequent, on the whole, rather closed environments. Several reasons may account for this difference. First, it may be more difficult to maintain group cohesion in a dense forest where visibility is poor and calls are muffled by the vegetation than in the open, especially since predators need much space in which to find sufficient prey. Second, a pride would find it difficult to subsist on the sparse prey populations that are characteristic of such forests as those of the Congo basin, but leopards readily survive there. When prey is scarce and members of a social unit are scattered, as is the case with lions, the mere logistics in keeping many animals in frequent contact with each other in, let us say, 500 sq km of forest might be so formidable that selection would favor either small cohesive groups or a solitary existence. Third, cooperation in hunting prey is undoubtedly of advantage to a species in open terrain. On the other hand, several leopards or tigers alone can probably find more of the scattered prey that tend to inhabit dense forests than if a group hunted together; a communal stalk in such a habitat is possibly no more successful than a solitary one. If this line of reasoning is correct, then lions which inhabit rather dense habitats should occur in smaller groups than those in open terrain. I have no information on this point, but Guggisberg (1961) noted that "in general, it can be said that on the open veld prides tend to be considerably larger than in dense bush." Cheetah are essentially solitary yet they inhabit open environments, probably because the way in

which they capture prey, by singling out an animal and sprinting after it, requires good visibility and because a brief, very fast run is not a hunting method suited to a group effort.

(2) Several lions stalking together are generally twice as successful in catching their prey as a solitary lion, in part because they employ such cooperative methods as encircling their quarry. It would seem that a social leopard would accrue the same benefit, but prey size probably has an influence on group size too. Lions prefer wildebeest, zebra, and similar prey that weighs at least 100 kg. Such quarry provides food for several lions, whereas a leopard subsists largely on such small animals that it would not be advantageous to have to share the spoils with others. If this is so, then the size of lion groups should be correlated with prey size. An average group consists of four to six animals, just the right number to enable each individual to gorge itself on a wildebeest, topi, zebra, or similar prey. But during the dry season when Thomson's gazelle was the principal prey around Seronera, lions often hunted alone and average group size was only about half as large as at other times of the year. In this context, it is interesting to note that when tigers kill small prey, such as an axis deer, they usually eat it alone, but when a domestic buffalo or other large animal is captured, several adults may join and share.

(3) Lions are able to increase their food resources by hunting together. A solitary lion usually hesitates to attack a buffalo or giraffe but a group does so more readily and presumably more successfully, a point of particular importance in the Serengeti, where some three-fourths of the resident prey biomass consist of these two species. Solitary species of cats, on the other hand, are unable to utilize most of the available resources because much of the prey is too large for them to kill.

(4) A lion group uses its food more fully than an individual. Wildebeest and zebra may be consumed within an hour or two by several lions but a solitary lion must guard the remains against other predators and vultures. A single lion is not always able to do this, and hyenas, for example, often drive it from a kill. Leopards circumvent this problem by storing the kill in a tree, but cheetah merely abandon the carcass after one meal, which considering their timid nature is a logical solution.

(5) A division of labor is possible in a group. For instance, one lion may protect the carcass while others seek shade, fetch cubs, or are otherwise employed. A lioness often rests near small cubs, thereby

functioning as guard, without the need to hunt because other pride members may kill something and provide meat for all.

(6) Group existence is a form of life insurance. Sick or crippled lions, unable to kill for themselves, can subsist for months by joining others on their kills. Cubs whose mothers fail to produce enough milk can supplement their diet by suckling on other lactating females, and in the event of a lioness's death her cubs remain with the pride. In contrast, the young of solitary cats are wholly dependent on their mother.

(7) Young lions typically remain dependent on the pride for at least the first $2\frac{1}{2}$ years of their life. In contrast, cheetah young become fully independent at the age of about 16 months and leopards and tigers between about 20 and 24 months. This prolonged period of dependence suggests that group life confers some advantage on young lions which has no relevance to other cats. Various ways of hunting cooperatively are the most complex tasks in a lion's behavioral repertoire and it is possible that cubs need several extra months of dependence to learn these ways well.

(8) "The number of individuals in the group determines to some extent the status of the species in the general predator dominance hierarchy. Interspecific rank is based largely on size, with the lion at the top, followed by the hyena, the leopard, and probably the wild dog, cheetah, and jackal, in that order. In an encounter between a solitary lion and a solitary hyena, the latter always retreats, but a dozen hyenas can readily chase a lion from its kill. Similarly, one hyena usually waits while a pack of wild dogs feed, but several may rush and grab the carcass" (Schaller and Lowther, 1969). Groups can thus increase their food resources at the expense of solitary animals; conversely, groups are better able to defend their meat against a larger but solitary opponent.

The evidence indicates that a close relation exists between the exploitation of a food resource and the type of social organization, a point also made by Crook (1965) with respect to birds. Lions are adapted for hunting large prey cooperatively in open terrain, but their social organization has retained a certain plasticity which enables individuals to change from solitary to social hunting while retaining the benefits of life in a pride. Although a social existence confers advantages on lions, there are at times also disadvantages. When, for example, prey is smaller than the optimum, competition for meat may be so fierce that cubs obtain little and starve. At such times it would benefit lions to be solitary. It is probable that the limited social life of most cats, as well as of some canids, is related primarily to the small prey that is hunted. For instance, jackals which subsist on mice and other small animals occur in pairs, no doubt a compromise between the solitary existence that one would expect from such a diet and the necessity to collect enough food for a litter.

The intricate relation between food resource and social organization is well exemplified by the hyena. In Ngorongoro Crater, where prey is abundant, each hyena clan maintains a territory. On the other hand, clans on the Serengeti plains disintegrate when the migrating herds move to the woodlands. Some hyenas then move to dens along the edge of the plains and commute to prey concentrations; others become nomadic (Kruuk, 1966). Other selective forces which influenced lion society may also have had an effect on hyenas. For example, several hyenas are about four times as successful in catching a wildebeest calf as a solitary hyena is, and they increase their food resources by cooperatively hunting zebra and other large prey which they are unable to kill alone (Kruuk, 1972).

Cooperative hunting, efficient use of food resources, a division of labor, protection against competitors, and other factors already mentioned with respect to lions also are of advantage to wild dog packs. Dog society differs from that of lions and hyenas by being relatively inflexible; it does not adjust its size seasonally to the available prey. When the principal quarry consists of Thomson's gazelle, a large pack may have to kill several animals to satiate every member, at a cost of more individual effort than if small groups hunted only once. Average pack size is about 10, of which 3 or 4 animals are usually juveniles. This size may well reflect both the optimum number of adults needed as a hunting unit and the number of animals which can obtain an adequate portion of meat from an average kill.

Social Structure

Many integrative mechanisms operate in a society, including for example, the birth of young and a preference for familiar faces, but to function smoothly, to survive the disruptive forces that occur, a structuring is necessary. This structure varies considerably between species.

Lion

The core of a pride consists of a closed society of lionesses all of which are directly related to each other. Some means of preventing excessive inbreeding is advantageous to such a society, for, in the final analysis, selection is for reproductive success. Genes flow through the population in several ways: (1) With rare exceptions, young males leave the pride in which they are born by the age of $3\frac{1}{2}$ years thereby reducing the probability of incestuous mating; (2) pride lionesses, though antagonistic toward females that do not belong to the pride, may court with nomadic males or males from neighboring prides; (3) adult males are only temporary members of prides, staying from a few months to as long as six years before either leaving on their own volition or being forcefully replaced by new males.

Male lions are larger than females and possess a voluminous mane which contrasts with the sleek pelage on other parts of their body. In many polygynous societies the male has conspicuous secondary sexual characters that are used in intraspecific display, indicating strong selection for such visual signals. Other cat species are also polygynous but they lack adornments (although male tigers may have a ruff), suggesting that the social existence of lions somehow favored the evolution of a mane. The mane makes the male look impressive; it enhances his appearance, especially during a strutting display in front of another lion which may be either intimidated or impressed depending on its sex. It is possibly significant that I never saw a lioness present herself sexually to a subadult male. In a social system in which females form a closed society and males are transients, it is of obvious advantage to lions to be able to distinguish sexes at a distance. A strange male that looked like a female might be attacked by lionesses before he could establish his identity. Pride males are antagonistic toward male intruders into their pride area. In this context the dichotomy between the sexes is useful in preventing a male from erroneously mauling a female and in enabling him to avoid such close contact with a male stranger, before recognizing him, that a fight is inevitable. The distinctive colors and sizes of manes may also help animals to identify individuals. An important secondary function of the mane is one of protection during fights. The dense mat of hair absorbs blows and harmlessly tangles claws in a part of the body toward which most social contact is directed; bites, too, may leave an opponent with a mouth full of hair rather than skin. A mane is, however, of disadvantage to a male when he tries to hunt in daytime: he looks like a moving haystack and prey spots him easily in open terrain. Yet it is possible that a mane is not essential in order for a male to function adequately in lion society. In Tsavo Park normal adult males often lack manes (A. Root and S. Trevor, pers. comm.).

The large size of males probably features importantly in their defense of the pride area. But males use a pride's food resources uneconomically, not only because they eat a disproportionately large share but also because they capture little prey themselves, relying instead on the lionesses. This being so, the uneven pride sex ratio favoring females by at least 2:1 is adaptive: less food is required than if the ratio were 1:1. Besides, many males are not needed in a polygynous society. In the population as a whole the male biomass is about equal to the female one.

Some cohesive forces must be operating on a pride to keep the animals together. Small cubs often serve this function among females especially when several litters are born at the same time and their care becomes a communal affair. Although young seem to have no special relationship with their mother after she has a new litter, contact with her for those that remain in the pride endures for life. Familiarity with and preference for the sight, sound, and smell of individuals they have grown up with, and an aversion to strange lionesses, may be a main cohesive influence on lionesses.

Pride males are often brothers or pride mates born at the same time. Even if they are not related they had become companions before establishing themselves in the pride. They thus have an intimate relationship, a bond based on the rather vague criteria of familiarity and congeniality. In areas with a sparse lion population one male may be able to retain jurisdiction over a pride without being replaced by rivals, but in the Serengeti most males need to have at least one companion in order to remain in the pride. To form a companionship is therefore advantageous for a male assuming that pride life has advantages which a nomadic existence does not.

(1) Being unable to run as fast as lionesses and finding it difficult to stalk because of their size and mane, males are inefficient hunters and let the lionesses kill whenever possible. Some 85% to 90% of the hunting in a pride is done by lionesses. Nomads, on the other hand, often have to kill for themselves. In effect, the pride becomes a meal ticket. Sick or injured males in a pride might recover more readily than a nomad in similar condition, giving them extra years to procreate.

(2) In theories of social cohesion, sex is often mentioned as an important factor. About four lionesses in the Seronera and Masai prides came into varying degrees of estrus every month, providing each of the resident males with the opportunity for roughly one sexual contact. Several males may mate with the same female and a male occasionally courts a nomad, but this does not appreciably change the

frequency of sexual contacts. Nomadic males have ready access to nomadic lionesses and to unattended residents, and it was my impression from observing such males as No. 134 and Tailless that they mated as often as pride males. However, it is necessary to think in terms of the population, not the individuals: do residents or nomads sire more viable offspring? Pride lionesses are more successful at raising cubs than are nomads so that most of the annual increment to the population emanates from prides and hence from pride males.

At first glance a lion pride gives the impression that males and females are poorly integrated. They often lie apart when resting, they may wander off by themselves for several days without maintaining contact, and interactions such as social licking and head-rubbing are more frequent between members of the same sex than between those of the opposite one. Males either ignore cubs or rebuff them, and their gluttonous and irascible behavior at kills is proverbial. Although social facilitation is apparent in many activities from drinking to chasing away strangers, both males and females seem to prefer to associate with their own sex. But the pride is perhaps best analyzed in terms of the social roles of males and females.

Females, as mentioned earlier, do most of the hunting, thereby providing the pride with sustenance. Males usually do not take part in the initial stages of a communal hunt. Instead they trail behind the lionesses, watching and waiting, and when prey has been caught they run up and try to claim rather more than their share. While males have been denigrated in the popular press for such behavior, it serves two useful functions: (1) males are so conspicuous that their participation in the hunt would increase the chance of the group being detected by the prey; (2) with the males in the rear any cubs there are protected from marauding hyenas and other dangers. The contribution of males in providing food is slight even though they can kill buffalo and other large prey more easily than lionesses can. In fact their presence may be detrimental because they appropriate small kills so persistently that in some seasons they hasten the death of cubs through starvation.

Females also take the major share of raising cubs, suckling, guarding, and transporting them for the first few weeks of their life, and later, after they eat meat and until they are almost independent, providing kills and a certain amount of protection. Males affect cubs directly in one minor but interesting way. After a kill has been largely consumed, a male often takes the remains and prevents lionesses from eating. But he does allow cubs to feed, and this may be their only meat if they arrive late at a kill. Lionesses, on the other hand, are less generous than males in sharing and even cuff starving cubs away.

The main role of a male is to maintain the integrity of the pride area either by escorting intruders out or by inducing them to move on by his mere presence. Although lionesses take part in territorial displays, males are the most active in patrolling, scent-marking, and roaring. The Seronera pride illustrated the importance of males to the pride. After one of the two pride males had been killed by neighbors, males from three other prides penetrated deep into the Seronera pride area, killed several cubs, and drove out the remaining male. The lionesses remained but their success in keeping cubs alive was much lower (19%) than that of a neighboring pride which was under the jurisdiction of vigorous males (46%). It required two years after the death of the male for the pride to settle down to a normal social life. While this may have been an extreme case, hunters who shoot pride males on the assumption that this has no effect on the population might well consider the long-term disruption and lowered reproductive success their act might cause.

Competitive interactions in societies are usually viewed in terms of dominance, a descriptive term which may lead to oversimplification of analysis as Gartlan and Brain (1968) have pointed out. Nevertheless, males of some primate and ungulate societies establish a hierarchy with high rank giving priority to food and estrous females. A rank order is not evident among pride males or among females; a limited resource is either shared with members of the same sex, or the animal which has first taken possession may be able to retain it without dispute. Considering the steep size gradient in lion society, assertion of strength is inevitable and some form of rank based on age and sex could be expected. Yet there is none in the sense of one animal giving another undisputed rights—for example, to a portion of meat. Each animal fights vigorously to retain its share, behavior which tends to inhibit an aggressor from asserting itself fully. In a pride every individual has learned the fighting potential of every other one, and the severity of interactions are tempered by this.

Aggression at a kill (and in other circumstances) is adaptive, although the vigor of it in disputes over a scrap may seem somewhat excessive, because an individual must first assert itself to obtain a share and then to keep it. Disputes at buffalo kills, which provide enough meat to satiate every member, are not essential and indeed seemed to me less severe than at small kills, except when one animal crowded another. Most carcasses have enough meat for only a few pride members. A male wildebeest provides about 120 kg of food, or 8 kg for each member of an average pride of 15, assuming that all are together; this is not enough to gorge each animal. It might be argued that a hierarchy would lessen the amount of strife at kills—around which most aggression occurs. Given the physical characteristics of lions, such a hierarchy would be based on size: the males would eat first, followed by females and finally the cubs. Such a system would lead to the dissolution or at least complete social reorganization of the pride, for those low in the hierarchy would seldom obtain a meal.

Hyena

Hyena clans contain an equal number of males and females. Although a stranger that enters a clan territory may be killed, hyena society is more open than that of lions. In Ngorongoro Crater, some males and an occasional female switch from one clan to another. Clans in the Serengeti dissolve in times of food scarcity, and the hyenas may then establish themselves with new members in temporary clans near prey concentrations (Kruuk, 1966). Such a plasticity in social structure is adaptive insofar as adults are able to remain in the vicinity of prey. But reliance on a distant food resource creates problems for a female when she has cubs in the den. The young develop slowly. They are almost wholly dependent on milk for the first 8 months of life and they do not accompany the hunt until a year old. Because hyena females do not permit cubs of other litters to suckle on them, the young obtain no food as long as their mother is away. A female often does not carry meat back to her cubs if she has to travel far, nor does she regurgitate any on her return. Consequently cubs sometimes starve (Kruuk, 1972). In some respects hyenas seem less well adapted to a nomadic existence than lions, which, for example, suckle cubs communally.

Female hyenas are larger than males, an unusual situation among carnivores which Kruuk relates to the cubs' need of protection against hyena strangers with a propensity to eat them. The social roles among hyenas are distributed somewhat more equitably between the sexes than among lions. Both sexes are active in territorial defense and in hunting, with females tending to take the lead in hunting. Only the females actively feed and protect cubs. Although females are dominant over males, aggression at kills is limited to occasional disputes for a bone; instead of fighting, each animal bolts food as rapidly as possible.

Wild dog

Dog packs differ from lion prides and hyena clans in two striking respects: (1) many packs contain more males than females; (2) packs are cohesive, possibly because they roam so widely that on logistic grounds alone a scattering of individuals would lead to a disintegration of the social unit. The sexes are of about the same size. The most conspicuous attribute of dog society is the lack of aggression even around a kill. When two dogs are confronted by a potentially aggressive situation, both often behave submissively, a system which represents the behavioral opposite of that used by lions. Since a pack has the ability to obtain daily enough food to satiate every member, at least in the Serengeti, it is not imperative for a dog to fight for a portion. Actual sharing in the sense of one animal regurgitating food for another is a conspicuous trait among wild dogs. Such behavior may help maintain the cohesiveness of the pack not only by strengthening social bonds but also by insuring that every member is at about the same physiological level of hunger, a necessary requisite for a cooperative hunt during which each animal must contribute. Lack of a rigid hierarchy, cooperative hunting by both sexes, and sharing of food almost presuppose a blurring of social roles. Indeed both males and females act as leaders, regurgitate meat for pups, and guard pups while other pack members search for prey. The role of the female consists mainly of suckling her young and of transporting them to another den.

Each of the three predator societies has evolved its own system of group organization, many facets of which have been shaped by the availability and vulnerability of the prey. One main trend is apparent: the greater the size difference is between the sexes the more aggression is shown within the group and the greater is the distinction between the social roles of males and females.

COMMUNICATION

There is a constant interchange of signals in a society. The frequency and form of expression of the communicatory patterns are presumably related to the complexity of the social system, a supposition that can be checked by comparing the signal repertoires of solitary and social carnivores.

Head-rubbing is a common tactile pattern in both *Panthera* and *Felis*, but for unknown reasons is rare in cheetah (table 73). The gesture functions as a greeting, as a means of establishing amicable relations, particularly after a separation, and of maintaining group cohesion before a communal endeavor such as a cooperative hunt. Wild dogs have an analogous gesture in a similar situation in that they nuzzle each other's mouth, to name only the most conspicuous component of a complex pattern. Hyenas greet by licking each other's genitalia.

Social licking is common in the big cats, even between females and cubs of the solitary species, as well as in hyenas. Wild dogs seldom lick each other possibly because their social system is so well integrated that another pattern promoting contact would be superfluous. However, from a utilitarian point of view it would seem that some social licking would benefit the scruffy appearance of their pelage.

The facial expressions of cats are remarkably similar, regardless of the extent of their social life, but it was my impression that cheetah have less mobile lips than the others, presumably because of their short face. In addition, all the cats that I studied have prominent markings on the ears which enhance the expressions. Wild dogs and hyenas use facial patterns which closely resemble those of cats and are elicited by the same types of situation. Although the ears of these two species are not marked, the same effect is achieved through the conspicuous size of the ears. Gartlan and Brain (1968) noted that as group structure in monkeys tightens there is an increased reliance on facial rather than vocal signals, a trend not apparent in cats.

The five species of predators discussed in this report all have either special color markings or long hairs on the tail, an indication that this appendage functions importantly in communication. All, too, crouch or roll on their back as a defensive gesture which signifies appeasement and tends to inhibit attack; in addition, hyenas and possibly cheetah indicate submission by walking with their forequarters lowered on their elbows. Lion, leopard, and cheetah threaten with a head-low posture, in contrast to the house cat, which raises its head and tucks in its chin (Leyhausen, 1960). Male lions make themselves look impressive by strutting, and Leyhausen told me that he has seen captive cheetah males behave similarly (see table 73); an analogous display among hyenas appears to be the parallel walk (Kruuk, 1972).

Vocalizations show some interesting similarities and differences among the four big cats (table 73). Those used in agonistic encounters occur in all species, except that cheetah also moan at such times, and miaowing too is widespread. Roaring is confined to the three *Panthera*. The pooking and prusten of tigers have no counterpart in the African cats. The omission from the lion's repertoire of prusten or a similar vocalization to signify friendliness is surprising, for at night such a sound would convey an animal's friendly intentions better than a visual signal. Hyenas emit vocalizations as varied as those of lions; Kruuk (1972) recognized eleven different calls. I have heard wild dogs produce six sounds but one of these is graded. Although dogs have a loud, hooting contact call, it is never used in the communal concerts so characteristic of lions and wolves. Their bark is a distinct vocalization which serves as a warning of danger. Cats lack such a call, merely borrowing the agonistic growl for that purpose.

Except for minor variations, the four species of cats mark their environment in the same manner—all spray scent, all make scrapes. In tigers and cheetah the feces, too, are an integral part of the marking system but are never covered with soil as is customary among house cats (table 73). Hyenas paste secretions on grass stalks and defecate communally at latrine areas, behavior more typical of viverrids than cats. Marking is not a prominent activity among wild dogs.

This brief overview of communicatory patterns makes several points. In spite of differences in social structure, habitat utilization, and land tenure systems, the signals emitted by the four species of cats are similar in form and complexity. There are, to be sure, differences in the presence or absence of specific vocalizations, but the size of the repertoire is about the same for the solitary tiger and social lion; cheetah have fewer calls than *Panthera* mainly because they lack graded roars. With minor variations the marking system is also the same whether a species is territorial or nomadic. However, some trends are noticeable, especially when hyenas and wild dogs are also considered. Social species which are territorial and in which the members of a group are scattered, namely, lion and hyena, need a long-distance signal that not only keeps individuals in contact with each other but also maintains intergroup distance. A loud sound is an obvious solution. To function adequately, such a sound must begin spontaneously, be contagious, and be most prevalent when the animals are active (Marler, 1969); these indeed are characteristics of the lion's roar and hyena's whoop. The solitary leopards and tigers also emit roars, though less often than lions, and these serve to space out individuals as well as bring them together in order to court or to share a kill, in the case of tigers. Nonterritorial species which exist in cohesive groups (wild dog) or are solitary (cheetah) have little use for a long-distance signal.

Scent-marking serves the same function as a loud call, and its value lies in the fact that it remains effective for many days. All cats and hyenas have a complex system of marking, but wild dog packs seem to depend more on visual than olfactory communication, having no need to maintain contact with others.

Land Tenure System

Land tenure patterns may not only affect all aspects of a predator's behavior but also influence prey populations by regulating the density of predators in an area. Concepts relating to land tenure, such as territoriality, are often difficult to apply to predators. As McNab (1963) has pointed out, carnivores tend to range over an area four times as large as that of herbivores of comparable body weight. Range size tends to increase with the size of the predator and at a rate greater than do energy requirements with size, a relationship which probably reflects the decreasing food density which is available to a predator as it becomes larger (Schoener, 1968). As ranges increase in size, boundaries are difficult to maintain and confrontations between groups become rare, making it difficult to distinguish defense of an area from mere intergroup antagonism. As with other behavior patterns, land tenure may be influenced by ecological conditions. Many Serengeti cheetah and hyena have, for example, modified their land tenure pattern in order to take advantage of the migrating herds, and caution must therefore be exercised in generalizing about a species on the basis of having studied it in one area.

Lion

Pride lionesses confine their wanderings to a limited area some 20 to 400 sq km in size, all or part of which constitutes a territory in the sense that other lions tend to avoid it or are driven away by the owners. Lionesses are, on the whole, more antagonistic toward intruding lionesses than toward males, behavior which is of adaptive value in view of the temporary residency of pride males. Persistent rebuffs of new males would make it difficult for them to become integrated, and this in turn would deprive lionesses of the necessary assistance in maintaining the pride area. When males join a pride they generally assume jurisdiction of the area that is used by the lionesses, although some also annex new terrain and a few associate with two prides. Pride areas may overlap so extensively that some prides retain no portion for their exclusive use. However, each pride has a focus of activity, an area within which it spends much of the year, and these foci tend to be discrete. Incursions into the focus of activity may occur, usually in the absence of the owners.

Nomads roam widely, some over 4,000 sq km and more of terrain. Their contacts with other nomads tend to be amicable. A few males, however, establish temporary territories in the plains and with the acquisition of property become intolerant of others, a trait also rather conspicuous in man.

When adults are lean and cubs are starving in one area while another nearby is inundated with wildebeest, the question obtrudes as to just why prides do not follow the migratory herds. Wolf packs in Algonquin Park, which has a large resident population of deer, confine their movements to an area of 100 to 300 sq km (Pimlott et al., 1969), whereas some packs in the Northwest Territories follow migrating caribou for distances of up to 800 km (Kelsall, 1968). What survival value does retention of a territory have in the case of the lion? Given the tendency of pride members to scatter widely, the pride would not be able to maintain itself in its present form if it assumed a nomadic existence: the animals would either have to remain together like wild dogs or wolves, or limit themselves to shifting and casual contacts in the manner of nomadic lions. Several advantages accrue to a resident. Territorial behavior is a potent spacing mechanism. It restricts the number of animals that may prey in a particular area and thus allows for a more uniform and socially less disruptive use of food resources than if most of the population competed for meat in a few places. Detailed knowledge of the location of waterholes, river crossings, ambush places, and other localities probably increases a lion's hunting success, as does, perhaps, a cooperative hunt with pride members rather than strangers. Familiarity with the environment permits an efficient harvesting of the resources. Although nomadic lionesses seem to give birth as often as resident ones, they manage to raise few cubs, indicating that communal care of young in a pride, as well as the protection afforded by the pride males, increases reproductive success. This fact alone would favor selection of a stable land tenure system for, as Brown (1969) stressed, the survival value of territoriality must be viewed in terms of the effects on gene frequencies.

I have tacitly assumed so far that the size of the pride area is correlated with prey abundance. Kühme (1966) speculated that Serengeti lions maintain large territories whose size is not dependent on the food supply but on a species-specific adaptation for areas that have little prey. Availability of food has an obvious effect on the distribution and size of pride areas and on the movements of lions within them. The plains are almost devoid of prides, no doubt because during the dry season the prey biomass drops below 100 kg per sq km. The woodlands, on the other hand, are wholly occupied by prides which have some 1,000 to 7,200 kg of prey per sq km available even during the leanest season (see table 35). Where resident prey is abundant, as in Manyara Park, a pride may subsist within an area of only 20 sq km in size. In 1966–67, the Masai pride remained in its focus of activity from June until March while the prey biomass there was at least 216 kg per sq km, but from April to June, when the biomass dropped to 115 kg per sq km and below (table 34), the animals moved eastward to a part of their pride area they had not used for many months. The effort to find prey, much less to find vulnerable prey, had become so great that hunting became uneconomical. These and other examples suggest that the pride area represents the minimum amount of space needed for lions to survive during periods of greatest food shortage. Such a shortage may occur only once every few years and, since the size of a pride area does not change constantly with the food supply, a brief study might give the impression that lions occupy more terrain than they need.

The number of pride members may oscillate considerably over the years, a fact which could be used to argue against the correlation between pride area size and food supply. The nucleus of the Seronera pride, for example, consisted of about seven lionesses, but there were once ten additional lionesses, all of them subadult, until emigration and death reduced the population to its former level. Being long-lived and occupying the same locality for many generations, lions have possibly adjusted the size of the pride area to take into account fluctuations in their number and in the prey; boundaries reflect the actual needs of the pride only at intervals of years. Lions possibly have a minimum space requirement, regardless of prey abundance, past which it is not possible to compress a population without affecting social structure. I have no evidence for this other than to note that densities over large areas never exceeded one lion per 2.6 sq km (1 sq mile).

Solitary Cats

Contacts between solitary cats are infrequent, not only because individuals avoid each other but also because various forms of scent-marking may tend to space animals out, thereby further reducing the probability of an encounter. Consequently my descriptions of land tenure in these cats is based largely on inference. Leopard females occupy overlapping ranges about 50 sq km in size, all or part of which may also be shared with a male. Females do not seem to be territorial but the fact that the range of the male at Seronera seemed to be discrete raises the possibility that males may be territorial. A similar system has been described for tigers (Schaller, 1967); puma males also maintain ranges without overlap whereas females share some areas (Hornocker, 1969). Such a system provides males with ready access to estrous females and reduces the chance of strife in competition over mates. Territorial defense by females would seem to have little survival value as long as food is not difficult to obtain. With respect to puma, Hornocker (1969) speculated that spacing of individuals may also afford "greater success in securing large prey animals. Mountain lions must employ stealth to place themselves within striking distance. The chances of success in an area already hunted or being hunted by another individual are much less than in an area where prey animals are undisturbed."

The question arises to what extent the size of a leopard's range is correlated with food supply. Table 34 shows that at certain times of the year only a few gazelle, reedbuck, and other hoofed animals of a size preferred by leopards remain around Seronera. The starved condition of one litter during the rainy season in 1968 indicated that this family had no surplus of food within its range. My impression was that the range size of Seronera leopards reflected their needs during lean seasons. On the other hand, Hornocker (1970) noted that the puma population he studied remained stable during a period of five years whereas deer and wapiti in the area increased. "These data offer evidence that lion numbers in the Big Creek area are determined by factors other than the food supply." While this may be so, it is also possible that range sizes in puma, as in lions, are adapted to periods of relative prey scarcity, especially since deer populations may fluctuate rapidly in numbers within a short time-span, and seasonal shifts in movements are common. A resident puma may not expand or contract its range with changes in prey density, although it may limit its activity to a portion of the range for long periods. On the death of that animal the area would be occupied by a newcomer whose movements are regulated by the surrounding occupants to conform to the range of the former owner. The carrying capacity of predators in a particular area possibly evolves over the years in such a way that a stable population is maintained as long as the environment remains fairly constant.

Cheetah show no evidence of being territorial. In Nairobi Park and other areas with a stable prey population, cheetah are resident, but on the Serengeti plains most are nomadic for at least part of the year, following the movements of their principal prey, the Thomson's gazelle, a way of life which provides them with an ample and constant supply of food. Animals space themselves out by avoiding contact, using both sight and scent to do so. Such a system can operate successfully only if populations remain at a low level, and in this context it is of note that cheetah tend to be less abundant than lions even in areas of prey abundance.

Hyena

According to Kruuk (1966), hyenas in Ngorongoro Crater maintain permanent territories whose boundaries they patrol and defend. On the other hand, in the Serengeti some hyenas leave their territory and commute to prey concentrations as far as 30 km away when the migratory herds leave the plains. By virtue of such mobility these hyenas are able to reside in areas with insufficient prey to support them, something lions and leopards are unable to do. However, Kruuk (1972) noted that a seminomadic existence is suboptimal for hyenas, judging by the high death rate of cubs in the Serengeti as compared to Ngorongoro Crater, a finding which is similar to mine with respect to cubs of resident and nomadic lions. In the Crater, hyenas are controlled by their food supply, and from this it might be inferred that the size of the eight clan territories there is correlated with the available and vulnerable prey (Kruuk, 1966, 1972).

Wild Dog

Packs roam over vast areas, except for about three months each year when they are localized around a den, without evidence that they are territorial. Such movement occurs even in the presence of much suitable prey. Considering the heavy though temporary disturbance that a pack creates among prey and the preference of dogs for young and sick animals, mobility may be a necessary adaptation to insure a constant food supply with minimal effort in areas with a sparse prey population. In the Serengeti there are few dogs and much prey, and it may simply be uneconomical in terms of energy for dogs to maintain a territory. In contrast, wolf packs tend to confine themselves to definite ranges which may vary from 60 to more than 5,000 sq km (Burkholder, 1959; Joslin, 1967), with size depending on the available food, judging by such published sources as Mech (1966) and Pimlott et al. (1969).

One obvious consequence of territorial behavior in lions and hyenas is that it spaces out groups. Such spacing is not immediately apparent in leopards and tigers and cheetah, but its existence may be inferred from the fact that population levels tend to remain constant. Similarly, pumas avoid contact with each other and "this spacing, brought about without apparent conflict, acts to limit population size" (Hornocker, 1969). Solitary species may have a communal organization and each individual probably recognizes its neighbor by sight and scent. New animals, notified by the calls and olfactory markings of the owners that an area is occupied, are unlikely to attempt to settle there. As Lack (1954) noted: "dispersion is primarily due to the avoidance of occupied or crowded ground by potential settlers, not to aggressive behavior of those in occupation." Such a method of spacing is economical in that little effort is required to mark the environment and the chances of conflict are reduced. An injury which to a lion might be an inconvenience could be fatal to a solitary cat which has to do its own hunting. Thus spacing appears to be the most important function of the land tenure of these species, and from it other benefits accrue such as reduced competition for available food. This generalization does not apply to wild dogs, but packs, being rare, automatically are widely spaced.

Population Dynamics

"It is becoming increasingly understood by population ecologists that the control of populations i.e. the ultimate upper and lower limits set to increase, is brought about by density-dependent factors, either within the species or between species. The chief density-dependent factors are intra-specific competition for resources, space or prestige; and inter-specific competition, predators or parasites" (Elton, 1949). The ways in which the large predators regulate their population is of particular interest, for they have few natural enemies, and the density they maintain has relevance from the standpoint of management and conservation.

Lion

Since about four-fifths of the lions are resident, density regulating factors have to work most effectively on that segment of the population rather than on the nomads. Lionesses have a high reproductive potential but for one reason or another few of them raised the average of 1.2 young per year of which they are capable: (1) about 80% of all sexual contacts did not result in young; (2) some 15% of the adult lionesses were barren; (3) the average interval between the death of a litter and birth of a new one was over 9 months rather than the expected 4 months; (4) cub mortality was at least 67% in the two prides that were studied. Only 23% of the expected total of cubs was raised. Because litter size is not influenced by normal oscillations in food abundance, regulatory factors have to work on the cubs, a finding similar to that made by Macpherson (1969) on arctic fox. Abandonment, violence by adult lions, predation, and starvation are the main causes of death among Serengeti cubs. In Manyara Park, where starvation is not a factor, the death rate of cubs is also high, suggesting that death rates may be somewhat intercompensatory in a stable population: victims escaping one factor are lost to another so that, on the average, about half the cubs die before independence.

In spite of a high death rate in cubs, the annual increment was about 11%. Pride size, as was noted earlier, tends to remain constant and the factors maintaining this stability must be explained. The annual death rate of pride adults was about 5.5%, due to old age, violence, and occasional illness. All male cubs leave the pride and enter the nomadic segment of the population, not wholly as surplus animals but as potential replacements of pride males. Some subadult lionesses may also become nomadic. Thus mortality of adults and dispersal of subadults are the main mechanisms that regulate the size of prides.

All suitable terrain, that is, terrain with enough prey to support lions throughout the year, is taken up by pride areas within the park, leaving nomads no place in which to settle. Nomads can become resident in one of several ways: (1) males can replace pride males but this has no influence on the size of the transient population; (2) nomads can remain within a pride area but this is at best a temporary arrangement judging by some which did so; (3) nomads can lay claim to a portion of the plains but those that attempt this are ultimately forced to retreat from lack of prey; (4) nomads can enter areas peripheral to the park where poaching continues to deplete the resident population. Neither tolerated nor desirous of remaining in occupied terrain, many nomads seem to choose the last of these four ways and are then killed by man. This mortality, together with that of the resident adults in the park, probably accounts for the surplus raised by prides with the result that the park population is relatively stable. The reports of Guggisberg (1961) and others indicate that the lions in Nairobi Park have maintained a stable population of about 20 to 25 animals for at least 18 years.

After observing Serengeti lions, Kühme (1966) stated that "the factor limiting the population density of lions might be the size of their territories, but it is most certainly not the food supply." He failed to realize that only about 37% of the biomass in the ecological unit is available to residents during the rains. Territory size appears to be correlated with food abundance during the leanest season, as noted earlier, and from this it follows that lion density in general is too. Several authors (for example Lamprey, 1964; Bourlière, 1965) have expressed the relation of lion density and prey abundance by a simple ratio; for example 1 lion per 300 ungulates in Kagera Park. Such a figure has a limited usefulness, for the actual biomass available to lions may have little relation to the number of prey animals (table 72). Prey biomass shows fairly good correlation with lion density in Manyara, Nairobi, and Kruger parks. The Serengeti has notably few lions when compared to the size of its prey biomass, an important finding which shows the effect of the movements of the migratory herds and to a lesser extent the depletion of the lion population through poaching outside of the park. If the existing prey were resident and poaching were stopped, the ecological unit could support at least three times the number of lions it does now. Ngorongoro Crater, too, has a smaller lion population than expected, but in that area the hyena is extraordinarily abundant and competes with lions for the same food resource more than it does in the other reserves. When total predator biomass is considered, there is good correlation with prey biomass (table 75). From this evidence I conclude that food is a main controlling factor of the lion population.

Biologists differ in their opinion about the density-dependent mechanisms which control or limit the number of animals within a population. Some feel that populations are limited by certain resources, particularly by the amount of food, and others speculate that animals control their own numbers by physiological and behavioral means, maintaining a population level below the carrying capacity of the resources. Although lion prides tend to remain stable in size, maturing cubs may double the number of animals in a pride for several years without affecting the amount of terrain that is occupied. This indicates that prides use larger areas than they need for at least part of their existence. On this basis, I speculated that the size of the pride area and hence the lion density is adapted to the amount of available prey not only during lean seasons but also during lean years when fewer than the normal number of animals are present and the number of pride members is above the optimum. While the correlation between prey biomass and lion density is reasonably good, some factor or factors other than food must determine and maintain pride size at its optimum level when prey is abundant, a level below the one which is based on carrying capacity of the prey. A self-limiting behavioral mechanism appears to regulate pride size to some extent. Size is most readily adjusted by the emigration of subadult females, but, as was noted in chapter 4, I was unable to determine what caused some animals to leave the pride and others not. Thus the data suggest that lion density is regulated both by the food supply and by a behavioral mechanism, an idea which includes the contrasting points of view expressed by Lack (1954) and Wynne-Edwards (1962).

Solitary Cats

The leopard population around Seronera remained stable during the three years of the study even though one adult is known to have died and several grown cubs left the area. Mortality of adults and dispersal of young appear to regulate the leopard population as they do the lion in spite of the different social systems of these two cats. With respect to tigers, Schaller (1967) noted that "the tiger population also appears to be self-limited, with perhaps a social spacing mechanism keeping the number of animals in a particular area relatively constant regardless of the abundance of prey. This was evident at Kanha when the wild prey concentrated during the hot season but there was no increase in the number of tigers. . . ." After studying puma for five years, Hornocker (1970) wrote: "Territoriality appears to be the primary factor regulating numbers of lions in the Idaho Primitive Area. The number of adults, well established on territories, remained unchanged from year to year despite the fact that 4 to 6 kittens were born into the population each year. . . . The actual regulating mechanisms operating in the mountain lion population appear to be dispersal and mortality, both primarily affecting young individuals." While land tenure patterns have a great influence on the dynamics of these species, it is not yet clear to what extent behavioral factors and the food supply determine how much terrain the predators occupy.

Cheetah differ in several respects from the other cats. Their rate of reproduction, as well as population turnover, is higher than that of leopard and lion: cheetah cubs become fully independent at the age of 15 to 17 months as compared to about 20 to 22 months for leopard and over 2 years for lion; cheetah females conceive for the first time at 21 to 22 months as compared to about 3 years in leopard and $3\frac{1}{2}$ to 4 years in lion; and they have average litters of 3 to 4 cubs as compared to 2 for leopard and 2 to 2.5 for lion. Litters break up and disperse when they reach adulthood, and after that contact is avoided. This spaces individuals, but in view of the low population level and indeterminate land-tenure system it is difficult to see how numbers are actually regulated. Reproduction is adequate, judging by the fact that one-third of the females were accompanied by cubs, and an ample supply of food seems to be available and vulnerable. Some unknown factor or factors keep the population depressed and seemingly stable at a low level.

Hyena

The hyena populations in Ngorongoro Crater and the Serengeti appear to be stable. In the former area, recruitment in clans is equal to mortality of adults, according to Kruuk (1972), making dispersion unimportant as a regulating factor, but whether it features in the Serengeti is not clear. The Serengeti hyenas contain a larger proportion of old animals and have about one-third as much annual recruitment as the Crater ones (5% versus 13.5%), indicating that their rate of population turnover is slow. Mortality of cubs in the Serengeti is often due to starvation and that of adults to violence such as attacks by man, lions, and other hyenas. Starvation of cubs would be less prevalent if the young would follow their mother to a kill at an early age, as is characteristic of cats. Instead they remain near the den and are almost wholly dependent on milk for about 8 months, an adaptation different from any felid or canid and one whose survival value has not been adequately explained. "From the present evidence, it seems likely that both the hyena populations are to some extent controlled by food supply, the hyenas in the crater by food for the adults and the hyenas in the Serengeti by food for the cubs" (Kruuk, 1972). Territoriality is not a primary factor regulating the density of Crater hyenas, and it has even less of an influence on density in the Serengeti, where the distance that hyenas must travel for food during the dry season seems to be critical.

Wild Dog

The selective forces that have shaped the reproductive biology of wild dogs seem to have been operating in opposite directions at the same time. On the one hand are factors which tend to decrease the rate of population growth, among them an unbalanced sex ratio favoring males, a large number of reproductively inactive females, and a high death rate of litters; on the other hand are factors which favor a rapid increase such as litters of up to 16 pups, males that help to feed and guard pups, and adults that give pups priority at a kill, the last being in striking contrast to the behavior of lions and hyenas. About one-third of the population consisted of large young less than a year old, a higher percentage than for any other large predator in the park and an indication of good reproductive success in spite of various depressants. Yet mortality of adults was of similar magnitude, judging by the fact that the population seemed to be stable. Canine distemper affected one pack, and probably others as well, and this disease alone could have caused a serious decline in the population. Yet with its high reproductive potential, the population could have rebounded rapidly, perhaps to be affected again, but, since there have been no oscillations, some other factor or factors must have stabilized it at its present low level. Dogs being rare and highly mobile, it is difficult to visualize how land tenure can have an influence. Slow, long-term fluctuations in the number of wild dogs have been noted in Kruger Park, as well as in Alaskan wolves (Murie, 1964), and these seem to be independent of the food supply. The Serengeti population may well increase some day; when it does so, it would be important to determine the cause.

Social Carnivore Behavior and the Study of Early Hominids

When trying to deduce the social system used by *Australopithecus* and other early hominids, anthropologists have usually looked for clues among nonhuman primates. This is logical on phylogenetic grounds but not on ecological ones. Social systems are so strongly influenced by the ecological conditions under which an animal lives that even the same species may behave differently from area to area as Rowell (1967) has shown for several primates and Kruuk (1970) for hyenas. Monkeys and apes are essentially vegetarians living in groups which confine themselves to small ranges. Man and his precursors, on the other hand, have been widely roaming scavengers and hunters for perhaps two million years, a way of life that has diverged so drastically from the nonhuman primates that similarities in the social systems of the two may well be accidental. More can probably be learned about the genesis of man's social system by studying phylogenetically unrelated but ecologically similar forms than by perusing nonhuman primates. The social carnivores provide an obvious choice. It may be conjectured that some of the same selective forces which had an influence on the social existence of lion, hyena, and wild dog also had an effect on hominid societies (Schaller and Lowther, 1969).

A nonhuman primate derives several advantages from living in a group, including the learning of traditions and protection from predators. While a carnivorous or omnivorous hominid might benefit in the same way, additional selective forces favored its group existence. These forces were mentioned earlier with respect to lions: several individuals hunting together are more successful than a solitary one, they can kill larger prey, and they can use their food resources more fully by being able to share them and protect any excess from other predators. The descriptions of cooperative hunting in lion and wild dog show that hominids at whatever stage of mental evolution could have employed relay races, encircling of prey, and other techniques without recourse to an advanced system of communication such as language. Group life also makes a division of labor possible, with, for instance, several adults protecting young at the home site while the rest hunt and bring back the spoils to share with the others—a system used by wild dogs but not by nonhuman primates. Before setting out on a hunt, hominids probably indulged in some activity analogous to the greeting ceremony in wild dogs which strengthened social bonds and synchronized behavior. The size of the hunting group was most likely flexible and depended on the prey that was to be pursued, as is the case with hyenas. If sexes were of about equal size and each individual assumed several social roles, then strife within a hominid group at kills may have been slight, but aggression toward other predators, including other hominid species, would have been marked. To store excess meat, hominids could either have guarded it, hung it in a tree, or submerged it in water, all methods used by today's predators. In areas where prey was abundant throughout the year, hominid groups probably hunted within limited ranges whose boundaries they may have defended, judging by the behavior of lions and hyenas, whereas in areas with a migratory prey population groups no doubt were seasonally nomadic. The size of a hominid's range, like that of most predators, was probably regulated by the food supply. Throughout his evolution, man most likely ate vegetal matter too, but the distinct requirements of a hunting existence were probably responsible for shaping many aspects of his society. Thus, knowledge of carnivore behavior, as well as of primate behavior, can provide the spectrum of possibilities that were open to the early hominids. "Human social systems are the product of the selective forces operating on man the primate and man the carnivore, and it might be useful to try to find out how each of those forces contributed to his way of life and that of his precursors" (Schaller and Lowther, 1969).

14 The Dynamics of Predation

Previous chapters describe some of the factors that influence predation, particularly the size of the predator population, the way in which prey escapes predation, and the food habits, hunting efficiency, and so forth of the predators. The question that remains to be answered is to what extent predators affect prey populations; that is, to what extent do they constitute an effective limiting factor in the sense of outweighing all other factors that tend to reduce the rate of increase? This is a difficult question to answer, for predators constitute only one of several forces operating on a population. The reproductive potential, the conditions of the habitat, intraspecific strife, disease, and others—the "opposing forces of procreation and destruction" as Craighead and Craighead (1956) phrased it—all interact simultaneously to keep a population in check.

In a classic review of vertebrate predation, Errington (1946) stated: "A great deal of predation is without truly depressive influence. In the sense that victims of one agency simply miss becoming victims of another, many types of loss—including loss from predation—are at least partly intercompensatory in net population effect. Regardless of the countless individuals or the large percentage of population that may annually be killed by predators, predation looks ineffective as a limiting factor to the extent that intra-specific self-limiting mechanisms basically determine population levels maintained by the prey." He noted, however, that predators can have "a truly significant influence on population levels of at least some wild ungulates." In 1956 he added: "The distinction to be kept in mind is that predation centering on essentially doomed surpluses or wastage parts of prey populations is in a different category from predation that cuts right into a prey population and results in the prey's reaching or maintaining a significantly lower level than it would if it did not suffer predation." Several detailed studies of the effects of carnivore predation on mammalian prey populations have been made and of these I mention only a few to indicate the widely varying results.

After measuring predation on a mouse population by feral house cats, gray foxes, raccoons, and skunks, Pearson (1964) concluded: "Nevertheless, I think it is important to document the extent of predation on mouse populations because many ecologists have convinced themselves that predators subsist on 'surplus' prey animals, or on sick or weakened individuals. Undoubtedly this is often true .... But the carnivores had not merely skimmed the cream of the *Microtus* population, they had taken almost everyone." Similarly, weasels almost exterminated a moderate lemming population in northern Canada (Maher, 1967); in other studies (Golley, 1960) predation had a less drastic influence on rodent populations.

Cowan (1947) found that in the Rocky Mountain National Parks a wolf population of one animal per 259 sq km of summer range or one per 30 sq km of winter range was not large enough to remove diseased and starving mule deer and wapiti, much less the annual increment of these species. On the other hand, with respect to wolves preying on Dall sheep in Alaska, Murie (1944) noted that "predation on lambs is the most important limiting factor in stabilizing sheep numbers." The twenty or so wolves on Isle Royale were thought by Mech (1966) to be controlling the moose population which in late winter numbered 600 animals. Pimlott et al. (1969) were uncertain if wolf predation was the primary limiting factor on the deer population in Algonquin Park, but they showed that it was a major one. From the available evidence, Mech (1970) generalized that wolves may have a limiting influence at prey densities of 3,363–10,909 kg (7,400–24,000 lbs) per wolf but not at higher ones.

After studying puma predation, Hornocker (1970) concluded "that elk and deer populations were limited by the winter food supply, and that predation by lions was inconsequential in determining ultimate numbers of elk and deer. Predation by lions, however, is a powerful force acting to dampen and protract severe prey oscillations and to distribute ungulates on restricted, critical range." Tiger predation "has a strong depressive influence on the number of wild hoofed animals" in Kanha National Park, India, in spite of the fact that domestic cattle provide a buffer of alternate prey. The barasingha deer declined as a result of predation on a small localized population whose annual recruitment was poor, possibly as the result of disease, yet axis deer and gaur were increasing in number (Schaller, 1967).

The results of these studies are so variable that generalizations about the effects of predation on mammalian populations are premature except as hypotheses to be tested. Wolves may limit prey on Isle Royale but not in the Rocky Mountain National Parks. I have tried to evaluate the influence of predation on prey in the Serengeti in spite of the fact that data are limited and some conclusions are based more on inference than on fact. But enough information is available that realistic conservation and management practices with respect to predators can be initiated should there be a need to do so. Of course, prey populations are never static for long, indeed some have been changing rapidly, and this may alter the relations between predators and their food supply in coming years. A baseline of information against which such changes can be compared is provided here.

Ecological Separation of Predators

When several predatory species hunt in the same area, competition in the sense of a joint demand for a limited prey resource is likely to occur. One way in which such competition can be reduced is for predators to occupy different habitats or to use the same one at different times. Although all habitats are used by the five large predators in the Serengeti, some are more abundant in one than another. Lions occupy mainly the woodlands, hyenas and cheetah the plains and woodlands-plains border, leopards the thickets and riverine forests, and wild dogs both woodlands and plains equally, depending on availability of prey (table 79). Cheetah and wild dog are diurnal, hunting predominantly in the morning and afternoon (see fig. 44), whereas the others are nocturnal except in special circumstances such as when lions stalk prey at waterholes.

It would seem to be a truism that predators occupy the ecological niche to which they are best adapted: the stalking cats are in dense cover and the coursing predators are in open terrain which enables them to keep their quarry in sight. But other factors also influence distribution, notably the availability of prey. Lions cannot permanently colonize the plains due to lack of prey during the dry season. Wild dogs and cheetah circumvent this problem by being nomadic and some hyenas do so by commuting. Kruuk (1972) noted that hyenas abhor tsetse flies and this may influence their habitat preference. There is also the possibility that direct antagonism between predators affects their distribution. In India, leopards tend to be scarce where tigers are abundant and vice versa (Schaller, 1967); a similar relationship has been noted for coyotes and wolves in North America (Mech, 1970). To leopards a riverine forest may not be a preferred habitat but a refuge from lions.

The various predators are obviously not separated completely in space and time. Cheetah and wild dogs hunt during the same hours, but both are so scarce that they seldom meet. Attracted by each other's kills, lions and hyenas often come into conflict, particularly in the plains where each appropriates kills from the other. In general, predators tend to be intolerant of each other, even to the extent of killing without provocation—and not just at kills. However, a clear distinction must be made between hunting behavior and aggression, between predators killing each other for food and for other reasons. Leopards commonly catch small carnivores such as jackals and servals and eat them as any other prey. On the other hand, lions may pursue hyena, leopard, and cheetah, using not the inexpressive facial features of a hunt but the bared teeth and vocalizations typical of intraspecific strife; they treat other predators as they would other lions. In this context it is of interest to note that man, too, is usually attacked like another predator rather than like a prey item. Such interspecific intolerance is particularly striking in lions and leopards and less so in hyenas and wild dogs. Indeed these last two species have an armed truce which usually remains in effect while they are not at kills, except when dogs have pups, even to the extent of both species resting side by side in the same mud hole. The various predators seem to view each other as competitors, and it is of crucial importance to find out to what extent they actually use the same resources.

It has been customary to divide the large African carnivores into predators such as lions and wild dogs, which kill their own food, and scavengers such as hyenas and jackals, which subsist mainly on the remains left by predators or on any other meat they can find. This distinction is not justified, for hyenas in the Serengeti kill two-thirds of their own food and jackals over four-fifths of theirs. Estes (1967) therefore termed these species scavenger-predators to contrast them to "pure" predators. But lions, too, scavenge much of their food in some areas. For example, of 63 carcasses on which lions were eating in Ngorongoro Crater, 81% had been killed by hyenas (Kruuk, 1972). Thus lions can be considered pure predators in one park and not in another, an awkward semantic distinction. Obviously the predators, with the possible exception of cheetah, take whatever food is easiest to obtain. The amount that they can scavenge depends on the relative abundance and absolute density of each predator species and on the number of prey animals dying from disease and starvation. Where lions are abundant but hyenas few in number, the latter can subsist on lion kills; where hyenas outnumber lions the hyenas must capture their own prey in order to survive. Hyenas are predators. Their powerful jaws and massive teeth, as well as the digestive system which enables them to assimilate organic matter in bone, might be viewed more as adaptations enabling them to utilize their own kill fully rather than solely as an adjustment to a scavenging existence. Jackals lack such specializations.

Based on their predominant hunting methods, the predators can be divided into stalkers and coursers, with the former characteristic of felids and the latter of canids. The cats are specialized to capture prey through stealth: "The long body; the powerful, quick-reacting (but quickly exhausted) 'white' muscles; the relatively short, thick limbs with padded feet and hooked, retractable claws; frontal eyes for good binocular vision; and the short jaws armed with long, stabbing canines are the main specializations that enable cats to stalk so stealthily, to gauge distances accurately, to make a lightning and overpowering surprise attack, and to grip and kill the prey" (Estes, 1967). The canids in contrast, have fairly long and slender legs and a deep chest, designed for running fast and far. Lacking the curved claws and powerful arms of cats, they have to hold and pull down their prey with the teeth alone.

Similar ecological requirements have led to evolutionary convergences in these family patterns of hunting. The hyena, which is most closely related to cats and viverrids (Thenius, 1966), has essentially taken up a canid mode of life, and in few respects the cheetah has too. Convergence is apparent not only in hunting methods, which probably are most strongly affected by selection pressures, but also in various social patterns. In table 79 certain types of behavior are compared on the assumption that lion and leopard are typical felids and wild dog and jackal typical canids. Each species has, of course, evolved specializations which may not occur in all family members. For example, wild dogs do not store small pieces of meat whereas wolves and jackals do. The resemblance between hyenas and canids lies particularly in their hunting and killing behavior, in their use of burrows, and in their method of fighting, whereas mating and parental-care patterns of hyenas show a felid affinity; in addition, they paste scent on stalks in a manner reminiscent of viverrids (Kruuk, 1972). Their communal defecation at latrine sites has no counterpart among felids, nor, as far as is known, among canids, but the members of a banded mongoose group may defecate together by the termite mounds in which they live. Cheetah remain cats in spite of their dog-like build, and even their coursing is limited to brief runs. However, they have been observed to regurgitate food for small young, a typical canid trait.

All large cats have a tawny pelage marked with stripes, rosettes, or spots, as the case may be, except for the lion which lacks a conspicuous pattern. While pelage color unquestionably helps to camouflage a stalking cat, all patterns seem to serve this function equally well. The variations in markings, or the lack of markings, may function as species-specific recognition signals which prevent contact and hybridization, especially in areas like India where tiger, lion, leopard, and cheetah were once found in the same general habitat. Hyena and wild dog, neither of which make an effort to conceal themselves when approaching prey, have a splotched, dark coat, one that is easily visible at a great distance, and it is possible that the coat helps to keep the scattered pack members in visual contact during the hunt.

A large predator has a greater variety of hoofed animals available as prey than a small one. All five of the large predators capture prey weighing less than 100 kg, whereas animals scaling more than 250 kg are killed only by lions, except on rare occasions when hyena attack them. Communal hunting also has an effect on the size of prey that can be killed by predators. For example, judging by its size, a solitary wild dog would subsist on animals weighing less than 20 kg, but packs readily kill prey of 100 kg. With the spectrum of predators ranging from jackal to lion, competition for the same food resource is reduced and predation pressure is distributed more evenly over the prey population.

Speed and endurance of a predator have an obvious effect on the frequency with which it can obtain a meal and the kind of prey it can catch. Lions and leopards are fairly slow animals (table 79), easily exhausted too, and this limits their pursuit to a short rush, a method whose success depends as much on the vulnerability of the prey as on the skill of the hunter. Stalkers may, however, increase their chances of success by hunting at night and near cover, though they do not consider wind direction. Hyenas have stamina but only moderate speed and agility. These attributes are reflected in the hunting success of lion, leopard, and hyena: only about one-third of their pursuits after

Thomson's gazelle succeed (table 79). Two or more days may pass without their being able to obtain prey, and when some is finally available they gorge themselves, with hyenas being able to ingest one-third of their body weight in one meal (Kruuk, 1972) and lions one-fourth—over 35 kg of meat in the case of a male. Food passes rapidly through the stomach which means that they can soon eat more. Excess meat may be stored, leopards doing so in a tree, hyenas sometimes in water. Wild dogs can maintain a high speed for a considerable distance, and their success in capturing gazelle is higher than that of hyenas. The cheetah's sprint is so fast that its success rate of 70% for gazelle exceeds that of other species. Both of these predators tend to capture prey daily and neither makes an effort to save excess meat.

It is interesting to speculate, as Schaller and Lowther (1969) have done, how a carnivorous hominid might have subsisted in the Serengeti, especially in view of the fact that man has been a member of this predator community since his evolutionary beginnings. A hominid could have obtained meat in one of four ways: by scavenging animals dead from disease, malnutrition and other causes; by driving predators off their kill; by capturing newborn young, sick individuals, and other vulnerable prey; and by capturing healthy large mammals. All large predators in the Serengeti obtain meat by all four methods, with the exception of cheetah which do not appropriate kills. Some anthropologists (Leakey, 1967) visualize an evolutionary progression from vegetarians to scavengers to hunters. The fact that no large mammalian predator in the Serengeti subsists solely by scavenging suggests that hominids would have found it difficult to do so unless they supplemented their diet with vegetal matter. There is no ecological room for a total scavenger, a conclusion which Lowther and I supported by a simple field experiment. After searching on foot for kills and dead animals in an area of great prey abundance for several days, we concluded that "under similar conditions a carnivorous hominid group could have survived by a combination of scavenging and killing sick animals" but that where prey was sparse it would have to hunt in order to survive. Like all predators, hominids probably obtained their meat in the easiest possible way, by scavenging and by killing the young and sick when possible, and by pursuing healthy animals when nothing else was available.

> The means by which scavenging and hunting hominids might fit into the ecological community without competing too extensively with other predators pose a number of questions. Their primate heritage

suggests that they were diurnal, and selection pressures from both their primate and carnivore way of life undoubtedly favored a social existence. The only other diurnal social carnivore is the wild dog, which hunts at dawn and dusk, and favors prey weighing 60 kg or less. An ecological opening exists for a social predator hunting large animals and scavenging during the day, an opening some early hominid may well have filled, assuming that none of the saber-toothed cats did so. A certain amount of competition undoubtedly existed between hominids and other predators, just as it exists among carnivores today (Schaller and Lowther, 1969).

ANTIPREDATOR BEHAVIOR

Arrayed against the predators are the various prey species, each with antipredator patterns, some morphological and physiological in nature and others behavioral, but all directly or indirectly designed to reduce the chance of the animals being eaten. Such antipredator devices can only be understood if they are related to the food habits and hunting techniques of each predator. The possibility should also be considered that some kinds of behavior represent anachronisms, patterns once useful for avoiding saber-toothed cats and various *Hyaenidae* which are not among the impoverished predator fauna of today. The responses of hoofed animals to a predator are so finely balanced between the need for avoidance and the endeavor to maintain a status quo, and between attempts of an animal to defend a member of the group and efforts to save its own life, that the adaptiveness of some antipredator patterns are not always apparent. Prey animals know the potential of each predator intimately and react to such small nuances of behavior that my rather casual observations revealed only the most obvious signals to which they responded. For example, a gazelle has the smallest flight distance in response to a jackal and progressively larger ones in response to hyena, lion and leopard, cheetah, and finally wild dog, which may be avoided as soon as a pack moves into view a kilometer or more away (Walther, 1969). The flight distance is correlated with the danger that each predatory species represents.

Similarly, wildebeest and zebra may permit wild dog and cheetah to approach to within 20 m or less without fleeing whereas a lion is usually avoided at 40 m. If, however, the herd contains vulnerable young, it may retreat when these predators are still 100 m away. Wildebeest and zebra are quite casual about the proximity of hyenas, permitting approach to within 10 m or less. Probably the animals can detect from the behavior of the hyena whether it is hunting.

The defensive actions of the prey are also adapted to each predator. A female Thomson's gazelle may butt a jackal vigorously when defending her fawn, but against a hyena she limits herself to distraction displays, and with the others she makes no effort to save her young because she may be killed herself. Wildebeest and zebra sometimes defend their offspring against cheetah, wild dog, and hyena but not against lion. Buffalo and rhinoceros readily attack lions when their young are threatened. The size relation between predator and prey is a critical factor in determining whether an animal will defend its young. If prey outweighs a predator by a ratio of at least 3:1 then it may feel secure enough to attack, if only briefly. Of course defense of a young when it is being attacked by several predators may not be adaptive, and Kruuk (1972) mentions a female wildebeest which was killed by a pack of wild dogs in such a situation. Defense of young is usually limited to the mother. Among zebra, however, several mares and the stallion may attack a predator. A communal effort is particularly evident among buffalo, and a solid phalanx of horns sometimes greets a lion in search of a meal. A cohesive social unit seems to promote communal defense, whereas an amorphous social organization, such as is typical of wildebeest, does not.

Although an animal may defend itself against a predator when cornered or caught, it usually does nothing to retaliate, "looking less the victim than the witness of its own execution" as Estes and Goddard (1967) phrased it with respect to a wildebeest being killed by wild dogs. Defense does have survival value, as I witnessed when a buffalo attacked a lion and a wildebeest a pack of wild dogs, and it is surprising that such weapons as horns and teeth, which are used consistently in intraspecific strife, are wielded infrequently by an animal in defense of its life.

Many antipredator patterns have relevance with respect to all predators, but some probably evolved in response to a particular hunting style. To escape from a stalker, for example, a gazelle needs only a moderate running speed, but a much greater one is required to elude a wild dog. Other patterns of particular value in deterring coursers include the bunching of herds, a zigzaging run, and the retreat of young to the center of the herd. Several antipredator devices have relevance mainly with respect to stalkers; among such devices are the tendency of animals to stay away from dense cover, to drink in daytime, to travel in single file, and to scatter and jump with twisting leaps when attacked. Some of these patterns may be culturally determined, transmitted to the young by the actions of adults in the herd. The tendency for some species to go to water only in daytime and to approach or be approached by a predator up to a certain distance probably belong to this category.

No species has a perfect defense against predators, for selection works in such a way that any increase in efficiency of the escape pattern brings about a refinement in the method of attack by the predator. In addition, the need to avoid being captured must be weighed against other daily requirements in the life of the animal, such as a supply of green forage which it might be able to obtain only by migrating, entering thickets, and in other ways making itself vulnerable.

A predator's hunting success reflects the efficiency of the antipredator patterns of a species. Once a lion has launched itself into the final rush, its chance of catching a reedbuck or topi is 13 to 14%, a gazelle, zebra, or wildebeest 26 to 32%, and a warthog 47% (table 60), figures which speak well for the vigilance, agility, and speed of these species. To compare the success of one predator with that of another is difficult because methods of selecting and catching prey are not directly comparable, and different antipredator patterns may contribute to the failure of the hunt. In addition, hunting success may differ from season to season depending on ecological conditions, as is, for instance, the case with lions capturing gazelle and with lynx pursuing snowshoe hare (Nellis and Keith, 1968). Lions may double their success and hyenas quadruple it by hunting cooperatively. Hyenas catch about 32 to 44% of the individuals they attempt to capture regardless of species (table 76). Since hyenas tend to select the weakest animal, the comparable success rate shows mainly their ability to accurately gauge their chances of securing their quarry (Kruuk, 1972). Wild dogs are highly efficient hunters, an average of 70% of their chases ending in a kill (table 70), and cheetah, too, have a success rate of similar magnitude. It is interesting to note, though possibly not significant, that the two predators most successful in obtaining a meal also kill most often and eat proportionately the greatest amount, yet are scarcer than the others.

Table 76 shows the hunting success of some North American carnivores, but since the results are based largely on spoor in snow, I have no idea how comparable they are to those of the African predators. Mech's (1966) percentage for wolves can not be compared to wild dogs and hyenas for he counted every individual the wolves "tested" whereas Kruuk and I tallied only chases. However, all data do indicate that most predators must make several vigorous attempts to secure a meal. In spite of this, a predator in areas where prey is abundant can fulfill its needs with relatively little effort, and it is characteristic of both stalkers and coursers that they spend three-fourths or more of each day resting.

Predators are generally more successful at catching small young than adults and a large percentage of the diet of several species consists of such animals. It has been speculated that birth peaks, which are typical of most Serengeti ungulates (fig. 36), represent an antipredator mechanism which makes young so abundant for a short period that all predators become satiated, only to have few or no young available the rest of the year, a system which is thought to result in a lower mortality than if births were equitably distributed. Estes (1966), for example, wrote that "seasonal breeding in equatorial Africa is obviously not dictated by climate. This and all the major features of the [reproductive] system have probably been shaped by predation, quite possibly by the one ranking predator of the young, the spotted hyena." On the other hand, Talbot and Talbot (1963), Gosling (1969), and others felt that climatic factors, specifically the growth of nutritious green forage resulting from the seasonal rains, affect the birth season of some species.

Three out of the four migratory species have a birth peak while they are on the short-grass plains between January and May, and Grant's gazelle have young then too. Birth at this season provides young with ample nutritious grass during the first months of life. The buffalo has a peak season then too, correlated, according to Sinclair (1970), with a high level of protein in the long grasses. Most woodlands species have birth peaks during the height of the dry season. While this may seem an inauspicious period for the young, there is usually some rain in late October or November providing green grass at the time the young are being weaned; in fact, it is tempting to correlate the earlier birth period of some species in the north and west of the park with the rains that reach there before they do Seronera. A fetus grows rapidly during the last three months of its development and adequate nutrition is important to the female then. Only those species with birth peaks early in the year have nutritious forage more or less regularly available during the last stages of pregnancy. However, topi, Thomson's gazelle, and possibly other females ready to give birth seem to concentrate on patches of green grass. A strong case can be made for the argument that births of most species are timed in such a way that young have good forage available at a critical time in their development and that peaks are caused by differential survival of young as a result of changes in the availability of food. But it seems likely that predation influences the magnitude of the peak in those species in which young are particularly vulnerable either because they are conspicuous or the social system does not provide for their protection adequately. Wildebeest are a good example. Hyena, wild dog, and cheetah all select newborn calves, making those born early and late in the season highly vulnerable. Predation in this case maintains the sharp birth peak, but there is no evidence that the peaks of zebra, topi, buffalo and warthog are similarly affected.

SEX, AGE, AND HEALTH OF PREY KILLED BY PREDATORS

Several species are preyed on by a variety of predators each with the tendency to kill animals of a certain sex, age, and health, but the question remains if these predators compete or if each concentrates on a distinct segment of the population.

Wildebeest

Hyenas and lions kill more adult males than females, the former at a ratio of 1.8:1, the latter at 2:1. The ratio in the living population is thought to be 1:1 (Watson, 1967), and if this is so then both predators find males more vulnerable than females, lion possibly because males are often alone, are inattentive and so forth, and hyena perhaps because there may be proportionately more sick males in the population.

The age of a wildebeest greatly influences its selection by a predator. Leopard and cheetah with rare exceptions kill only calves, and wild dogs also specialize on this age class. Figure 46 compares wildebeest killed by hyenas and lions. Unfortunately Kruuk (1970) and I used different aging criteria, but the data nevertheless show that hyenas take a much higher proportion of wildebeest less than one year old and a significantly lower proportion of adults ($p = <.05$) than do lions. Both predators captured more elderly animals than expected. Hyena seldom preyed on prime wildebeest, in contrast to lion, and it is interesting that a sample of 31 adult wildebeest which died of disease and malnutrition near the Mara River showed an age distribution similar to that of the hyena kill (see Kruuk, 1970). This suggests that hyenas take a segment of the population that is doomed anyway.

Being pragmatists, predators may kill yearlings or adults in poor condition because they are easier to capture than healthy ones, but the proportion of such animals in the diet varies considerably between

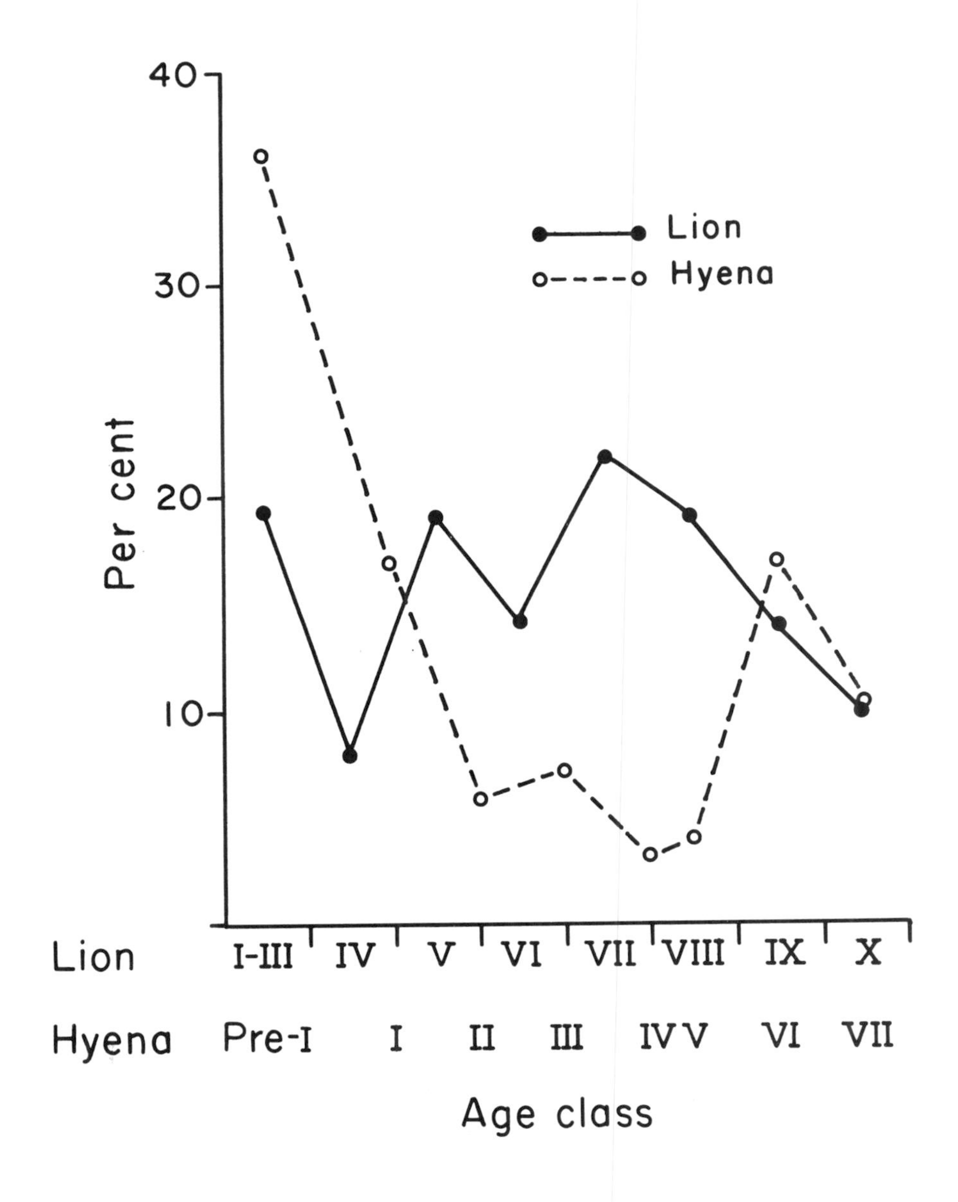

Fig. 46. The ages of wildebeest killed by lions and hyenas. (The hyena data are from Kruuk, 1970.)

species. Wild dogs take only sick yearlings and adults, as far as I was able to ascertain; hyenas probably select animals in poor condition (see fig. 39); and lions predominantly kill wildebeest in good health, only about 20% of those in classes III to X being below par physically.

Zebra

Lions prey on adults in the expected sex ratio of 1:1 except in the oldest classes, in which males predominate significantly. More females than males die of disease in middle age, making it likely that a disproportionately large segment of the population consists of old males. Kruuk (1972) found that hyenas kill zebra at a ratio of 0.6 males to 1 female, and he ascribes this disparity to antipredator mechanisms. While he probably is right, the effect of disease on the population may also have an influence.

Predators select the various age classes of zebra much as they do wildebeest. Leopard, cheetah, and wild dogs kill mostly foals, but the last-named also attacks yearlings and adults in poor condition. Lions capture young animals in roughly the same proportion as they occur in the population, middle-aged ones about as often as expected but old ones in significantly large numbers, a fact that may be attributed to the old zebra's lack of agility and speed for want of a better explanation. Hyenas kill zebra of all ages too, but no further conclusions can be drawn on the basis of Kruuk's small sample. About 20 to 25% of the adult zebra that were killed by lions may have been in poor condition; Kruuk (1970) noted that hyenas appear to select sick animals if present.

Thomson's Gazelle

Because of their size and abundance in certain parts of the park, Thomson's gazelle are preyed on by all large predators. Counts at Seronera showed that adult males are outnumbered by adult females in a ratio of about 0.8:1. All predators except cheetah kill proportionately more males than females. There are several reasons for this. Lions capture only as many animals of each sex as expected in normal seasons, but in dry years many males, particularly nonterritorial males, congregate along river courses where they become vulnerable, especially since their reaction time to danger is slower than that of females. Leopards also take a high proportion of such males. Kruuk (1972) suggested that the preponderance of males in the hyena kill sample is due to a greater susceptibility of males than females to disease, making the former more vulnerable to predation—a valid conjecture, considering the fact that I found gazelle dead or dying of disease at a ratio of 2.4 males to 1 female. The wild dog data show a slight preponderance of male kills in my sample but a considerable one in Kruuk's, an unexplained difference which emphasizes the need for caution in drawing firm conclusions.

Fawns up to the age of one year (classes I to IV) constituted roughly 35% of the population. All predators capture fawns but eat them so quickly that usually they are not fully represented in a kill sample. Lions caught fawns in about the expected proportions, the percentage being 39% in the total sample and 44% in the observed killings only (see table 49). According to Kruuk (1972), 43% of the hyena kills comprised fawns, all two months old or less (my class I), with the older animals being almost immune to predation. Of the cheetah kills, 55% were fawns, but it is probable that the figure in an unbiased sample would be at least 66%. Forty-five percent of the wild dog kills were fawns, two-thirds of them in class I; in contrast, Kruuk (1972) found that dogs "took only 17% fawns out of gazelle killed," a difference which I am unable to explain, for we collected data in the same area at the same seasons.

Table 78 compares a small randomly shot sample of gazelle in classes IV to X with predator kills. The lion, cheetah, wild dog, and mass-killed hyena samples are similar to the one from the living population in that the percentages of animals killed in classes IV, V, and VI are low, the one in class VII is high and is followed by fairly low percentages in classes VIII and IX, only to show a rising one in class X. The distribution of leopard kills does not quite conform to this pattern, possibly because the sample is small. The predators harvest gazelle of all age classes in about the same proportion as they occur in the population except that coursers take notably few large fawns and yearlings.

It appeared to me that lions, cheetah, and wild dogs killed primarily healthy animals; at least I could seldom detect a difference between an individual that was selected and one that was not. Predators are more astute in perceiving a minor debilitation than I, and it may be that wild dogs in particular often kill animals in poor condition. Kruuk derived the same conclusion with respect to hyena hunting gazelle.

In sum, leopard, cheetah, and wild dog kill mainly young wildebeest and zebra, and competition between lions and hyenas for the adult segment of the population is reduced by the fact that the former captures mainly animals in good condition and the latter those in poor condition. On the other hand, all predators kill subadult and adult

Thomson's gazelle. Although some capture a higher percentage of fawns or sick individuals than others, it is obvious that there is competition for the same resource, except insofar as the predators are ecologically separated in time and space. My findings for wildebeest possibly apply also to topi, hartebeest, and other prey of that size and those for Thomson's gazelle may be relevant with respect to predation on impala and Grant's gazelle, among others. Lions have no competitors for large prey, such as buffalo and giraffe. On the basis of these data a number of generalizations about prey selection and competition in Serengeti predators may be advanced.

(1) The hunting methods of predators result in somewhat different prey selection processes: coursers choose a specific individual and pursue it whereas stalkers often depend on prey that selects against itself, so to speak, by becoming vulnerable in some way. Consequently, in areas where both types of predator prey on the same species, their impact tends to be different.

(2) When stalkers and coursers prey on the same species, the age structure of animals captured by the former tends to be more similar to that found in the living population than the age structure of animals killed by the latter, which take a disproportionately large number of young and sick individuals. In other words, stalkers select fairly randomly from the population and coursers kill whatever is easiest to catch.

That coursers tend to take the young, weak, and old has also been shown in several wolf studies. Mech (1966) noted that of 51 moose kills, 33% consisted of calves, none were 1 to 6 years old, and 67% were over 6 years old; furthermore 39% of the adults showed symptoms of some debilitating condition. Pimlott et al. (1969) found that wolves in Algonquin Park showed "a marked selectivity for deer 5 years of age or older during the winter months; predation on fawns was greatest during the summer."

Turning to stalkers, Schaller (1967) concluded with respect to tiger predation that it "is not confined to the young, old, sick, and surplus animals, but that prime ones are also readily taken." According to Hornocker (1970), 75% of the wapiti killed by puma were less than 1.5 years old or more than 9.5 years old; for mule deer the corresponding figure was 62%, a pattern more typical of coursers than stalkers. This raises the point that under certain conditions the method of hunting is less important in determining the class of animals killed than the characteristics of the prey population. Since predators obtain their meals in the easiest possible way, the percentage of young and sick in the kill sample depends largely on the number of such animals in the population. Hornocker conducted his study on a range overstocked with deer where there were no coursing predators to weed out the unfit individuals, which then fell prey to puma. Conversely, in areas where herds are healthy and well within the carrying capacity of their habitat, coursers may have to subsist mainly on prime individuals, as Kruuk (1972) suggested when he wrote: "The hyenas' selection and hunting methods cause them to select the least physically fit from the populations, at least in wildebeest and gazelle; in zebra, the interaction is complicated by the active defense of the family-group by the stallion. This by no means signifies that hyenas eat only very young, old, or diseased animals but if these are available, they will be selected. In the Ngorongoro the latter categories are hardly present at all, and the hyenas have to be content there with what are probably perfectly healthy prey in the prime of life." In such a situation, hyenas and lions are likely to compete.

(3) Direct competition between predators for the same food resource increases as the size of the prey species decreases. Only lions can subdue healthy adult buffalo but five kinds of predators capture adult Thomson's gazelle. Competition is to some extent held in abeyance by a preference for large prey by large predators.

The Impact of Predation on Prey Populations

To measure the impact of predation is a difficult task, for numerous factors influence the results. The number of predators, the killing rate, and the degree of carcass utilization, to mention just three aspects, influence the number of animals harvested, as do such factors as the density of the prey populations, the extent of their movement, and so forth. Table 75 compares the predator density and biomass in five African reserves. Lions contribute the highest biomass in all areas except Ngorongoro Crater where hyenas occupy the top position. Ngorongoro Crater, Nairobi Park, and Kruger Park each supports about 1 kg of predator per 100 kg of prey; in Manyara Park the ratio is 1:174, possibly because much of the biomass consists of buffalo, which are almost invulnerable to a solitary lion. The Serengeti has about 1 kg of predator per 250–300 kg of prey, a dearth of predators readily understood if it is remembered that most lions and leopards are resident in the woodlands, whereas the migratory herds, which constitute 62% of the total biomass, spend the rainy season on the plains; that is, the three most abundant species escape the full impact of lion and leopard predation by migrating. Cheetah, wild dog, and hyena circumvent this problem by being nomadic for at least part of the year.

By observing predators continuously for several days at various seasons, it is possible to obtain an idea about how much food they ingest and how often they kill. I estimated the amount of prey killed and scavenged annually by several predators, and Kruuk (1972) did the same for hyenas. While these figures appear to be precise, it should be remembered that the calculations are based on several unproven assumptions. At best, the figures represent only the right order of magnitude. Hyenas generally consume almost 100% of a carcass but the other predators leave 25 to 35% of it as wastage, some of which is also taken by hyenas. The predators removed roughly 9–10% of the estimated prey biomass of 107.7 million kg, a great proportion of it by lion and hyena.

	Amount of Prey Killed and Scavenged Annually
Lion	4,935,671–5,922,807 kg
Hyena	3,285,000
Leopard	720,000–1,080,000
Cheetah	373,985–450,692
Wild dog	182,500–273,750
	9,497,156–11,012,249

Kruuk and Turner (1967) presented a useful method by which an average killing rate of a predator can be roughly calculated. In brief, the percentage of each species in the kill sample is multiplied by a common factor which is derived by dividing the total weight of prey in the sample into the average weight of prey killed annually by the predator. According to this method, each lion in the Seronera and Masai prides acquired 32 prey animals a year (table 74). Because gazelle are not represented fully in the kill sample, the total should be raised somewhat. Most other prides and nomads subsist on fewer gazelle and more on such species as topi and zebra. For example, if the same method is applied to the kill record from the edge of the woodlands (column 3, table 36), the calculated total of prey animals taken by each lion is 16. Small kills are again not represented fully. Lions, of course, also scavenge, but most obtain relatively little meat in this fashion. If a killing rate of 20–30 animals per year per lion (including cubs) is taken as an arbitrary average, then lions in the ecological unit account for 40,000 to 72,000 animals, a higher estimate than the 24,500 to 36,400 given by Talbot and Talbot (1963).

Kruuk (1970) calculated that hyenas take 23,749 to 32,356 animals annually, or 8 to 10 prey per hyena per year excluding small young. I suspect that these figures are too low, especially when the propensity of hyenas to capture gazelle fawns and wildebeest calves is considered; in fact Kruuk expressed his dissatisfaction with the data when he first calculated a wildebeest calf kill of 679 to 1,116 animals and then noted that in one year it might be as high as 19,000.

A solitary cheetah kills at the rate of 150 animals a year and a mother with cubs at least double that number, for a total of 26,964 to 35,052 animals. However, these figures are based on observing cheetah hunt gazelle, and I am not sure that cheetah which prey, for example, mainly on impala kill at the same rate. I have no way of assessing how many animals are captured by leopard. Those around Seronera kill gazelle for part of the year at an estimated rate of slightly more than one animal a week, but when such preferred prey is unavailable they undoubtedly take many such small animals as dik-dik, hyraces, and guinea fowl. It was my impression that cheetah kill proportionately about twice as much in terms of biomass as leopard. Each wild dog was estimated to catch 55 prey a year, for a total of 9,125 to 10,950 animals. As a whole, the five predators may account for 150,000 to 200,000 animals annually, including small prey for which no adequate population estimates are available.

Many animals are obviously taken by predators, but a critical question that must be answered is to what extent they represent a surplus destined to die anyway and to what extent the population level is actually affected by their removal.

Wildebeest

Thousands of wildebeest calves die each season—Kruuk (1970) calculated from Watson's (1967) data that 59,840 or 45% of the calves died in 1966—but the extent of loss to predators is mitigated in several ways. First, wildebeest have a sharp birthpeak which makes small young available for only a few months; second, traveling en masse, the population reduces the general availability of calves; and third, by calving on the plains, lion and leopard predation on newborn young has been reduced to a negligible amount. Although predators kill many calves, the loss is so great that other mortality factors must also be operating. Watson (1969) asserted that mortality of newborn calves is a result of their being unable to maintain contact with their mother in large and dense herds. He felt that this loss was density-dependent and the most important factor governing the rate of increase in the total population. Sinclair (1970) challenged this idea, pointing out that, to a calf, a crowd of 100,000 or 200,000 wildebeest can make little difference. He thought that abandonment occurred because the calf was weak from malnutrition and disease. In any event, abandoned calves are slated to starve, for no cow will accept them, unless a predator rescues them from this fate. As direct observation indicates, however, predators do not subsist only on such doomed animals: they usually capture young which are with their mother. In spite of heavy calf mortality, an average of 14% of the population consisted of yearlings between 1963 and 1966 (Watson, 1967) and 10% did so in 1969 (Sinclair, 1970).

Talbot and Talbot (1963) estimated that the average annual death rate of adults was 8% between 1959 and 1961, and Watson (1967) calculated that it varied from 4.2% in 1962–63 to 15.7% in 1965–66. Disregarding the few sick individuals taken by wild dogs, only hyena and lion affect adult wildebeest. Hyena remove 5,592 to 9,286 yearlings and adults annually (Kruuk, 1970). If in mid-year the population of 410,000 wildebeest consisted of 80% yearlings and adults, or 328,000 animals, then hyenas took 1.7 to 2.8% of these. The estimate includes many sick and starving individuals as well as those dead from one cause or another. In addition, hyenas prey heavily on males, many of which are not essential to a polygynous society. Thus hyenas harvest a large proportion of surplus animals.

Most lions have wildebeest available for about one-third of the year, during which time at least half of their food consists of this species. Some prides and many nomads prey on wildebeest for more than four months a year, but others live away from migration routes and seldom have any available. Wildebeest may contribute 20% to 25% to the annual diet of a lion, or about 987,034 to 1,480,702 kg. Taking 95 kg as the average weight of a wildebeest, this amounts to 9,139 to 13,710 animals (including calves), or 2.2% to 3.3% out of a population of 410,000. The wildebeest population numbered well over 500,000 animals in 1970. Since the lion population showed no proportional increase, the percentage killed that year was even less than indicated. While lions kill many healthy adults, they also account for a significant number of animals in poor condition and other surplus individuals.

Sinclair (1970) calculated that the following number of yearling and adult wildebeest died from various causes between 1967 and 1969: 1967, 61,400; 1968, 52,200; 1969, 34,700. Hyenas and lions accounted, on the average, for roughly a third to a half of these, many of them doomed anyway. These calculations are crude and based on several assumptions which may not be valid. But they do show that predation by itself has little impact on the population, even when poaching by man is taken into account. Many thousands of individuals die from other causes each year, most probably from a combination of malnutrition and disease. In the final analysis, the fact that the number of wildebeest has increased dramatically in recent years is the most potent argument in support of the statement that predation is not an important limiting factor.

The impact of predation is even less significant when the actual potential of the population is considered. Watson (1967) found that in Ngorongoro Crater 75% of the cows produced a calf in their second year. This population is a vigorous and healthy one. Hyenas, and to a lesser extent lions, prey on wildebeest so heavily that mortality is of the same magnitude as recruitment (Kruuk, 1970). In the Serengeti, on the other hand, only 35% of the cows produced a calf in their second year (Watson, 1967), a figure which, with the increase in the population, dropped to 4% (1 cow in a sample of 27) according to Sinclair (pers. comm.). Increased predation is compensated for by increased fecundity, a characteristic also reported for North American deer (Klein, 1965).

In contrast to the Serengeti, where predation has little effect, and to Ngorongoro Crater, where it is the main limiting factor, lion predation in Nairobi Park may be partly responsible for the decline of wildebeest in recent years. In the past, several thousand wildebeest concentrated in the park each dry season, but in 1961 a drought killed many animals. The survivors continued to decrease steadily as several counts show: 1961, 1,780; 1962, 956; 1963, 691; 1966, 253 (Foster and Kearney, 1967). Wildebeest feature conspicuously in the lion's diet. When they "were abundant (1961) they were captured by lion at a rate which was more than twice that to be expected from the percentage of wildebeest in the population .... As the wildebeest declined in numbers, this discrepancy increased, until by 1966 it is four times that to be expected" (Foster and Kearney, 1967). No such disproportionate selection was shown for zebra and hartebeest. Possibly wildebeest make themselves vulnerable by frequenting certain habitats, a suggestion made by Hirst (1969) to account for the large proportion of wildebeest kills in the Timbavati Reserve, or they may be particularly susceptible to predation as a result of disease. Whatever the reason, Foster and Kearney calculated that the annual decrease of 100 or so wildebeest was due largely to lion predation. However, the decline halted in 1967. With wildebeest numbers low, lions turned increasingly to other prey such as eland (Foster and McLaughlin, 1968).

Zebra

Without information to the contrary, I assume that the population of about 150,000 animals remained stable during my study. Kruuk (1970) calculated that hyenas accounted for 4,750 to 6,271 zebra annually, 475 to 614 of them being foals, or a total of 3.2% to 4.2% of the population. At least a few zebra are available to most lions for two-thirds of the year, and my guess is that about 30% of their food consists of this species: that is, 8,967 to 10,834 zebra weighing an average of 164 kg each, or 5.9–7.2% of the population. It is difficult to determine to what extent these percentages represent an expendable surplus. Two-thirds of the foals disappear within the first few months of life (Klingel, 1969; Skoog, pers. comm.). Since neither lions nor hyenas prey to a marked degree on foals, the high death rate is probably attributable to disease and malnutrition. Many adults that are killed by hyenas are undoubtedly in poor condition, as are some of those that are taken by lions; in addition both predators find and eat some zebra that have died. Predation seems to act as a limiting factor to a greater extent on the zebra population than on the wildebeest, but the impact of disease and possibly malnutrition is also needed as a depressing influence to keep the population from fluctuating markedly.

Thomson's Gazelle

Gazelle have a higher reproductive potential than any other main prey species. Many females have young twice a year and most yearlings probably give birth at least once. Without adequate information on gazelle numbers and population dynamics, the impact of predation is difficult to assess, and my calculations are based on little more than guesses. The figures for the number of fawns and adults taken by the five large predators are based, among other factors, on known killing rates coupled with the availability of gazelle in various parts of the ecological unit. Taking the biomass of prey obtained by leopard and assuming that 10% of it consists of gazelle, of which 35% are fawns, then the kill can be calculated on the basis of an average weight of 12 kg per gazelle. Of course, Seronera leopard subsist on a much higher percentage of gazelle, but those in other parts may have few or none available. Only an estimated 2.5% of the lion's prey biomass consists of gazelle. One-third of the total kill of wild dogs probably consists of gazelle, 45% of it fawns, and for cheetah the respective figures are 60% and 66%. Data for hyenas were taken from Kruuk (1970); his figures include both Grant's and Thomson's gazelle. The total number of animals taken by each predator was roughly estimated to be as follows:

Predator	*No. fawns*	*No. yearlings and adults*
Lion	3,599–4,319	6,683–8,020
Leopard	2,100–3,150	3,900–5,850
Cheetah	10,677–13,880	5,501–7,251
Hyena	1,330–4,008	3,871–11,701
Wild dog	1,369–1,643	1,673–2,007
	19,075–27,000	21,628–34,829

More fawns, of course, are killed than these figures suggest. Aside from the fact that some estimates may be too low, jackals and other small predators catch many young. All large predators, with the possible exception of hyena, kill seemingly healthy adult gazelle and this, together with the total number taken, suggests that predation may be an important limiting factor on the population.

Buffalo

An estimated 15% of the prey biomass taken by lions consists of buffalo. This amounts annually to 1,762 to 2,115 buffalo weighing an average of 420 kg each. According to Sinclair (pers. comm.), 6,300 yearling and adult buffalo die each year, indicating that lions account for only about one-third of this total, especially if the fact that lions occasionally scavenge dead buffalo and often kill those in poor condition is taken into account. Disease possibly is a major cause of death in calves and malnutrition in adults (Sinclair, 1970), and these two factors affect the Serengeti population more than predation. The population has been increasing steadily in recent years, a further proof that predation is inconsequential in determining the level maintained by these buffalo.

The buffalo population in Manyara Park has remained stable in recent years. The local lions subsist largely on the buffalo, and it is tempting to correlate these two facts. However, the lions primarily kill bulls whose loss can have little effect on the population. Judging by the lean condition of many animals, the population has been held in check not by predation but by the carrying capacity of the restricted habitat, which in recent years has decreased markedly due to a rise in the lake level.

Other

During the rainy season, topi, hartebeest, impala, warthog, waterbuck, and several other resident and semimigratory woodlands species constitute the principal prey of lion and leopard, and other predators use them as well. My record of predator kills from the woodlands is inadequate for purposes of assessing the impact of predation on these species, but several points can be made. The woodlands prey, excluding those already discussed, comprise 156,500 animals, 16% of the total prey population and 13% of the biomass. Giraffe are only preyed upon by lions and then but sparingly. Yet this small fraction provides an estimated 27.5% to 32.5%, or 1,357,310 to 1,924,912 kg of the lion's food annually. Leopard and man, to mention just two other important predators, use it as well. Although detailed studies in the woodlands are needed to confirm these figures, it seems that predation is perhaps the most important limiting factor operating on these species, except on giraffe, and it is not coincidental that disease and malnutrition are less prevalent, or at least less conspicuous, among resident than migratory prey.

Predation may be particularly heavy on species with a restricted distribution whose habitat coincides with that of a predator. Around Seronera, leopard and to a lesser extent lion probably eliminate enough reedbuck along the rivers to keep the population depressed. Lion predation on warthog at Seronera may have a similar effect, in this instance because a disproportionately large number is killed when the pigs approach waterholes in the dry season. Poaching around the western periphery of the park has reduced resident populations to a low level.

To sum up, wildebeest and buffalo have been increasing steadily, probably as a result of improved range conditions, after successive seasons of good rain distribution, coupled with a reduced incidence of disease. The Thomson's gazelle, zebra, and eland populations have probably remained stable or at least have not fluctuated markedly, judging by the censuses of Talbot and Stewart (1964) and Watson (1967). In the absence of information to the contrary, I assume that the other species too have either been stable or fluctuated only moderately in number. Wildebeest and buffalo escape the full impact of predation, the former by migrating and the latter by being large. The remaining species are both available and vulnerable to predation throughout the year, and the evidence, though limited, indicates that this is a factor in keeping the populations depressed. I am not certain that predation actually regulates any of these populations in the sense of removing the total annual increment, but for some species this may be the case. The giraffe is an exception to the pattern in that predators seldom kill it, yet its population remains fairly low.

Predation, of course, works in conjunction with other factors, such as disease and malnutrition, but on the basis of various studies it has almost become axiomatic that as the intensity of predation increases the prevalence of disease decreases. It might then be asserted that if animals are not killed by predators, they would merely die from illness, starvation, and other factors and that, in the final analysis, predation has no effect on ungulate populations because only surplus individuals are taken. This argument has little relevance to the role of predation per se: predators can be a major force in keeping a population depressed, keeping it at a level below that which it would attain if predation did not exist. Under normal circumstances it seems unlikely that an ungulate species in the Serengeti can draw enough predation pressure to depress it below its threshold of security. But predation can hold species below their carrying capacity, below a level at which disease, starvation, and other regulatory forces associated with poor nutrition can take effect. Wildebeest and buffalo may have reached the carrying capacity of their range, and drastic forces may ultimately reduce their numbers. The new level might be considerably below that which they would have maintained if predation had been effective in limiting them, as has been noted repeatedly in studies of North American cervids (Russo, 1964). Whatever the ultimate effect, the behavior of these two species is typical of many ungulates; they tend to increase up to and usually beyond the carrying capacity of their range unless checked in some way. Predators constitute an important check. The most important influence of predation is this dampening of the tendency of populations to increase beyond the carrying capacity of their range, an effect that prevents severe oscillations. The same conclusion was reached by Hornocker (1970) with respect to puma predation. "The fact that predation operates as a steadily functioning force throughout the seasons and year after year, in spite of continually changing physical and biotic conditions of the environment, gives it a great advantage, in comparison with regulators that operate intermittently or only under special conditions, in harmoniously regulating animal populations with one another and with the rest of their environment" (Craighead and Craighead, 1956).

While predation may be a major factor limiting the size of populations, the primary factor which ultimately exercises control is the habitat. The habitat may affect animals by providing cover or the lack of it, whichever the species requires to reduce its vulnerability. And in the final control of numbers the food supply has always been directly or indirectly implicated (Leopold et al., 1947; Klein and Olson, 1960). In this context, Pimlott (1967) made an apt point when he wrote: "In considering the population dynamics of some big-game species . . . the question arises, as to why intrinsic mechanisms of population control have not evolved to prevent them from increasing beyond the sustained level of their food supply. It seems reasonable to postulate that it may be because they have had very efficient predators, and the forces of selection have kept them busy evolving ways and means not of limiting their own numbers but of keeping abreast of mortality factors."

Most prey species within the Serengeti Park maintain a tenuous balance with their environment. There is the habitat which determines the maximum level a population can reach, and the reproductive potential, delicately adjusted to population density, which influences the rate of increase. The food supply is affected mainly by the amount and distribution of rainfall, whose vagaries may cause drastic declines. Predation, disease, accidents, and other factors also act as controls, the impact of each one varying from year to year but all operating continuously to impose a population level which normally lies between a species' threshold of security and its carrying capacity. Conditions seldom remain constant; an improvement in the habitat, for example, may favor an increase in a population. Because of various self-limiting and other mechanisms, the predators are unable to adjust their numbers quickly to such a change. As a result, oscillations of varying magnitude in prey numbers may occur.

CONCLUSION

The Serengeti is a boundless region with horizons so wide that one can see clouds between the legs of an ostrich. It is a Pleistocene vision throbbing with the life of over half a million wildebeest and zebra, a stern yet lovely wilderness where man can renew his ancient ties with the predators that were once his competitors and the prey that gave him sustenance. There, too, in the silence and primeval surroundings, he can find spiritual solace from the complications of civilization. Tanzania possesses in the Serengeti one of the great national parks of the world. There are many reasons to preserve this natural heritage in perpetuity, not the least being the income the country can derive from tourism and the moral obligation it has toward future generations who will want to see and feel this earth as it once was. The aim of the Tanzania National Parks is "to conserve the present variety of the different habitats within the Park and of the different species of fauna using these habitats and to maintain them as nearly as possible in their present distribution and proportions" (Owen, n.d.). Conservation alone may not be sufficient to accomplish this objective, for the park is not a self-contained unit. Wandering freely back and forth across its borders, the animals are coming into increasing conflict with agriculturalists and pastoralists, so much so that the huge herds may some day have to be managed through judicious cropping—not only to reduce their interference outside of the park, but also to keep them within the carrying capacity of the range within it. Any scheme that tampers with the hoofed animals automatically involves consideration of the predators. Early in this report I asked "What effect does predation have on prey populations?" Equally if not more to the point is the question "What effect do prey numbers have on population levels of predators?" The predators are in many ways the park's most valuable resource for they are the ones which add excitement and authenticity to the natural scene, they are the symbol of this wilderness; tourists are drawn not to the impala and zebra but to the lion and leopard. This being so, hoofed animals should be managed in such a way that predator populations are maintained at the highest possible level. The previous pages contain many facts, some empirical conclusions, and occasional speculations about the relations between predators and prey, and some of these will have relevance in the event that active management of the park's resources becomes necessary.

The resident lion population in the woodlands of the park seems to have stabilized at a level commensurate with its food supply, except in peripheral areas where poaching is heavy. Any attempt to raise the number of animals by, for example, feeding starving cubs is a futile gesture. With pride size in part regulated by self-limiting mechanisms, excess animals become nomadic and often leave the park. The level of the resident population is determined by the amount of prey available during the leanest time of the year, the rainy season, when the migratory herds are on the plains. At that time lions depend more on topi, buffalo, hartebeest, and other resident species for survival than on wildebeest and zebra. The existence of leopards also is largely based on woodlands prey, particularly on impala and other species of that size. If some day a percentage of prey must be harvested by man for his own use, only wildebeest and those species whose influence on the population levels of lion and leopard has the least effect should be taken. Hyena are dependent on the migratory species, but it is unlikely that a moderate reduction in wildebeest would affect the population size of this predator. For unknown reasons both cheetah and wild dog are rare, and until further research clarifies the dynamics of their existence, little can be done to encourage their increase other than to protect them vigorously and insure that their principal prey, the Thomson's gazelle, continues to be an abundant source of food for them.

The Serengeti predators are an integral and essential part of the ecological community. They help maintain an equilibrium in the prey populations within the limits imposed by the environment, they prevent severe fluctuations in the number of animals and condition of the habitat. To this task they bring a discernment that can not be matched by man: predators are the best wildlife managers. It is not coincidental that species which have somehow escaped the effects of predation, such as elephant, pose the greatest management problems in reserves. The predators weed out the sick and old, they keep herds healthy and alert. The beauty of antelope, their fleetness and grace, their vital tension, are the evolutionary products of a constant predator pressure that has eliminated the stolid and slow. Man, one hopes, has gained enough wisdom from his past mistakes to realize that, to survive in all their vigor and abundance, the prey populations need the lion and other predators. Ecological and aesthetic considerations aside, predators should be allowed to survive in national parks without justification, solely for their own sake. Only by so doing can man show his good intentions and atone in a small way for the avarice and prejudice with which he continues to exterminate predators throughout the world.

APPENDIXES

Appendix A

Common and scientific names of wild mammals, birds, and reptiles mentioned in the text.

*Mammals**

Order Marsupialia	
Flying opposum	Petaurus breviceps
Order Primata	
Howler monkey	Alouatta palliata
Olive baboon	*Papio anubis
Vervet monkey	*Cercopithecus aethiops
Gray langur	Presbytis entellus
Chimpanzee	Pan troglodytes
Order Pholidota	
Pangolin	*Manis temmincki
Order Lagomorpha	
Hare	*Lepus capensis
Order Rodentia	
Springhare	*Pedetes surdaster
Cane rat	*Thryonomys sp.
Porcupine	*Hystrix cristata
Order Carnivora	
Wolf	Canis lupus
Coyote	Canis latrans
Wild dog	*Lycaon pictus
Black-backed jackal	*Canis mesomelas
Asiatic jackal	*Canis aureus
Side-striped jackal	*Canis adustus
Bat-eared fox	*Otocyon megalotis
Gray fox	Urucyon cinereoargenteus
Arctic fox	Alopex lagopus
Raccoon	Procyon lotor
Coati	Nasua narica
Short-tailed weasel	Mustela erminea
Striped skunk	Mephitis mephitis
Ratel	*Mellivora capensis
Civet	*Viverra civetta

* An asterisk indicates a mammal occurring in the Serengeti Park.

Genet	*Genetta genetta and *G. tigrina
White-tailed mongoose	*Ichneumia albicauda
Banded mongoose	*Mungos mungo
Spotted hyena	*Crocuta crocuta
Striped hyena	*Hyaena hyaena
Lynx	Felis lynx
Puma (mountain lion)	Felis concolor
Caracal	*Felis caracal
Serval	*Felis serval
Wild cat	*Felis libyca
Jaguar	Panthera onca
Tiger	Panthera tigris
Leopard	*Panthera pardus
Lion	*Panthera leo
Cheetah	*Acinonyx jubatus
Order Tubulidentata	
Aardvark	*Orycteropus afer
Order Proboscidea	
Elephant	*Loxodonta africana
Order Hyracoidea	
Hyrax	*Procavia habessinica and *Heterohyrax brucei
Order Perissodactyla	
Zebra	*Equus burchelli
Black rhinoceros	*Diceros bicornis
White rhinoceros	Ceratotherium simum
Order Artiodactyla	
Bushpig	*Potamochoerus porcus
Warthog	*Phacochoerus aethiopicus
Hippopotamus	*Hippopotamus amphibius
Giraffe	*Giraffa camelopardalis
Caribou	Rangifer arcticus
Wapiti (elk)	Cervus canadensis
Barasingha	Cervus duvauceli
Mule deer	Odocoileus hemionus
Whitetail deer	Odocoileus virginianus
Axis deer	Axis axis
Moose	Alces americana
Kudu	Tragelaphus strepsiceros
Bushbuck	*Tragelaphus scriptus
Nyala	Tragelaphus angasi
Eland	*Taurotragus oryx
Buffalo	*Syncerus caffer
Gaur	Bos gaurus
Bush duiker	*Sylvicapra grimmia
Waterbuck	*Kobus defassa
Lechwe	Kobus leche
Puku	Kobus vardoni
Bohor reedbuck	*Redunca redunca
Mountain reedbuck	*Redunca fulvorufula
Impala	*Aepyceros melampus
Roan	*Hippotragus equinus

Sable	Hippotragus niger
Oryx	*Oryx gazella
Topi	*Damaliscus korrigum
Tsessebe	Damaliscus lunatus
Hartebeest	*Alcelaphus buselaphus and A. lichtensteini
Wildebeest	*Connochaetes taurinus
Klipspringer	*Oreotragus oreotragus
Dikdik	*Madoqua kirki
Steinbok	*Raphicerus campestris
Grysbok	Raphicerus melanotis
Oribi	*Ourebia ourebi
Blackbuck	Antilope cervicapra
Thomson's gazelle	*Gazella thomsoni
Grant's gazelle	*Gazella granti
Dall sheep	Ovis dalli

Birds

Ostrich	Struthio camelus
European stork	Ciconia ciconia
Saddle-bill stork	Ephippiorhynchus senegalensis
Marabou stork	Leptoptilos crumeniferus
Egyptian goose	Alopechen aegyptiacus
Secretary bird	Sagittarius serpentarius
Rüppell's griffon	Gyps rüppelli
White-headed vulture	Trigonoceps occipitalis
Lappet-faced vulture	Torgos tracheliotus
White-backed vulture	Pseudogyps africanus
Hooded vulture	Necrosyrtes monachus
Egyptian vulture	Neophron percnopterus
Martial eagle	Polomaetus bellicosus
Tawny eagle	Aquila rapax
Coqui francolin	Francolinus coqui
Grey-breasted spurfowl	Pternistis rufopictus
Helmeted guineafowl	Numida mitrata
Yellow-throated sandgrouse	Eremialector gutturalis

Reptiles

Monitor lizard	Varanus niloticus
Crocodile	Crocodilus niloticus
Leopard tortoise	Geochelone pardalis
Python	Python sebae
Spitting cobra	Naja nigricollis

Appendix B
Tables 1 to 79

Table 1
Monthly rainfall (mm) and number of days with rain (in parentheses) at Seronera, 1963–69

	1963	1964	1965	1966	1967	1968	1969
January	85.1	97.1	23.6	85.0	21.6 (7)	31.0 (7)	156.3 (11)
February	99.3	104.1	166.1	112.5	37.1 (6)	140.0 (16)	133.3 (9)
March	75.7	104.1	74.4	76.2	112.0 (11)	122.4 (17)	47.5 (6)
April	252.2	4.5	226.6	73.4	211.9 (17)	192.7 (12)	44.4 (7)
May	140.5	38.1	82.6	54.3	56.0 (12)	97.1 (7)	51.6 (10)
June	14.2	51.8	0	4.3	34.8 (5)	69.2 (6)	12.7 (5)
July	0	40.4	0	0	22.3 (5)	.3 (1)	.3 (1)
August	1.5	20.3	30.5	40.9	22.3 (1)	17.6 (3)	2.8 (2)
September	1.5	32.8	11.4	32.5	27.1 (4)	28.9 (6)	39.1 (2)
October	8.1	19.3	96.8	12.2	34.8 (6)	44.3 (6)	66.5 (5)
November	215.4	41.9	24.1	31.5	131.8 (11)	53.8 (7)	74.3 (9)
December	137.9	65.8	85.1	98.0	190.2 (16)	149.0 (11)	108.6 (11)
Total	1,031.4	620.2	821.2	620.8	901.9 (99)	946.3 (99)	737.4 (80)

TABLE 2
Size and composition of some Serengeti lion prides

No.[a]	Name of pride	Date	Adult male	Sub-adult male	Adult female	Sub-adult female	Large cub	Small cub	Total
I	Masai	Oct. '66	3	0	6	1	0	10	20
II	Seronera	Sept. '66	2	0	6	5	0	7	20
III	Loliondo	July '67	2	0	11	2	0	22	37
IV	Nyaraswiga	July '67	2	0	2	0	0	0	4
V	Plains	Oct. '66	2	0	7	0	0	1	10
VI	Kamarishe	Dec. '67	2	0	6	0	0	7	15
VII	Magadi	Oct. '66	3	0	9	0	4	3	19
VIII	Mbalageti	Oct. '68	2	0	8	0	7	0	17
IX	Mukoma	Jan. '67	2	0	4	0	0	8	14
X	Nyaraboro	Nov. '68	4	1	8	1	11	1	26
XI	Banagi	Oct. '67	2	0	4	0	4	0	10
XII	Nanyuki	Sept. '66	2	0	4	0	0	4	10
XIII	East Moru	Jan. '67	2 (XIII and XIV together)	0	3	0	1	0	
XIV	Simba	Apr. '68		0	4	0	0	0	
Total			30	1	82	9	27	63	

[a] See figure 5 for location of prides.

TABLE 3
Degree of association between each of the Seronera pride lionesses in 1967 (right side of table) and 1968 (left side)

Female	A	B	C	D	E[a]	F	G	H	I	J	K	
A	—	.58	.19	.31	.57	.51	.15	.18	.38	0	.02	
B	.19	—	.20	.37	.75	.76	.16	.23	.14	.03	0	
C	.39	.15	—	.38	.15	.18	.61	.58	.31	.02	.05	
D	.21	.33	.38	—	.38	.37	.35	.34	.24	.05	.06	
E	—	—	—	—	—	.76	.16	.20	.13	.06	.03	
F[b]	.04	.07	0	0	—	—	.19	.19	.14	.03	.03	1967
G	.37	.19	.48	.43	—	.06	—	.70	.26	.02	.03	
H	.64	.25	.49	.31	—	.13	.33	—	.27	.04	.05	
I	.61	.41	.45	.33	—	.05	.45	.30	—	.02	.04	
J	.04	0	0	0	—	0	0	0	0	—	.42	
K	0	0	0	0	—	0	0	0	0	.38	—	

1968

[a] Died October, 1967.
[b] Died November, 1968.

TABLE 4
Degree of association between each of the Masai pride lionesses in 1968

Female	L	M	N	O	P	Q	R
L	—	.45	.34	.36	.40	.89	.49
M		—	.43	.45	.45	.51	.58
N			—	.64	.54	.33	.56
O				—	.61	.31	.63
P					—	.41	.68
Q						—	.54
R							—

TABLE 5
Ages of females in 4 prides, July 1967

Age Class	Pride				Total
	Seronera	Masai	Loliondo	Nyaraswiga	
Subadult 2–4 yrs.	0	0	2	0	2
Young adult 4–6 yrs.	5	1	5	1	12
Prime	1	3	4	1	9
Past prime	3	2	1	0	6
Old	2	1	1	0	4
Total	11	7	13	2	33

TABLE 6
Sex and age composition of 362 nomadic lion pairs on the plains

Composition	%
2 adult males	28
1 adult male, 1 subadult male	4
2 subadult males	17
1 adult male, 1 adult female	22
1 subadult male, 1 adult female	1
1 subadult male, 1 subadult female	6
2 adult females	10
2 subadult females	4
1 adult female, 1 cub	3
Others	5

Table 7

Relative frequency of occurrence of several social and body care patterns, based on 372 lion-hours of observation at a time of day when such activities were most prevalent

	Head-rubbing	Social licking	Self-licking	Scratching with hind paw
No. obs.	568	181	271	42
No. per lion-hour	1.8	.5	.7	.1

Note: 1 lion observed for 1 hour = 1 lion-hour.

Table 8

Frequency of head-rubbing by members of the Masai pride, November, 1968–January, 1969, based on a total of 780 interactions and 690 lion-hours of observation

Individual	No. interactions per lion-hour of observation	
	Initiates head-rubbing	Has head rubbed
3 adult males	.03 av.	.5 av.
Subadult male	.5	.4
Female L	1.6	2.6
M	1.1	2.2
N	.4	.6
O	1.1	2.2
P	.8	1.5
Q	1.3	2.8
R	.5	.7
6 subadult females	1.1 av.	.9 av.
2 cubs (10–12 mo.)	2.1 av.	.6 av.
2 cubs (3½–5½ mo.)	1.1 av.	.3 av.

TABLE 9

Number of times two litters of cubs initiated head-rubbing with 3 lionesses of the Masai pride

	Female		
	L	M	Q
2 cubs (10–12 mo.)	49	19	61[a]
2 cubs (3½–5½ mo.)	14	32[a]	16

[a] Mother of the cubs.

TABLE 10

Social licking by lionesses and cubs, showing parts of body licked, based on 703 observations

Part licked	Female licks female %	Female licks cub %	Cub licks female %	Cub licks cub %
Head and upper neck	76	42	67	82
Shoulder and chest	13	5	10	0
Back and sides	7	30	8	12
Abdomen	3	19	10	3
Others	1	4	5	3
No. observations	266	281	94	62

TABLE 11

Frequency with which lions licked various parts of their body, based on 895 observations

Part licked	%
Head	0
Shoulder, neck, and chest	9
Back and sides	4
Abdomen	4
Groin, rump, and thighs	27
Forepaw	44
Hindpaw	6
Wipe face with paw	6

Table 12

Frequency of social licking by members of the Masai pride, November, 1968–January, 1969, based on 314 interactions and 254 lion-hours of observation.

Individual	No. interactions per lion-hour of obs.	
	Initiates licking	Is licked
3 adult males	.03 av.	0
Subadult male	0	.1
Female L	1.1	1.1
M	1.0	.3
N	.2	.3
O	.1	.2
P	.4	.2
Q	.7	.6
R	0	.1
6 Subadult females	.2 av.	.3 av.
2 Cubs (10–12 mo.)	.5 av.	.5 av.
2 Cubs (3½–5½ mo.)	.03 av.	.4 av.

Table 13

Average amount of time spent by lions in various activities, based on 24-hour time samples

Lions	Months in which sampled	No. days sampled	Av. no. hours per day (hrs/min)		
			Walking	Feeding	Resting
Nomad male No. 159	VI, VII	16	2/25	/50	20/45
Nomad male No. 134	I, XII	10	3/10	/40	20/10
Other nomads	I, V, VI, XI	6	2/5	/50	21/05
Masai pride	II, III, V, IX, X, XI, XII	9	2/0	/40	21/20
Masai pride males only	IX, X, XII	5	2/0	1/30	20/30
Seronera pride	III, IV	8	2/5	/30	21/25
Masai and Seronera pride (dry season)	VI, VII, VIII	8	2/20	/45	20/55

TABLE 14
Activity of nomadic male No. 134 during 9 consecutive days of observation around the Gol kopjes

Date	Distance traveled (to nearest $\frac{1}{2}$ km)	Feeds on:	Time spent eating/min	Food obtained from:
Dec. 9–10[a]	7.0±	Wildebeest	65[b]	Hyenas
10–11	17.5	Wildebeest	110	Hyenas
11–12	12.0	Wildebeest	30	Hyenas
12–13	21.0	Wildebeest	15	Lion
13–14	10.0	Wildebeest	85	Lion
14–15	5.0	—		
15–16	5.0	Zebra	125	Lion
16–17	11.0	Zebra	30±	Lion
17–18	20.0	—		

[a] Lost contact at 0130 hours.
[b] Feeding interrupted when radio was placed around his neck.

TABLE 15

Activity of nomadic male No. 159 during 21 consecutive days of observation around Naabi Hill

Date	Distance traveled (to nearest ½ km)	Feeds on:	Time spent eating/ min	Food obtained from:	Drinks	Comments
June						
27–28	8.0	Thomson's gaz.	35	Leopard		
28–29	8.0					
29–30	21.5				×	
30–July 1	16.0				×	
1–2	19.0±					Lost contact for the night at 0400 hours
2–3	8.0	Zebra	60	Hyena		
3–4	5.0	Grant's gaz.	150±	Disease death?		Feeding periods estimated
		Wildebeest	60±	Lion		
4–5	4.0					
5–6	14.5±					Lost contact 1100–0020 hours
6–7	11.0±				×	
7–8	4.0					
8–9	4.0	Zebra	135	Hyena		
9–10	3.0	Wildebeest	195	Lion		
10–11	2.5	Wildebeest	110	Lion		
11–12	0					
12–13	17.5				×	
13–14	16.0				×	
14–15	1.0	Eland	75	Lion		
15–16	9.5±				×?	Lost contact for night at 0115 hours
16–17	9.5±					Lost contact for night at 0030 hours
17–18	17.5±				×	Lost contact for night at 0530 hours

NOTE: See figure 12 for social contacts.

Table 16
Tree-climbing by Serengeti lions

Probable reason for climbing	No. obs.	% male	% female	% cubs
Play	66	0	38	62
Resting	28	4	28	68
Look around	20	0	65	35
Pursuit of leopard	5	0	100	0
Escape buffalo	2	0	100	0
Escape car	5	40	40	20
Escape flies	1	0	100	0

Table 17
Head-rubbing interactions among Masai pride lionesses

		Animal rubbed						
		L	M	N	O	P	Q	R
Animal rubbing	L	—	32	0	10	5	29	2
	M	15	—	2	10	7	13	0
	N	0	2	—	3	3	1	1
	O	5	3	1	—	3	3	2
	P	3	2	1	5	—	6	2
	Q	22	14	2	2	2	—	1
	R	2	3	2	2	0	5	—

Table 18
Social licking interactions among the Masai pride lionesses

		Animal licked						
		L	M	N	O	P	Q	R
Animal licking	L	—	5	1	4	5	7	2
	M	10	—	4	3	3	8	1
	N	0	3	—	6	7	0	3
	O	4	3	8	—	2	1	1
	P	6	1	6	4	—	4	0
	Q	10	6	1	3	6	—	0
	R	0	1	3	2	0	0	—

TABLE 19
Frequency of occurrence of some behavior patterns in mating lions

Behavior		%
Copulation initiated by:	Male	39
(n = 300)	Female	61
Vocalizations emitted by:	Both male and female	99
(n = 273)		
Behavior of male toward nape of female:	Ignores nape	15
(n = 273)	Licks nape only	2
	Gapes over nape	12
	Touches nape with teeth	14
	Bites nape	57
Copulation terminated when:	Male dismounts	50
(n = 276)	Female jerks head at male	46
	Female jerks head and slaps	4
Behavior of female after copulation	Sits or lies	16
(n = 269)	Walks off	8
	Rolls on side	36
	Rolls on back	40

TABLE 20
Extent of social isolation of courting lion couples

	Courting male and female alone	One other male near courting couple	Other male(s) and female(s) near courting couple	Other female(s) near courting couple
Copulation observed (n = 103 days)	50%	20%	24%	6%
Copulation not observed (n = 170 days)	46%	25%	25%	4%

TABLE 21

The means by which cubs, 3 to 12 months old, obtained meat from their mother (Seronera and Masai prides)

Prey	Total No. obs.	Cubs with female when kill made	Female fetches cubs to kill	Female takes meat to cubs	Cubs not at kill, nor receive meat
Wildebeest, zebra, topi	44	57%	22.5%	2.5%	18%
Thomson's gazelle, reedbuck	54	48%	6.0%	13.0%	33%

TABLE 22

Frequency of various types of social play among females and cubs of the Seronera pride, November, 1968 to June, 1969.

Play group	Age of cubs	No. obs.	Frequency of play (in %)			
			Chasing	Wrestling	Pawing	Stalking
Cub–cub	3–5½ mo.	242	16	59	17	8
Cub–cub	5–10½ mo.	622	33	20	22	25
Female–cub	5–10½ mo.	160	24	3	54	19
Female–female	—	76	57	0	6	37

TABLE 23

Number of nomads on the plains based on the Lincoln index

Month	No. lions according to index	S.E.	% tagged lions seen
April–May 1967	267	±51	23
March, 1968	177	±60	16
April, 1968	164	±38	26
May, 1968	194	±42	22
Feb., 1969	123	±50	12

TABLE 24
Ratios of nomadic males and cubs to 100 females on the plains

Date	Sample size	Sample			No. different lions seen	Count		
		Males	100 females	cubs		Males	100 females	cubs
Jan. 1968	241	106	100	20	119	75	100	33
Feb. 1968	100	200	100	44	83	160	100	73
March 1–15, 1968	277	82	100	17	106	77	100	18
April 1968	198	107	100	27	99	93	100	27
Feb. 1969	172	56	100	64	119	65	100	51
Average		96	100	31		88	100	38
June, 1966–May, 1969 (from table 26)	3,111	109	100	42				

TABLE 25
Ratios of nomads and resident males and cubs to 100 females in the woodlands

Area	Sample size	Sample			No. different lions seen	Count		
		Males	100 females	cubs		Males	100 females	cubs
Woodlands–plains border[a]	2,287	82	100	61	99	83	100	91
Musabi–Ndoha	653	68	100	74	173	55	100	66

[a] Excludes main study prides.

TABLE 26
Composition of the lion population in several areas of the park, based on all animals tallied June, 1966–May, 1969

Area	Resident or Nomad	Sample size	% Adult male	% Sub-adult male	% Adult female	% Sub-adult female	% Large cub	% Small cub
Plains	Nomad	3,111	26	17	31	9	4	13
Woodlands–plains border	Both	2,287	20	13	30	12	14	11
Musabi-Ndoha area	Both	653	12	16	28	13	18	13
Woodlands outside study area	Both	101	16	10	37	11	14	12

TABLE 27
Sex and litter size of lion cubs

Location	Age of cubs/mo.	No. litters	No. cubs	Male	Female	Average litter size
Masai and Seronera prides	0–6	34	79[a]	38	40	2.3
Woodlands	0–6	34	82	40	42	2.4
Woodlands	6–12	29	66	31	35	2.3
Woodlands	12–18	32	61	24	37	1.9
Plains	0–6	16	31	17	14	1.9
Plains	6–12	15	26	12	14	1.7

[a] One was not sexed.

TABLE 28
Number of cubs born to and raised by each lioness in the Seronera and Masai prides, 1966–69

Female No.	No. litters	No. cubs born	No. cubs raised to age of 2 yrs.	No. cubs 1–2 yrs. old, Dec. 1969	No. cubs less than 1 yr. old, Dec. 1969
A	4	9	1	4	
B	3	5	1		
C	2	6	0		
D	3	9	0		
E	0	0	0		
F	0	0	0		
G	2	4	0	1	
H	2	5	0		
I	2	4	0	2	
J	2	5	0		2
K	0	0	0		
L	2	8	4		
M	1	2	0		
N	2	3	1		2
O	3	5	1		2
P	3	8	1		3
Q	2	3	1		
R	1	3	0		
Total	34	79	10	7	9

TABLE 29

Cross-correlations of variables related to lion conceptions and births

	Rain	Litters born	Sex contacts seen	No. zebra days	No. wildebeest days	Relative zebra abundance	Relative wildebeest abundance
Rain							
Litters born	−.167 (.0687)						
Sex contacts seen	.050 (.3264)	.156 (.0808)					
Number zebra days	.0425 (.3520)	−.0359 (.3745)	.217 (.0262)				
Number wildebeest days	.129 (.1230)	.116 (.1401)	.240 (.0162)	.581 (<.00003)			
Relative zebra abundance	.132 (.1210)	.012 (.4443)	.389* (<.00003)	.928 (<.00003)	.595* (<.00003)		
Relative wildebeest abundance	.119 (.2266)	.137 (.1038)	.228* (.0207)	.578 (<.00003)	.932 (<.00003)	.561 (.0007)	

The upper figure in each cell is the Kendall rank correlation coefficient, tau. The figure in parentheses represents the one-tailed probability of a chance occurrence of a tau-value as large as the value observed. Thus, in the cell treating rain versus litters born, a negative tau-value of .167 could be expected from the sample 6.87 percent of the time.

The asterisks mark correlations which are significant at the .05 level.

TABLE 30

Sex and approximate age of 29 subadult and adult lions found dead

Class	Av. diam. of wear surface of upper pm2 (in mm)	Male	Female	Total
Subadult	0–1	5	1	6
Young adult	1–2	3	1	4
Prime	2–4	7	5	12
Past prime	4–6	3	1	4
Old	6–8	1	2	3

TABLE 31
Causes of death of cubs in the Seronera and Masai prides

Cause	Age/months				Total
	0–6	6–12	12–18	18–24	
Killed by lion	4	5	2		11
Killed by leopard	1				1
Killed by hyena	1				1
Starvation	1	10	4		15
Unknown	17	4	2	2	25

TABLE 32
Number and biomass (in kg) of prey in the Serengeti Ecological Unit

	Species	No. animals in ecological unit	Av. weight per animal[c]	Total biomass in million kg
Migratory	Wildebeest	400,000	108	43.200
	Zebra	132,500	164	21.730
	Thomson's gaz.	165,000	12	1.980
	Eland	3,500	225	.788
	Total	701,000		67.698
Woodlands resident	Wildebeest	10,000	128	1.280
	Zebra	17,500	164	2.870
	Thomson's gaz.	10,000	12	.120
	Eland	3,500	225	.788
	Topi	27,000	82	2.214
	Hartebeest	18,000	95	1.710
	Grant's gaz.	4,000	32	.218
	Impala	65,000	32	2.080
	Buffalo	50,000	420	21.000
	Giraffe	8,000	716	5.728
	Warthog	15,000	40	.600
	Waterbuck	3,000	131	.393
	Ostrich	3,000	83	.249
	Other[a]	10,000	30	.300
	Total	244,000		39.550
Plains resident[b]	Grant's gaz.	6,000	32	.192
	Thomson's gaz.	5,000	12	.060
	Ostrich	2,000	83	.166
	Total	13,000		.418
Grand Total		958,000		107.666

[a] Roan, oryx, reedbuck, bushbuck, etc.
[b] Excluding a few topi, warthog, oryx, etc.
[c] Weights from Sachs (1967) and Wilson (1968); based on three-quarters of weight of average female; resident wildebeest are heavier than migratory ones.

TABLE 33
Number and biomass (in kg) of lion prey in 4 African reserves

Species	Manyara Park (91 km²) this study		Ngorongoro (260 km²) Kruuk (1972)		Nairobi Park (115 km²) Foster & Kearney (1967)		Kruger Park (19,084 km²) Pienaar (1966)	
	No.	Biomass	No.	Biomass	No.	Biomass	No.	Biomass
Buffalo	1,500	630,000	60	25,200			10,614	4,457,880
Giraffe	60	42,960			76	54,416	2,975	2,130,100
Zebra	50	8,200	4,500	738,000	488	80,032	14,400	2,361,600
Wildebeest			13,528	1,731,584	253	32,384	13,035	1,668,480
Eland			400	90,000	58	13,050	540	121,500
Impala	700	22,400			633	20,256	204,050[a]	6,529,600
Thomson's gaz.			3,500	42,000	344	4,128		
Grant's gaz.			1,500	48,000	501	16,032		
Waterbuck	15	1,965	60	7,860	90	11,790	4,085	535,135
Reedbuck	40	1,480	60	2,220	12	444	1,210	44,770
Hartebeest			100	9,500	1,095	104,025		
Bushbuck	25	550			20	440		
Warthog	25	900			154	6,160		
Kudu							6,875	1,182,500
Sable							1,236	196,524
Roan							351	79,677
Tsessebe							715	65,000
Nyala							980	53,454
Ostrich			no data		94	7,802	no data	
Others							17,950[b]	312,850
No./km²	26		91		33		14	
Biomass/km²		7,785		10,363		3,052		1,034

[a] R. Estes told me that the estimates for impala may have been too high. Another census in 1968 gave a total of 97,400 animals. I do not know which of the two estimates is the most accurate one. The figures for the other species were similar for the two years except that there were about 5,000 more buffalo in 1968 than in 1966.
[b] Warthog, bushpig, bushbuck, and other small antelopes.

TABLE 34

Resident prey numbers and biomass (in kg) in 115 sq km usually occupied by the Masai pride (based on counts by car)

Species	Sept. 23, 1966		Feb. 8, 1967		Feb. 22, 1967		March 6, 1967		May 24, 1967	
	No.	Biomass	No.	Biomass	No.	Biomass	No.	Biomass	No.	Biomass
Topi	105	8,610	70	5,740	108	8,856	113	9,266	68	5,576
Hartebeest	180	17,100	11	1,045	113	10,735	3	285	15	1,375
Grant's gaz.	873	27,936	88	2,816	96	3,072	158	5,056	6	192
Waterbuck	0	0	8	1,048	13	1,703	6	786	0	0
Bushbuck[a]	6	132	6	132	6	132	6	132	6	132
Reedbuck[a]	120	4,440	120	4,440	120	4,440	120	4,440	120	4,440
Warthog	26	1,040	8	320	10	400	9	360	10	400
Buffalo[b]	10	6,500	2	1,300	1	650	3	1,950	0	0
Giraffe[b]	8	7,200	12	10,800	1	900	1	900	0	0
Ostrich[b]	29	3,190	10	1,100	14	1,540	16	1,760	10	1,100
Total	1,357	76,148	335	28,741	482	32,428	435	24,935	235	13,215
Kg/sq km		662		250		282		216		115
No./sq km	12		3		4		4		2	

[a] Not counted, only estimated.

[b] No young animals present.

TABLE 35
Number of prey per sq km in several woodland sample plots of the park during the rainy season (based on aerial counts)

Species	Loliondo pride area (325 km²)	Musabi (130 km²)	Upper Mbalageti (221 km²)	Roakari River (182 km²)	Bololedi River (65 km²)	Kogatende (112 km²)	Total No. in sample	Sample extrapolated to park woodlands
Wildebeest	0	0	.004	0	0	0	1	11
Zebra	.02	.40	.07	.44	.94	.09	230	2,466
Thomson's gaz.	0	.28	0	.01	1.43	0	132	1,416
Eland	0	1.41	.04	.05	.12	0	210	2,252
Topi	1.24	4.00	.75	1.46	2.14	1.60	1,672	17,931
Hartebeest	1.61	.21	1.24	.51	3.46	.50	1,196	12,827
Grant's gaz.	.04	.22	.01	.27	3.00	0	293	3,142
Impala	3.47 ±	2.87 ±	3.75 ±	3.00 ±	7.78 ±	7.17 ±	4,180 ±	44,829
Buffalo	1.49 ±	6.67	.35	2.78	11.97 ±	14.44	4,334 ±	47,552
Giraffe	.24	.42	.70	.16	1.43	.57	480	5,148
Warthog	.02	.20	0	.01	.24	.91	156	1,673
Waterbuck	.09	.06	.10	0	0	.80	147	1,577
Oribi	—	—	—	—	—	1.13	127	—
Ostrich	.05	.33	.08	.02	.46	.07	123	1,319
Total no./km²	8.27	17.07	7.09	8.71	32.97	27.28		
Biomass (kg/sq km)	1,193	3,982	998	1,654	6,981	7,234		

TABLE 36

Food items (killed and scavenged) eaten by lions in various parts of the Serengeti Park (percentages are in parentheses)

Species	1. Plains	2. Masai and Seronera prides	3. Edge of Woodlands	4. Corridor	5 Northern Extension
Wildebeest	159 (56.7)	121 (22.0)	97 (37.3)	22 (32.8)	10 (47.6)
Zebra	81 (28.9)	87 (15.8)	63 (24.2)	21 (31.3)	3 (14.3)
Thomson's gaz.	21 (7.5)	276 (50.0)	31 (11.9)	3 (4.5)	
Buffalo		13 (2.4)	40 (15.4)	5 (7.5)	7 (33.3)
Topi	4 (1.4)	18 (3.2)	7 (2.7)	4 (6.0)	1 (4.8)
Warthog		12 (2.2)	5 (1.9)	4 (6.0)	
Eland	9 (3.2)		3 (1.2)	1 (1.4)	
Grant's gaz.	3 (1.1)	7 (1.3)	1 (.4)		
Hartebeest	1 (.4)	1 (.2)	4 (1.5)		
Giraffe		3 (.5)	1 (.4)	5 (7.5)	
Impala		1 (.2)	1 (.4)	2 (3.0)	
Reedbuck		6 (1.1)	1 (.4)		
Bushbuck			1 (.4)		
Waterbuck			1 (.4)		
Pangolin			1 (.4)		
Hare		1 (.2)			
Lion	1 (.4)	2 (.3)			
Hyena	1 (.4)	1 (.2)			
Ostrich			3 (1.1)		
Guinea fowl		1 (.2)			
Sand grouse		1 (.2)			
Saddle-bill stork		1 (.2)			
Total	280	552	260	67	21

TABLE 37
Food items eaten by lions in 4 African reserves

	Manyara park 1967–69 (this study)	Nairobi park 1954–66 (Foster and Kearney, 1967)	Kafue park 1960–63 (Mitchell et al., 1965)	Kruger park 1954–66 (Pienaar, 1969)
No. kills	100[d]	257	410	12,313
Species				
Wildebeest	2.0%	48.3%	6.1%	23.6%
Zebra	16.0	20.7	7.3	15.8
Impala	11.0	2.7	2.0	19.7
Waterbuck	1.0	trace	5.9	10.5
Eland		1.8	2.9	.5
Hartebeest		10.5	16.3	
Warthog		9.7	9.5	1.9
Giraffe	2.0	3.9		3.9
Buffalo	62.0		30.5	9.2
Bushbuck			.2	.3
Bushpig			2.0	trace
Duiker			.2	.1
Hippopotamus			1.4	trace
Kudu			1.0	10.9
Lechwe			.5	
Puku			1.0	
Reedbuck			2.0	.3
Roan			5.6	.3
Sable			5.1	1.5
Tsessebe				.4
Small antelope[a]				trace
Baboon	6.0			trace
Carnivores[b]				.4
Ostrich		2.3		.1
Porcupine			.5	.1
Others[c]				.3
No. prey species	7	9	19	38

[a] Steenbuck, grysbuck, klipspringer.
[b] Lion, leopard, hyena, cheetah, jackal, civet, ratel, caracal.
[c] Nyala, white rhino, aardvark, pangolin, crocodile, tortoise.
[d] Three persons who were eaten by lions just outside of the park are not included.

Table 38

Number of prey animals killed or scavenged (percentages in parentheses)

	Wildebeest	Zebra	Thomson's g.	Buffalo	Others	Total
Plains						
Killed	73 (46%)	42 (52%)	5 (24%)	0	11 (58%)	131 (47%)
Scavenged	53 (33)	22 (27)	13 (62)	0	5 (26)	93 (33)
Unknown	33 (21)	17 (21)	3 (14)	0	3 (16)	56 (20)
Seronera						
Killed	112 (93)	83 (95)	232 (84)	10 (77)	49 (89)	486 (88)
Scavenged	6 (5)	4 (5)	42 (15)	0	5 (9)	57 (10)
Unknown	3 (2)	0	2 (1)	3 (23)	1 (2)	9 (2)
Woodlands						
Killed	99 (77)	71 (82)	21 (62)	37 (71)	42 (91)	270 (78)
Scavenged	13 (10)	11 (13)	12 (35)	5 (10)	2 (4.5)	35 (12)
Unknown	17 (13)	5 (5)	1 (3)	10 (19)	2 (4.5)	35 (10)

TABLE 39
Sources of meat scavenged by lions

	Wildebeest	Zebra	Thomson's g.	Buffalo	Other	Total
Plains						
Hyena	40	8	1	0	2	51 (54.8)
Wild dog	6	2	1	0	0	9 (9.7)
Cheetah	0	0	2	0	0	2 (2.2)
Leopard	1	0	3	0	0	4 (4.3)
Disease/ malnutrition	1	9	2	0	2	14 (15.0)
Unknown	5	3	4	0	1	13 (14.0)
						93
Seronera						
Hyena	3	1	5	0	0	9 (15.8)
Wild dog	0	0	1	0	1	2 (3.5)
Cheetah	0	0	22	0	0	22 (38.6)
Leopard	0	0	3	0	1	4 (7.0)
Disease/ malnutrition	0	3	3	0	2	8 (14.0)
Drowned	2	0	0	0	0	2 (3.5)
Jackal	0	0	3	0	0	3 (5.3)
Unknown	1	0	5	0	1	7 (12.3)
						57
Woodlands						
Hyena	9	4	8	0	1	22 (51.2)
Wild dog	0	0	2	0	1	3 (7.0)
Cheetah	0	0	0	0	0	0
Leopard	0	0	1	0	0	1 (2.3)
Disease/ malnutrition	1	3	1	4	0	9 (20.9)
Drowned	1	2	0	0	0	3 (7.0)
Unknown	2	2	0	1	0	5 (11.6)
						43

TABLE 40
Average number of lions feeding on wildebeest, zebra, and Thomson's gazelle carcasses. The average number of subadult and adult lions (excluding cubs) is given in parentheses.

	Wildebeest	Zebra	Thomson's gaz.
Killed			
Plains	4.7 (3.7)	4.5 (3.7)	—
Seronera	13.1 (6.9)	12.1 (7.3)	3.1 (2.0)
Woodlands	6.6 (4.7)	5.2 (4.0)	2.8 (1.7)
Scavenged			
Plains	2.2 (2.2)	2.5 (2.1)	1.5 (1.5)
Seronera	—	—	2.2 (1.3)
Woodlands	3.0 (2.6)	4.0 (3.4)	1.2 (1.2)

TABLE 41
Number of lions feeding on various prey species

	Kills	Scavenged	Unknown	Total
Wildebeest	2,245 (38.3%)	219 (39.7%)	173 (45.8%)	2,637 (38.9%)
Zebra	1,660 (28.4)	151 (27.3)	75 (19.8)	1,886 (27.8)
Thomson's gaz.	831 (14.2)	122 (22.1)	8 (2.1)	961 (14.2)
Buffalo	596 (10.2)	27 (4.9)	110 (29.1)	733 (10.8)
Others	523 (8.9)	33 (6.0)	12 (3.2)	568 (8.3)
Total	5,855	552	378	6,785

Table 42
Frequency with which lions killed more than one animal of a species at the same time

	Total killed	No. in multiple kills	% in multiple kills
Wildebeest	284	96	34
Zebra	216	22	10
Thomson's gazelle	258	44	17
Buffalo	47	4	8

Table 43
Relative importance as lion food of the 7 most abundant prey species in the woodlands of the ecological unit

Species	Food items based on kill record[a] (table 36, columns 3–5)	Assumed annual importance rating	Assumed critical time importance rating during rains
Wildebeest	1	2	6
Zebra	2	1	5
Buffalo	3	3	1
Thomson's gazelle	4	7	7
Topi	5	4	2
Hartebeest	6	6	4
Impala	7	5	3

[a] 1 = most important prey.

TABLE 44
Relative age scale for adult wildebeest, based on tooth wear[a]

Class	Characteristic
VI	*Mandible:* Full permanent dentition. Infundibulum on pm3 usually open; posterior infundibulum on pm4 present and usually open or just closed; incisor length is greater than 2 × the width. *Maxilla:* m1 accessory infundibulum usually present; m2 accessory infundibulum present, and the m3 one present or not yet formed. Posterior infundibulum on pm4 usually present.
VII	*Mandible:* Anterior infundibulum of m1 chevron or u-shaped; infundibulum of pm3 often closed; posterior infundibulum of pm4 present or absent, and, if present, closed at posterior end; incisors oval, often touching, with length less 2 × the width. *Maxilla:* At least one accessory infundibulum present on m3, and usually one on m2. Accessory infundibulum absent on m1. Usually no posterior infundibulum on pm2 and pm3.
VIII	*Mandible:* Infundibulum on anterior part of m1 small, round or oval, on posterior part chevron or u-shaped; posterior infundibulum on pm4 absent; incisors oval or rectangular, often not touching. *Maxilla:* Infundibulum on anterior part of m1 oval or chevron-shaped, on posterior part u-shaped; accessory infundibulum on m3 present or absent; anterior infundibulum on pm4 present.
IX	*Mandible:* Infundibulum worn off anterior part of m1, and posterior part has small round or oval infundibulum. Usually one or both infundibuli worn off pm4; incisors: some non-functional, worn to gums, others round, not touching. *Maxilla:* Infundibulum worn off anterior part of m1 or rarely a small remnant left; posterior surface has small infudibulum. Anterior infundibulum of pm4 present or absent.
X	*Mandible:* Both infundibuli worn off m1, sometimes off m2; most incisors non-functional. *Maxilla:* Both infundibuli worn off m1 except for occasional remnant on posterior part; infundibulum worn off pm4.

[a] For tooth eruption stages of classes I to V see Talbot and Talbot (1963) and Watson (1967).

Table 45

Age and sex of wildebeest killed by lions

Age class	Age	Male	Female	Unsexed	Total
I	0–2 mo.			13	13 (5%)
II	2–6 mo.			21	21 (8)
III	6–12 mo.	1	5	11	17 (6.5)
IV	12–24 mo.	15	5	1	21 (8)
V	24–42 mo.	23	25	2	50 (19)
VI	3½–5½± yrs.	30	7	—	37 (14)
VII	—	14	7	1	22 (8.5)
VIII	—	14	3	2	19 (7)
IX	—	21	14	1	36 (14)
X	—	22	3	1	26 (10)

Table 46

Age distribution of adult wildebeest killed by lions (data calculated by Sinclair, pers. comm.)

Age Class	Male		Female	
	Observed	Expected	Observed	Expected
V	23	23.0	25	26.8
VI	30	27.5	7	29.5
VII	14	17.5	7	14.8
VIII	14	9.9	3	5.1
IX	21	5.8	14	3.2
X	22	4.4	3	3.1

TABLE 47

Age and sex of zebra killed by man, disease, and lions

	Age Class (after Klingel, 1966)	Approximate age	Sample shot (Skoog, pers. comm.)			Disease/malnutrition deaths				Lion kills			
			♂	♀	Total	♂	♀	Unsexed	Total	♂	♀	Unsexed	Total
Young	I–IV	0–9 mo.	5	4	9			10	10			15	15
Yearling	V–VI	1–2 yrs.	9	15	24			1	1			9	9
Subadult	VII–VIII	2–3	13	13	26	1	1	1	3			12	12
Young adult	IX–XI	3–4 yrs.	11	12	23 (9%)	1	3		4 (10%)	7	11		18 (13%)
Adult	XII–XIII	4–5	21	13	34 (13)		2		2 (5)	5	6		11 (8)
	XIV–XV	5–9	58	33	91 (36)	3	15		18 (44)	19	15		34 (25)
	XVI–XVII	9–13	48	23	71 (28)	2	3		5 (12)	7	13		20 (14)
	XVIII–XIX	12–16	23	8	31 (12)	3	3		6 (15)	16	7		23 (17)
	XX–XXI	15–20+	3	1	4 (2)	2	4		6 (15)	22	10		32 (23)
Total—all zebra			191	122	313	12	31	12	55	76	62	36	174
Total—young adults and adults			164	90	254	11	30	0	41	76	62	0	138

TABLE 48
Relative age scale for Thomson's gazelle, based on tooth eruption and wear in the lower jaw

Class	Approx. age at beginning of class (in months)	Character
I	0	Only premolars present; anterior cusp of m1 sometimes barely visible through gums.
II	2±	M1 erupting to erupted.
III	6±	M1 present; m2 erupting to erupted.
IV	9±	Deciduous pm4 still present; m3 erupting.
V	18±	Full permanent dentition; 3rd cusp of 3rd molar not worn; posterior infundibulum of pm4 open.
VI	24±	Posterior infundibulum of pm4 present and usually closed. Anterior infundibulum of m1 very small, round or oval, or just worn off; posterior infundibulum of m1 present; 3rd cusp of m3 worn.
VII		Both infundibuli off m1; posterior infundibulum of pm4 small and round, or gone.
VIII		Both infundibuli off m1 and off anterior part of m2; posterior infundibulum of pm4 gone.
IX		All infundibuli off m1 and m2.
X		Infundibuli off m1 and m2 and one or more off m3; some incisors non-functional.

NOTE: The ages were kindly provided by L. Robinette.

TABLE 49
Age and sex of Thomson's gazelle killed by lion

Age class	Male	Female	Unsexed	Total	% killings observed	% killings not observed
I			18	18 (8.8%)	17	3
II			31	31 (15.2)	14	16
III	4	4	10	18 (8.8)	9	9
IV	6	6	—	12 (5.9)	4	7
V	8	8	—	16 (7.8)	8	8
VI	13	7	—	20 (9.8)	13	7
VII	26	20	—	46 (22.6)	19	25
VIII	6	5	—	11 (5.4)	3	7
IX	7	5	—	12 (5.9)	4	7
X	7	13	—	20 (9.8)	9	11

TABLE 50
Age and sex of buffalo killed by lions[a]

Age (yrs.)	Male Observed	Male Expected	Female Observed	Female Expected
2–3	2	7.6	2	4.2
4–5	3	6.9	5	3.8
6–7	3	5.7	3	3.3
8–9	3	4.2	1	2.6
10–11	3	2.9	1	1.8
12–13	7	1.5	2	.9
14–15	4	.45	2	.34
16–17	3	.06	1	.005
18–19	1	0	0	0
Total	29		17	

[a] Prepared by A. Sinclair from his data and mine; 4 calves (aged 0–1 yr.) have not been included; 3 buffalo kills collected by Sinclair were not included in table 36.

TABLE 51
Relative age of 7 species of prey killed by lions

Class	Topi	Hartebeest	Eland	Grant's gaz.	Reed-buck	Impala	Giraffe
I (newborn)	1				1	1	1
II (subadults: not full dentition)	2	2	1		2		
III (young adults: full dentition, little wear)	2		1	2	2	1	2
IV (prime adults: mod. wear)	7	1	5	2	2		2
V (past prime: anterior infundibulum worn off 1st molar)	6	1				1	2
VI (old: both infundibuli worn off 1st molar)	4	1				1	1
Total	22	5	7	4	7	4	8

TABLE 52
Cover types in which wildebeest and zebra were killed by lions

	Seronera	Woodlands	Plains
Open-groundcover < .3 m high	25%	27%	23%
Groundcover .3 to .6 m high	18	10	46
Groundcover .6+ m high	12	6	10
Thicket	1	20	—
In or within 60 m of river	41	30	—
Along road bank or erosion terrace	1	6	8
By kopje	2	1	3
No. of kills	179	89	91

TABLE 53
Composition of group to which lion belonged when hunting alone (stalking and running only)

Group composition	Thomson's Gazelle No. Hunts	Wildebeest and Zebra No. Hunts	Other No. Hunts	Total No. Hunts
Male alone	0	0	0	0
Female alone	39 (25%)	8 (26%)	14 (48%)	61 (28%)
Two or more males	2 (1)	1 (3)	1 (3)	4 (2)
Two or more females	60 (38)	13 (42)	7 (25)	80 (37)
Female(s) and male(s)	55 (36)	9 (29)	7 (24)	71 (33)
Total	156	31	29	216

TABLE 54
Number of females, males, and large cubs using various hunting methods

	Hunting method				
	Stalking and running	Unexpected hunting	Driving	Ambushing	Other
Female	1115 (92%)	38 (72%)	19 (83%)	18 (100%)	6 (100%)
Male	37 (3%)	13 (24%)	0	0	0
Large cub	58 (5%)	2 (4%)	4 (17%)	0	0
Average number lions hunting	2.5 (1–14)	1.5 (1–5)	1.3 (1–5)	1.1 (1–2)	1.5 (1–3)

Table 55
Hunting success comparing day and night hunts (stalking and running only)

	Thomson's gazelle		Wildebeest and zebra		Other		Total	
	No. hunts	% success	No. hunts	% success	No. hunts	% success	No. hunts	% success
Day	338	23	73	23	49	14	460	21
Night	24	29	30	37	9	33	63	33

Table 56
Hunting success in various habitat types in daytime (stalking and running only)

Habitat	Thomson's gaz.		Wildebeest and zebra		Other		Total	
	No. hunts[a]	% success	No. hunts	% success	No. hunts	% success	No. hunts	% success
Little or no cover	43	7	10	10	12	33	65	12
No cover but near some	53	9	14	14	9	22	76	12
Scattered shrubs	20	15	1	0	4	25	25	12
Thickets or tall grass	175	26	15	7	13	8	203	24
Thickets along river	45	49	31	42	9	0	85	41

[a] Habitat type not recorded in 3 hunts.

Table 57
The relation of wind direction to daytime hunting success of Thomson's gazelle (stalking and running only)

	Lion moves downwind		Lion moves upwind	
	No. hunts	% success	No. hunts	% success
Single lion	28	7	27	18.5
Two or more lions	26	15	29	48

TABLE 58
The relation of hunting success to method of hunting

Hunting method	Thomson's gazelle		Wildebeest and zebra[b]		Other		Total	
	No. hunts	% success	No. hunts	% success	No. hunts	% success	No. hunts	% success
Stalking by single lion	86 (20.6%)	17	24 (20.9%)	17	23 (35.4%)	17	133 (22.3%)	17
Running by single lion	71 (17.0)	6	7 (6.1)	14	6 (9.2)	33	84 (14.1)	8
Stalking or running by single lion[a]	28 (6.7)	29	2 (1.7)	0	2 (3.1)	50	32 (5.4)	28
Stalking and running by several lions	177 (42.4)	32	70 (60.9)	23	27 (41.5)	11	274 (46.0)	30
Hunting unexpectedly	24 (5.8)	58	10 (8.7)	60	3 (4.6)	67	37 (6.1)	61
Driving	17 (4.1)	18	—	—	—	—	17 (2.8)	18
Ambushing	11 (2.6)	27	2 (1.7)	0	3 (4.6)	0	16 (2.6)	19
Grabbing prey in water	3 (.7)	100	—	—	—	—	3 (.5)	100
Digging	—	—	—	—	1 (1.5)	100	1 (.2)	100
Total	417		115		65		597	

[a] Ambiguous information supplied by informant.
[b] Wildebeest and zebra were also combined in tables 55, 56, and 59 because there were no differences between these species.

TABLE 59
The relation of hunting success to the number of lions stalking or running

No. animals hunting	Thomson's gazelle		Wildebeest and zebra		Other		Total	
	No. hunts	% success	No. hunts	% success	No. hunts	% success	No. hunts	% success
1	185 (51.1%)	15	33 (32.0%)	15	31 (53.4%)	19	249 (47.6%)	15
2	78 (21.5)	31	17 (16.5)	35	11 (19.0)	9	106 (20.3)	29
3	42 (11.6)	33	16 (15.5)	12.5	5 (8.6)	20	63 (12.0)	27
4–5	42 (11.6)	31	16 (15.5)	37	4 (6.9)	25	61 (11.9)	32
6+	15 (4.1)	33	21 (20.4)	43	7 (12.0)	0	43 (8.2)	33
Total	362		103		58		523	

Table 60
The relation of hunting success to species hunted (all hunts combined)

Species	Total No. hunts	No. incomplete hunts	No. complete hunts	% hunts completed	% hunts successful
Reedbuck	21	12	9	43	14
Topi	17	11	6	35	13
Warthog	17	7	10	58	47
Wildebeest	59	28	31	53	32
Zebra	56	32	24	43	27
Thomson's gazelle	417	153	264	63	26

Table 61
The relation of herd size of prey to hunting success of lion (all hunts combined)

Herd size	Thomson's gazelle		Wildebeest		Zebra	
	No. hunts	% successful	No. hunts	% successful	No. hunts	% successful
1	64	33	19	47	5	60
2–10	164	21	8	13	19	21
11–75	165	25	11	9	26	23
76+	24	33	20	50	6	33

Table 62
The relation of prey size to sex of lion feeding on it

	Total no. kills	Lions on kill		
		Male(s) only	Male(s) and female(s)	Female(s) only
Medium-sized prey				
Topi-Hartebeest	31	10%	32%	58%
Wildebeest	233	15	50	35
Zebra	161	15	53	32
Large prey				
Eland	8	37	50	13
Buffalo	31	22.5	64.5	13
Giraffe	6	17	66	17

TABLE 63
Prey killed by cheetah and leopard[a]

Species	Cheetah: This study No.	Cheetah: This study %	Leopard: This study No.	Leopard: This study %	Leopard: Kruuk and Turner (1967) No.	Leopard: Kruuk and Turner (1967) %
Thomson's gaz.	238	91.2	104	63.4	15	27.3
Grant's gaz.	6	2.3	10	6.1	2	3.6
Reedbuck	2	.8	19	11.6	6	10.9
Wildebeest	5	1.9	11	6.7	5	9.0
Topi	2	.8	3	1.8	1	1.8
Hartebeest	1	.4	2	1.2		
Zebra			2	1.2	4	7.2
Waterbuck			1	.6		
Dik-dik	1	.4				
Impala	3	1.1			9	16.3
Bushbuck					1	1.8
Warthog			1	.6		
Baboon			1	.6	2	3.6
Golden jackal			1	.6		
Black-backed jackal			1	.6	1	1.8
Bat-eared fox			2	1.2		
Serval			2	1.2		
Cheetah					1	1.8
Hare	3	1.1				
Rock hyrax					1	1.8
Springhare					1	1.8
Python					1	1.8
Secretary bird					1	1.8
European stork			4	2.4	2	3.6
Helm. guinea fowl					1	1.8
Vulture					1	1.8
Total	261		164		55	

[a] Scavenged meat not included.

Table 64
Age and sex of Thomson's gazelle killed by cheetah and leopard.

Age class	Cheetah				Leopard			
	Male	Female	Unident.	Total	Male	Female	Unident.	Total
I			42	42				
II			37	37				
III	8	8		16			7	7
Unaged young I–III			29	29			2	2
IV	2			2	3	1		4
V	1	1		2	4	1		5
VI	5	5		10	3	2		5
VII	12	12		24	4	1		5
VIII	5	3		8				
IX	1	8		9	1			1
X	3	10		13	1			1
Unaged adult VI–X	9	27		36	42	16		58

Table 65
Size and composition of some wild dog packs

	Male		Female		Total
	Adult	Large young (6 to 12 mo.)	Adult	Large young	
	1		1		2
	2		2		4
	4				4
			5		5
	3		2		5
	4		1		5
	3		5		8[a]
	7		3		10[b]
	4	6	1		11[c]
	4	1	2	4	11
	5	1	4	2	12
	4	6	1	3	14
	3	4	2	5	14
	3	3	5	5	16
	11	8	8	5	32
Total	58	29	42	24	153

[a] 1 female with 10 small young
[b] 1 female with 16 small young
[c] 1 female with 14 small young

TABLE 66
Wild dog kills, Serengeti Park and Ngorongoro Crater

Species	Serengeti (this study) Jan.–June[a]	July–Dec.	Total	Serengeti (Kruuk and Turner, 1967)	Ngoronoro[b] (Estes and Goddard, 1967)
Wildebeest	75 (56.8%)	1 (1.5%)	76 (38.4%)	8 (19%)	18 (36%)
Thomson's gaz.	32 (24.2)	52 (78.8)	84 (42.4)	27 (64)	27 (54)
Grant's gaz.	9 (6.8)	0	9 (4.5)	4 (10)	4 (8)
Zebra	11 (8.3)	0	11 (5.6)		
Topi	0	2 (3.0)	2 (1.0)	2 (5)	
Hartebeest	0	1 (1.5)	1 (.5)		1 (2)
Warthog	3 (2.3)	3 (4.5)	6 (3.0)		
Bushbuck	0	1 (1.5)	1 (.5)		
Impala	1 (.8)	1 (1.5)	2 (1.0)	1 (2)	
Hare	1 (.8)	0	1 (.5)		
Mouse	0	1 (1.5)	1 (.5)		
Ostrich	0	4 (6.1)	4 (2.0)		
Total	132	66	198	42	50

[a] Wild dog pups eaten by a female are not included.
[b] Springhare and hare were mentioned as being killed but they were not added to the kill record by these authors.

TABLE 67
Age and sex of Thomson's gazelle and wildebeest killed by wild dogs

Age Class	Thomson's gazelle Male	Female	Unsexed	Total	Wildebeest Male	Female	Unsexed	Total
I			23	23			21	21
II			6	6			41	41
III			1	1			2	2
Unaged I–III			4	4				
IV	2			2	1	5	2	8
V	1	1		2	1	1		2
VI	2	1		3				
VII	8	6		14				
VIII	1	1		2				
IX	2	1		3				
X	1	4		5				
Unaged adult VI–X	6	9		15		2		2
Total				80				76

NOTE: The Thomson's gazelle and wildebeest kills are not directly comparable except for classes I to III, which are of the same age.

Table 68

Killing frequency and weight of prey killed by a pack of 11 wild dogs in a 20-day period, 1967

Date	Morning Hunt		Evening Hunt		Kg. killed per dog per day
	Species killed	Est. wgt/kg	Species killed	Est. wgt/kg	
April 6	No hunt	0	Warthog F	53	4.8
7	Wildebeest F	163	Wildebeest yg	40	18.4
8	2 wildebeest yg	70	Unsuccessful	0	6.3
10	Zebra F	219	Unsuccessful	0	19.9
11	Zebra F	219	Wildebeest yg	45	24.0
12	No hunt	0	Wildebeest yg	45	4.1
13	Wildebeest yg	45	Wildebeest yg	45	8.2
14	Wildebeest yg	30	2 wildebeest yg	75	9.5
17	Zebra yg	40	Unsuccessful	0	3.6
18	Thomson's g. yg	3	Unsuccessful	0	.3
20	Zebra yg	20	No hunt	0	1.8
21	Thomson's g. yg	3	Zebra F	219	20.2
22	No hunt	0	Grant's g. yg	8	.7
23	No hunt	0	Warthog M	86.5	7.8
26	Unsuccessful	0	Wildebeest F	163	14.8
27	Thomson's g. F	16	Wildebeest yg	45	5.5
May 9	Wildebeest yg	30	Wildebeest yg	45	6.8
10	No hunt	0	Warthog M	86.5	7.8
11	No hunt	0	Wildebeest yg	45	
			Wildebeest yearl.	85	
			Hare	2	12.0
19	Wildebeest yg	45	No hunt	0	4.1

Table 69

Killing frequency and weight of prey killed by wild dogs in a 5-day period, 1967

Date	Morning Hunt		Evening Hunt		No. dogs in pack	Kg. killed per dog per day
	Species killed	Est. wgt/kg	Species killed	Est. wgt/kg		
Nov. 6	4 ostrich yg	8				
	Thomson's g. yg	2	Unsuccessful	0	16	.6
10	Thomson's g. F	16	Unsuccessful	0	16	1.0
11	Thomson's g. M	20.5	Thomson's g. F	16	16	2.3
12	Impala M	57	No hunt	0	15	2.8
15	Unsuccessful	0	Thomson's g. M	20.5	7	2.9

TABLE 70
Hunting success of wild dog

Species	Age	No. chases observed (includes crouched gazelle fawns)	Successful chases	
			No.	%
Wildebeest	Young (0–6 mo)	36	27	75
	Others	6	4	66
Thomson's gaz.	Young (0–2 mo)	22	21	95
	Others	47	23	49
Grant's gaz.	Young (0–2 mo)	7	7	100
Warthog	Yearling and adult	7	4	57
Others (excluding zebra)	All ages	8	7	87
Total		133	93	70

TABLE 71
Percentage of prey killed and scavenged by hyenas in the Serengeti (after Kruuk, 1972)

	No. obs.	% killed	% scavenged
Wildebeest adult	136	62.5	37.5
Wildebeest calf	48	68.8	31.2
Zebra adult	38	50.0	50.0
Zebra foal	12	33.0	67.0
Thomson's and Grant's g. adult	77	41.6	58.4
Thomson's and Grant's g. fawn	37	81.0	19.0

TABLE 72
Prey abundance and lion density in 5 African reserves

Area	Size of area km^2	Biomass[a] kg/km^2	No. lions	No. km^2 per lion	Lion per no. prey
Ngorongoro Crater	260	10,363	70	3.7	1/338
Manyara Park	91	7,785	35	2.6	1/70
Serengeti Unit	25,500	4,222	2,000–2,400	10.6–12.7	1/397–1/477
Nairobi Park	115	3,052	25	4.6	1/152
Kruger Park	19,084	1,034	1,120	17.0	1/249

[a] See table 33 for sources.

TABLE 73

Occurrence of some communicatory patterns in 4 species of cats

Pattern	Lion	Tiger	Leopard	Cheetah
Tactile				
Head-rubbing	+	+	+	(+)
Social licking	+	+	+	+
Visual				
Head-low posture	+	?	+	+
Head-twist posture	+	+	+	+
Strut	+	?	?	+
Elbow-walk	−	−	−	+
Ear strikingly marked	+	+	+	+
Tail tip strikingly marked	+	−	+	+
Large mane in male	+	−	−	−
Vocalizations				
Puffing	+	?	+	?
"Prusten"	−	+	−	−
Purring	(+)	(+)	(+)	+
Bleating	+	?	?	+
Pooking	−	+	−	−
Humming	+	+	+	−
Woofing	+	+	?	?
Grunting and soft roaring	+	+	+	−
Loud roaring	+	+	+	−
Miaowing	+	+	+	+(?)
Chirring	−	−	−	+
Moaning	−	−	−	+
Coughing, hissing, growling, snarling	+	+	+	+
Marking				
Scraping with hindpaws	+	+	+	+
Urinating on scrape	+	+	?	(+)
Defecating on scrape	−	+	−	(+)
Defecating in prominent locations	−	+	−	+
Defecating at one site	−	+	−	−
Spraying scent	+	+	+	+
Tree clawing	+	+	+	+
Dung rolling	+	+	?	?
Pawing dirt onto kill	+	+	(+)	(+)

+ present
(+) rare
? Not heard or seen, possibly present
− absent

TABLE 74
Average number of prey consumed annually by one lion in the Seronera area

Species	a % kills (from table 36, column 2)	b Av. weight of prey in kg	c = a × b Total kg in sample of 100 prey	d = a × f[a] Calculated no. animals eaten
Wildebeest	22.0	108	2,376	7.04
Zebra	15.8	164	2,591	5.06
Thomson's gaz.	50.0	12	600	16.00
Buffalo	2.4	550	1,320	.77
Topi	3.2	82	262	1.02
Warthog	2.2	40	88	.70
Grant's gaz.	1.3	32	42	.42
Reedbuck	1.1	37	41	.35
Hartebeest	.2	95	19	.06
Impala	.2	32	6	.06
Giraffe	.5	800	400	.16
Hare	.2	.5	trace	.06
Lion	.3	10	3	.10
Hyena	.2	40	8	.06
Birds	.6	2	1	.19
Total			7,757	32.05

[a] f = 2,500 kg/7,757 kg.
NOTE: 2,500 kg represents the approximate amount killed and scavenged by an average lion (including cubs) annually.

TABLE 75

Predator biomass (in kg/km²) in 5 African reserves

Reserve	Lion		Leopard		Cheetah		Spotted Hyena		Wild Dog		Total predator biomass	Total prey biomass	Kg. prey per 1 kg of predator
	No.	Biomass	No.	Biomass	No.	Biomass	No.	Biomass	No.	Biomass			
Ngorongoro	70	25.3	20 [a]	2.3	v.o.	—	479	68.1	v.o.	—	95.7	10,363	108
Manyara	35	37.7	10 [a]	3.3	—	—	10 [a]	4.0	—	—	44.7	7,785	174
Serengeti Unit	2,000–2,400	7.6–9.2	800–1,000	.9–1.2	200–250	.3–.4	3,500	5.1	250–300	.1–.2	14.0–16.1	4,222	262–301
Nairobi	25	21.3	10 [a]	2.6	15	4.9	12 [a]	3.8	v.o.	—	32.4	3,052	94
Kruger	1,120	5.7	650	1.0	263	.5	1,500 [a]	2.9	335	.3	10.4	1,034	100

[a] = my guess; v.o. = visitor only.

Kruger Park data from Pienaar (1969) who stated with respect to hyenas that they are probably "the most abundant predator in the Park," but the actual figure represents my arbitrary guess.

Hyena data from Ngorongoro is from Kruuk (1972); for the Serengeti he gives a figure of 3,000 meat-eating hyenas. I have added 500 to represent small cubs.

The predator biomass figures are based on three-quarters of the average weight of an adult female, except in hyenas where the male is used: lion 98 kg; leopard, 30 kg; cheetah, 38 kg; hyena, 37 kg; wild dog, 15 kg.

TABLE 76
Hunting success of various carnivores[a]

Predator	Prey	No. attempts	% successful	Source	Comments
Lynx	Snowshoe hare	98	16	Nellis and Keith (1968)	Based on spoor in snow
Lynx	Snowshoe hare	43	42	Saunders (1963)	Based on spoor in snow
Lynx	Ruffed grouse	40	12.5	Nellis and Keith (1968)	Based on spoor in snow
Puma	Deer and wapiti	45	82	Hornocker (1970)	Based on spoor in snow; final attack only
Coyote	Snowshoe hare	40	10	Ozoga and Harger (1966)	Based on spoor in snow
Wolf	Moose	77	8	Mech (1966)	All moose "tested"
Jackals	Thomson's g. yg	?	33	Wyman (1967)	Direct observation
Hyena	Thomson's gaz.	43	33	Kruuk (1972)	Direct observation
Hyena	Wildebeest calf	108	32	Kruuk (1972)	Direct observation; chase of at least 50 m
Hyena	Wildebeest adult	49	44	Kruuk (1972)	Direct observation; chase of at least 50 m
Hyena	Zebra	47	34	Kruuk (1972)	Direct observation; chase of at least 100 m
Cheetah	Thomson's gaz.	87	70	This study	Only fast chases tallied

[a] For wild dog see table 70; for lion see table 60.

TABLE 77
Sex ratio of Thomson's gazelle killed by various predators in the Serengeti[a]

Predator	Size of sample	Ratio of ♂:♀	Source
Lion	178	1.1:1.0	This study
Leopard	79	2.8:1.0	This study
Cheetah	104	.6:1.0	This study
Hyena	54	3.2:1.0	Kruuk (1972)
Wild dog	67	1.0:1.0	This study
Wild dog	46	1.5:1.0	Kruuk (1972)

[a] My samples include age classes IV–X.

TABLE 78

Percentage of Thomson's gazelle of classes IV to X killed by predators compared with population samples

Class	Shot randomly[a]	Lion kills	Leopard kills	Cheetah kills	Wild dog kills	Mass kill by hyenas[b]	Hyena kills (Kruuk, 1972)
IV	14.8	8.8	18.8	2.9	6.5	3.2	} 0
V	14.8	11.6	25.0	2.9	6.5	19.3	
VI	14.8	14.6	18.8	14.7	9.7	3.2	19.6
VII	26.0	33.6	25.0	35.3	45.2	35.5	} 42.9
VIII	11.1	8.0	—	11.8	6.5	16.1	
IX	7.4	8.8	6.2	13.3	9.7	9.7	} 37.5
X	11.1	14.6	6.2	19.1	16.1	13.0	
No. in sample	27	137	16	68	31	31	56

[a] Collected by A. deVos and H. Hvidberg-Hansen.
[b] Collected by H. Kruuk.

TABLE 79
Some behavioral comparisons of Serengeti predators

Characteristic	Lion	Leopard	Cheetah	Hyena[a]	Wild dog	Jackal (2 species)
Av. weight of adult (kg)	110–180	35–55	35–55	45–60	17–20	5–9
Most commonly used habitat	Open woodlands	Thickets and riverine forest	Plains	Plains	Woodlands and plains	Woodlands (1 species) Plains (1 species)
Main activity time	Nocturnal	Nocturnal	Diurnal	Nocturnal	Diurnal	Nocturnal
Fastest running speed km/hr	60±	60±	95±	65±	70±	60±
Main hunting method for ungulates	Stalk, then fast brief rush	Stalk, then fast brief rush	Often stalk; very fast run of 200–300 m	No stalk; long chase of 1+ km	Stalk rare; long chase of 1+ km	Stalk infrequent; moderately long chase
Usual size of hunting group	Solitary or groups of 2–15	Solitary	Usually solitary	Solitary or groups of 2–30±	Groups of 2–30±	Solitary or groups of 2
Usual maximum prey size (kg)	900±	60±	60±	300±	250±	5±
Average hunting success for Thomson's gaz.[b]	26%	?	70%	33%	57%	33%
Main killing method	Usually neck-bite; often strangulation	Usually neck-bite; often strangulation (?)	Strangulation	Evisceration	Evisceration	Evisceration
Maximum no. meals from prey	Several	Several	One	Usually one; occasionally several	One	Usually one
Method of caching meat	Lies by it; drags it into thicket	Hangs it in tree; drags it into thicket	None	Occasionally submerges it in water	None	Hides small pieces of meat

Method of food transport to young	Usually leads young to kill; no regurgitation	Usually leads young to kill; no regurgitation	May regurgitate to small young; usually leads them to kill	Usually leads young to kill	Regurgitates	May carry food to den; regurgitates
Claws retractile	Yes	Yes	No	No	No	No
Climbs trees readily	Yes	Yes	Yes	No	No	No
Sharpens claws on trees	Yes	Yes	Yes	No	No	No
Uses burrows	No	No	No	Yes	Yes	Yes
Claws used in fighting	Yes	Yes	Yes	No	No	No
Lifts leg to urinate	No	No	No	No	Occasionally	Yes
Male helps in care of young	No	No	No	No	Yes	Yes
Copulation brief	Yes	Yes	?	Yes	?	Prolonged
Usually carries young with neck grip	Yes	Yes	?	No	No	?

[a] Some hyena data from Kruuk (1972).
[b] Excludes crouched fawns.

Appendix C

After this report had gone to press, I came across a recent paper by Hendrichs (1970) in which he estimated animal numbers and biomass in the Serengeti Park based on repeated car transects over the same route in three areas during 1967 and 1968. His estimates for several species are as follows

Prey		*Predator*	
Wildebeest	370,000	Lion	1,650
Zebra	193,000	Leopard	500
Thomson's gazelle	980,000	Cheetah	500
Grant's gazelle	3,100	Spotted hyena	6,000
Eland	7,200	Wild dog	1,100
Topi	26,000	Black-backed jackal	13,500
Hartebeest	20,000	Golden jackal	5,000
Impala	75,000		
Waterbuck	3,200		
Giraffe	8,000		
Warthog	17,000		
Buffalo	38,000		

Many of these figures are roughly similar to those I used, but some, notably the ones for Thomson's gazelle, wild dog, cheetah, and hyena, are considerably higher (see tables 32 and 75).

For the woodlands portion of the park, Hendrichs calculated an average biomass of 5,000 kg/km^2, including elephant and rhinoceros. "The average population weight was taken to be three quarters of the maximum female weight for species over 100 kg body weight, and four fifths of the maximum female weight for those under 100 kg."

References cited

Adamson, G. 1963. Observations on lions in the southern portion of Serengeti National Park, mid-June to mid-November 1962. Manuscript in the Tanzania National Parks files.

——— 1964. Observations on lions in Serengeti National Park. *E. Afr. Wildl. J.* 2:160–61.

——— 1968. *Bwana game*. Collins, London.

Adamson, J. 1960. *Born free*. Collins, London.

——— 1961. *Living free*. Collins, London.

——— 1966. Pippa kehrte in die Freiheit zurück. *Das Tier.* 6(12):4–7.

——— 1969. *The spotted sphinx*. Collins, London.

Adler, H. 1955. Some factor of observational learning in cats. *J. Genetic. Psych.* 86:159–77.

Akeley, C. and M. Akeley. 1932. *Lions, gorillas and their neighbors*. Dodd, Mead and Co., New York.

Ali, S. 1927. The Moghul emperors of India as naturalists and sportsmen. *J. Bombay. Nat. Hist. Soc.* 31(4):833–61.

Altmann, S., ed. 1967. *Social communication among primates*. University of Chicago Press, Chicago.

Anderson, G. and L. Talbot. 1965. Soil factors affecting the distribution of the grassland types and their utilization by wild animals on the Serengeti Plains, Tanganyika. *J. Ecol.* 53:33–56.

Anon. 1960. Annual report of the biologist, 1958–1959. Kruger National Park, *Koedoe* 3:1–205.

Asdell, S. 1964. *Patterns of mammalian reproduction*. Cornell University Press, Ithaca.

Backhaus, D. 1959. Experimentelle Untersuchungen über die Sehschärfe und das Farbsehen einiger Huftiere. *Z. Tierpsych.* 16:445–67.

Baker, J. 1960. A trypanosome of *T. congolense* group in African lion and leopard. *Trans. roy. Soc. trop. Med. Hyg.* 54:2.

——— 1968. Trypanosomes of wild mammals in the neighbourhood of the Serengeti National Park. *Symp. Zool. Soc. London* 24:147–58.

Baumann, O. 1894. *Durch Massailand zur Nilquelle*. Dietrich Reimer, Berlin.

Bell, R. 1967. Annual report, 1966–1967. Manuscript in the Serengeti Research Institute files.

Bere, R. 1966. *The African elephant*. Golden Press, New York.

Beyers, C. de. 1964. Lions versus buffalo. *Animals* 5(8):220–21.

Bolwig, N. 1964. Facial expression in primates with remarks on a parallel development in certain carnivores. *Behaviour* 22(3–4):167–92.

Bothma, J. du. 1965. Random observations on the food habits of certain carnivora. *Fauna and Flora* (Transvaal) 16:16–22.

Bourlière, F. 1963. Specific feeding habits of African carnivores. *African Wild Life* 17(1):21–27.

——— 1965. Densities and biomasses of some ungulate populations in eastern Congo and Rwanda, with notes on population structure and lion/ungulate ratios. *Zool. Afr.* 1(1):199–207.

Bourlière, F. and Verschuren, J. 1960. *L'écologie des ongulés du Parc National Albert.* Fasc. 1. Inst. des Parcs Nat. du Congo Belge, Brussels.

Brand, D. 1963. Records of mammals bred in the National Zoological Gardens of South Africa during the period of 1908 to 1960. *Proc. Zool. Soc. Lond.* 140(4):617–59.

Bridge, B. 1951. Animal strategy. *African Wild Life* 5(2):121–25.

Brocklehurst, H. 1931. *Game animals of the Sudan.* Gurney and Jackson, London.

Brooks, A. 1961. *A study of the Thomson's gazelle* (*Gazella thomsonii* Gunther) *in Tanganyika.* Colonial Res. Publ. No. 25, London.

Brown, J. 1969. Territorial behavior and population regulation in birds. *Wilson Bull.* 81(3):293–329.

Burkholder, B. 1959. Movements and behavior of a wolf pack in Alaska. *J. Wildl. Mgmt.* 23(1):1–11.

Burt, W. 1943. Territoriality and home range concepts as applied to mammals. *J. Mammal.* 24:346–52.

Cade, C. 1967. Notes on breeding the Cape hunting dog *Lycaon pictus* at Nairobi Zoo. In *Int. zoo yearbook*, vol. 7, ed. C. Jarvis, pp 122–23. Zool. Soc. London, London.

Carpenter, C. 1965. The howlers of Barro Colorado Island. In *Primate behavior*, ed. I. DeVore, pp. 250–91. Holt, Rinehart, and Winston, New York.

Carr, N. 1962. *Return to the wild.* Collins, London.

Carvalho, C. 1968. Comparative growth rates of hand-reared big cats. In *Int. zoo yearbook*, vol. 8, ed. C. Jarvis, pp. 56–59. Zool. Soc. London, London.

Clough, G. 1969. Some preliminary observations on reproduction in the warthog, *Phacochoerus aethiopicus* Pallas. *J. Reprod. Fert.*, Suppl. 6 : 323–37.

Cooke, H. 1963. Pleistocene mammal faunas of Africa with particular reference to southern Africa. In *African ecology and human evolution*, ed. F. Howell and F. Bourlière, pp. 65–116. Viking Fund Publ. Anthr. No. 36.

Cooper, J. 1942. An exploratory study on African lions. *Comp. Psychol. Monogr.* 17(7):1–48.

Cott, H. 1940. *Adaptive coloration in animals.* Methuen and Co., London.

Cowan, I. 1947. The timber wolf in the Rocky Mountain national parks of Canada. *Can. J. Res.* 25:139–74.

Cowie, M. 1961. *Royal National Parks of Kenya.* Annual Report. Nairobi.

——— 1966. *The African lion.* Golden Press, New York.

Craighead, J. and F. Craighead. 1956. *Hawks, owls, and wildlife*. Stackpole Co., Harrisburg.
Crandall, L. 1964. *The management of wild mammals in captivity*. University of Chicago Press, Chicago.
Crofton, M. 1958. Nematode parasite populations in sheep on lowland farms. IV. Sheep behaviour and nematode infections. *Parasitology* 48:251–60.
Crook, J. 1965. The adaptive significance of avian social organization. *Symp. Zool. Soc. London* 14:181–218.
Darling, F. 1960. *An ecological reconnaissance of the Mara Plains in Kenya Colony*. Wildl. Monogr. No. 5. The Wildlife Society.
Davis, A. 1954. *Let's eat right to keep fit*. Harcourt, Brace and World, New York.
Dekker, D. 1968. Breeding the Cape hunting dog at Amsterdam Zoo. In *Int. zoo yearbook*, vol. 8. ed. C. Jarvis, pp. 27–30. Zool. Soc. London, London.
Demmer, H. 1966. Beobachtungen über das Verhalten verschiedener wildlebender Tiere beim Fang in freier Wildbahn in Afrika. *Zool. Garten.* 32(5): 210–15.
Denis, A. 1964. *Cats of the world*. Constable and Co., London.
DeVore, I., ed. 1965. *Primate behavior*. Holt, Rinehart, and Winston, New York.
DeVore, I. and K. Hall. 1965. Baboon ecology. In *Primate behavior*, ed. I. DeVore, pp. 20–52. Holt, Rinehart, and Winston, New York.
Dinnik, J. and R. Sachs. 1969. Zystizerkose der Kreuzbeinwirbel bei Antilopen und *Taenia olngojinei* sp. nov. der Tüpfelhyäne. *Z. Parasitenk.* 31: 326–39.
Eaton, R. 1969. The cheetah. *Africana*. 3(10): 19–23.
———. 1970a. Group interactions, spacing and territoriality in cheetah. *Z. Tierpsych.* 27(4): 481–91.
———. 1970b. The predatory sequence, with emphasis on killing behavior and its ontogeny, in the cheetah (*Acinonyx jubatus* Schreber). *Z. Tierpsych.* 27(4): 492–504.
———. 1970c. Hunting behavior of the cheetah. *J. Wildl. Mgmt.* 34(1): 56–67.
———. 1970d. Notes on the reproductive biology of the cheetah. In *Int. Zoo yearbook*, vol. 10, ed. J. Lucas, pp. 86–89. Zool. Soc. London, London.
Edey, M. (photographs by J. Dominis). 1968. *The cats of Africa*. Time-Life Books, New York.
Eisenberg, J. 1970. A splendid predator does its own thing untroubled by man. *Smithsonian* 1(6): 48–53.
Eloff, F. 1964. On the predatory habits of lions and hyaenas. *Koedoe* 7: 105–12.
Elton, C. 1949. Population interspersion: an essay on animal community patterns. *J. Ecol.* 37: 1–23.
Encke, W. 1960. Birth and rearing of cheetahs at Krefeld zoo. In *Int. zoo yearbook*, vol. 1, ed. C. Jarvis and D. Morris, pp. 85–86. Zool. Soc. London, London.

Errington, P. 1946. Predation and vertebrate populations. *Quart. Rev. Biol.* 21(2):144–77; (3):221–45.

———. 1956. Factors limiting higher vertebrate populations. *Science* 124(3216):304–7.

Estes, R. 1966. Behaviour and life history of the wildebeest (*Connochaetes taurinus* Burchell). *Nature* 212:999–1000.

———. 1967. Predators and scavengers. *Nat. Hist.* 76(2):20–29; 76(3):38–47.

Estes, R., and J. Goddard. 1967. Prey selection and hunting behavior of the African wild dog. *J. Wildl. Mgmt.* 31(1):52–70.

Ewer, R. 1968. A preliminary survey of the behaviour in captivity of the Dasyurid marsupial, *Sminthopsis crassicaudata* (Gould). *Z. Tierpsych.* 25:319–65.

———. 1969. The "instinct to teach." *Nature* 223:698.

Eyssen, J. 1949. Rogue elephant's dramatic charge into van. *Afr. Wild Life* 3(1):7–16.

Florio, P. and L. Spinelli. 1967. Successful breeding of a cheetah in a private zoo. In *Int. zoo yearbook*, vol. 7, ed. C. Jarvis, pp. 150–52. Zool. Soc. London, London.

Fosbrooke, H. 1948. An administrative survey of the Masai social system. *Tanzania Notes and Records* 26:1–50.

———. 1963. The stomoxys plague in Ngorongoro, 1962. *E. Afr. Wildl. J.* 1:124–26.

Foster, J. 1966. The giraffe of Nairobi National Park: home range, sex ratios, the herd, and food. *E. Afr. Wildl. J.* 4:139–48.

Foster, J. and M. Coe. 1968. The biomass of game animals in Nairobi National Park, 1960–66. *J. Zool., Lond.* 155(4):413–25.

Foster, J. and D. Kearney. 1967. Nairobi National Park game census, 1966. *E. Afr. Wildl. J.* 5:112–20.

Foster, J. and R. McLaughlin. 1968. Nairobi National Park game census, 1967. *E. Afr. Wildl. J.* 6:152–54.

Fox, M. 1969. The anatomy of aggression and its ritualization in Canidae: a developmental and comparative study. *Behaviour* 35(3–4):242–58.

———. 1970. A comparative study of the development of facial expressions in canids; wolf, coyote and foxes. *Behaviour* 36(1–2):49–73.

Frädrich, H. 1965. Zur Biologie und Ethologie des Warzenschweines (*Phacochoerus aethiopicus* Pallas), unter Berücksichtigung des Verhalten anderer Suiden. *Z. Tierpsych.* 22(3–4):328–93.

Gartlan, J. 1968. Structure and function in primate society. *Folia primat.* 8:89–120.

Gartlan, J. and C. Brain. 1968. Ecology and social variability in *Cercopithecus aethiops* and *C. mitis*. In *Primates; studies in adaptation and variability* ed. P. Jay, pp. 253–92. Holt, Rinehart and Winston, New York.

Geist, V. 1966. The evolution of horn-like organs. *Behaviour* 27(3–4):175–214.

Goddard, J. 1967. Home range, behaviour, and recruitment rates of two black rhinoceros populations. *E. Afr. Wildl. J.* 5:133–50.

Golani, I. and H. Mendelssohn. 1971. Sequences of precopulatory behavior of the jackal (*Canis aureus* L.). *Behaviour*. 38(1–2):169–92.

Golley, F. 1960. Energy dynamics of a food chain of an old-field community. *Ecol. Monogr.* 30:187–206.

Gosling, L. 1969. Parturition and related behaviour in Coke's hartebeest, *Alcelaphus buselaphus cokei* Günther. *J. Reprod. Fert.*, suppl. 6:265–86.

Grafton, R. 1965. Food of the blackbacked jackal: a preliminary report. *Zool. Afr.* 1(1):41–53.

Graham, A. 1966. East African Wild Life Society cheetah survey; extracts from the report by Wildlife Services. *E. Afr. Wildl. J.* 4:50–55.

Grzimek, M. and B. Grzimek. 1960. A study of the game of the Serengeti plains. *Z. Säugetierk.* 25:1–61.

Guggisberg, C. 1961. *Simba*. Howard Timmins, Capetown.

Haas, G. 1958. 24-Stunden-Periodik von Grosskatzen im Zoologischen Garten. *Säugetierk. Mitt.* 6:113–17.

v. Haffner, K., G. Rack, and R. Sachs. 1969. Verschiedene Vertreter der Familie Linguatulidae (*Pentastomida*) als Parasiten von Säugetieren der Serengeti (Anatomie, Systematik, Biologie). *Mitt. Hamburg. Zool. Mus. Inst.* 66:93–144.

Hall, K. and I. DeVore. 1965. Baboon social behavior. In *Primate behavior*, ed. I. DeVore, pp. 53–110. Holt, Rinehart and Winston, New York

Haltenorth, T. and W. Trense. 1956. *Das Grosswild der Erde und seine Trophäen.* Bayerischer Landwirtschaftsverlag, Bonn.

Hanks, J., M. Price, and R. Wrangham. 1969. Some aspects of the ecology and behaviour of the Defassa waterbuck (*Kobus defassa*) in Zambia. *Mammalia* 33(3):471–94.

Hanström, B. 1949. Cheetahs provide an unusual experience. *Afr. Wild Life* 3(3):203–9.

Hays, W. 1963. *Statistics for psychologists.* Holt, Rinehart and Winston, New York.

Hendrichs, H. 1970. Schätzungen der Huftierbiomasse in der Dornbusch-savanne nördlich und westlich der Serengetisteppe in Ostafrika nach einem neuen Verfahren und Bemerkungen zur Biomasse der anderen pflanzenfressenden Tierarten. *Säugetierk. Mitt.* 18(3):237–55.

Hirst, S. 1969. Populations in a Transvaal lowveld nature reserve. *Zool. Afr.* 4(2):199–230.

Holling, C. 1965. The functional response of predators to prey density and its role in mimicry and population regulation. *Mem. Ent. Soc. Canada* 45:1–60.

Holzapfel, M. 1956. Über die Bereitschaft zu Spiel and Instinkthandlungen. *Z. Tierpsych.* 13:442–62.

Hoogstraal, H., K. Wassif, I. Helmy, and M. Kaiser. 1966–67. The cheetah, *Acinonyx jubatus* Schreber, in Egypt. *Bull. Zool. Soc. Egypt.* 21:63–68.

Hornocker, M. 1969. Winter territoriality in mountain lions. *J. Wildl. Mgmt.* 33:457–64.

———. 1970. *An analysis of mountain lion predation upon mule deer and elk in the Idaho Primitive Area.* Wildl. monogr. No. 21, The Wildlife Society.

Howell, A. 1944. *Speed in animals*. University of Chicago Press, Chicago.
Hunter, J. 1960. The rapacious wild dog. *Wild Life* 2(1): 42.
Jaeger, F. 1911. *Das Hochland der Riesenkrater und die umliegenden Hochländer Deutsch-Ostafrikas*. Mittler u. Sohn, Berlin.
Jarman, P. and M. Jarman. 1969. *Impala ecology and behaviour*. Serengeti Research Institute Annual Report 1969, pp. 26–29. Tanzania National Parks.
Jarvis, C., ed. 1966. *The international zoo yearbook*, vol. 6. Zool. Soc. London, London.
Jearey, B. 1936. *Pride of lions*. Longmans, Green and Co., London.
Jewell, P. 1966. The concept of home range in mammals. In *Play, exploration and territory in mammals*, ed. P. Jewell and C. Loizos, pp. 85–109. Symp. Zool. Soc. London, no. 18, London.
John, E., P. Chesler, F. Bartlett and I. Victor. 1968. Observation learning in cats. *Science* 159 (3822): 1489–91.
Johnson, M. 1929. *Lion*. G. P. Putnam's Sons, New York.
———. 1935. *Over African jungles*. Harrap, London.
Jordan, P., P. Shelton, and D. Allen. 1967. Numbers, turnover, and social structure of the Isle Royale wolf population. *Am. Zool.* 7: 233–52.
Joslin, P. 1967. Movements and home sites of timber wolves in Algonquin Park. *Am. Zool.* 7: 279–88.
Jouvet, M. 1967. The states of sleep. *Scient. American* 216(2): 62–72.
Kaufmann, J. 1962. Ecology and social behavior of the coati, *Nasua narica* on Barro Colorado Island, Panama. *Univ. Calif. Pub. Zool.* 60(3): 95–222.
Kearton, C. 1929. *In the land of the lion*. Arrowsmith, London.
Keay, R. 1959. *Vegetation map of Africa*. Oxford Univ. Press, Oxford.
Kelsall, J. 1968. *The caribou*. Canadian Wildlife Service Monographs no. 3, Ottawa.
Kiley-Worthington, M. 1965. The waterbuck (*Kobus defassa* Rüppel 1835 and *K. ellipsiprimnus* Ogilby 1833) in East Africa—spatial distribution: a study of the sexual behaviour. *Mammalia* 29(2): 177–210.
Klein, D. 1965. *Ecology of deer range in Alaska*. Ecol. Monogr. 35: 259–84.
Klein, D. and S. Olson. 1960. Natural mortality patterns of deer in Southeast Alaska. *J. Wildl. Mgmt.* 24(1): 80–88.
Klingel, H. 1965. Notes on the biology of the plains zebra *Equus quagga boehmi* Matschie. *E. Afr. Wildl. J.* 3: 86–88.
———. 1967. Soziale Organisation und Verhalten freilebender Steppenzebras. *Z. Tierpsych.* 24: 580–624.
———. 1969. The social organisation and population ecology of the plains zebra (*Equus quagga*). *Zool. Afr.* 4(2): 249–63.
Klingel, H. and U. Klingel. 1966. Tooth development and age determination in the plains zebra (*Equus quagga boehmi* Matschie). *Zool. Garten.* 33(1–3): 34–54.
Klopfer, P. 1964. *Behavioral aspects of ecology*. Prentice-Hall, New Jersey.
Kovach, J. and A. Kling. 1967. Mechanisms of neonate sucking behavior in the kitten. *Animal Behaviour* 15(1): 91–101.

Krampitz, H., R. Sachs, G. Schaller, and R. Schindler. 1968. Zur Verbreitung von Parasiten der Gattung *Hepatozoon* Miller, 1908 (Protozoa, Adeleidae) in ostafrikanischen Wildsäugetieren. *Z. Parasitenk.* 31:203–10.

Kruuk, H. 1964. Predators and anti-predator behaviour of the black-headed gull, *Larus ridibundis* L. *Behaviour,* suppl., no. 11.

———. 1966. Clan-system and feeding habits of spotted hyaenas (*Crocuta crocuta* Erxleben). *Nature* 209:1257–59.

———. 1967. Competition for food between vultures in East Africa. *Ardea* 55(3–4):171–93.

———. 1970. Interactions between populations of spotted hyaenas (*Crocuta crocuta* Erxleben) and their prey species. In *Animal populations in relation to their food resources,* ed. A. Watson, pp. 359–74. Blackwell, Oxford.

———. 1972. *The spotted hyena.* University of Chicago Press, Chicago.

Kruuk, H. and M. Turner. 1967. Comparative notes on predation by lion, leopard, cheetah and wild dog in the Serengeti area, East Africa. *Mammalia* 31(1):1–27.

Kühme, W. 1965. Freilandstudien zur Soziologie des Hyänenhundes (*Lycaon pictus lupinus* Thomas 1902). *Z. Tierpsych.* 22(5):495–541.

———. 1966. Beobachtungen zur Soziologie des Löwen in der Serengeti-Steppe Ostafrikas. *Z. Säugetierk.* 31(3):205–13.

Kurtén, B. 1968. *Pleistocene mammals of Europe.* Weidenfeld and Nicolson, London.

Lack, D. 1954. *The natural regulation of animal numbers.* Clarendon Press, Oxford.

Lamprey, H. 1962. A study of the ecology of the mammal population of a game reserve in the Acacia savanna of Tanganyika, with particular reference to animal numbers and biomass. Ph.D. thesis. University Museum, Oxford.

———. 1964. Estimation of the large mammal densities, biomass, and energy exchange in the Tarangire Game Reserve and the Masai Steppe in Tanganyika. *E. Afr. Wildl. J.* 2:1–46.

Leakey, L. 1967. Development of aggression as a factor in early human and pre-human evolution. In *Aggression and Defense,* ed. C. Clemente and D. Lindsley, pp. 1–33. University of California Press, Los Angeles.

Ledger, H., R. Sachs, and N. Smith. 1967. Wildlife and food production. *World review of animal production* 3(11):13–37.

Leopold, A., L. Sowls, and D. Spencer. 1947. A survey of over-populated deer ranges in the United States. *J. Wildl. Mgmt.* 11(2):162–77.

Leyhausen, P. 1950. Beobachtungen an Löwen-Tiger-Bastarden mit einigen Bemerkungen zur Systematik der Grosskatzen. *Z. Tierpsych.* 7(1):48–83.

———. 1960: 2d edition. *Verhaltensstudien an Katzen.* Paul Parey, Berlin.

———. 1965a. The communal organization of solitary mammals. *Sym. Zool. Soc. London* 14:249–63.

———. 1965b. Über die Funktion der relativen Stimmungshierarchie. *Z. Tierpsych.* 22(4):412–94.

Leyhausen, P. and R. Wolff. 1959. Das Revier einer Hauskatze. *Z. Tierpsych.* 16(6):666–70.

Leyn, G. de. 1962. Contribution a la connaissance des Lycaons au Parc National de la Kagera. *Inst. Parcs Nat. Congo et du Rwanda*. Brussels.
Lindbergh, A. 1966. Immersion in life. *Life* 61(17):89–99.
Loizos, C. 1967. Play behaviour in higher primates: a review. In *Primate ethology*, ed. D. Morris, pp. 176–218. Aldine, Chicago.
Longhurst, W., A. Leopold, and R. Dasmann. 1952. *A survey of California deer herds*. Cal. Dept. Nat. Res., Div. Fish and Game, bull no. 6.
Lorenz, K. 1964. *Man meets dog*. Penguin Books, Baltimore.
Macpherson, A. 1969. *The dynamics of Canadian arctic fox populations*. Canadian Wildlife Service Report Series, no. 8, Ottawa.
Maher, W. 1967. Predation by weasels on a winter population of lemmings, Banks Island, Northwest Territories. *Can. Field-Nat.* 81(4):248–50.
Makacha, S. and G. Schaller. 1969. Observations on lions in the Lake Manyara National Park, Tanzania. *E. Afr. Wildl. J.* 7:99–103.
Mangani, B. 1962. Buffalo kills lion. *Afr. Wild Life* 16(1):27.
Manton, V. 1970. Breeding cheetah at Whipsnade Park. In *Int. zoo yearbook*, vol. 10, ed. J. Lucas, pp. 85–86. Zool. Soc. London, London.
Marler, P. 1957. Specific distinctiveness in the communication signals of birds. *Behaviour* 11(1):13–39.
———. 1965. Communication in monkeys and apes. In *Primate behavior*, ed. I. DeVore, pp. 544–84. Holt, Rinehart and Winston, New York.
———. 1969. *Colobus guereza*: territoriality and group composition. *Science* 163:93–95.
Mason, W. 1965. The social development of monkeys and apes. In *Primate behavior*, ed. I. DeVore, pp. 514–43. Holt, Rinehart and Winston, New York.
McDiarmid, A. 1962. *Diseases of free-living wild animals*. FAO Agricultural Studies, no. 57.
McLaughlin, R. 1970. Aspects of the biology of cheetahs *Acinonyx jubatus* (Schreber) in Nairobi National Park. M.Sc. thesis. University of Nairobi, Nairobi.
McNab, B. 1963. Bio-energetics and the determination of home range size. *Am. Nat.* 97:133–40.
Mech, L. 1966. *The wolves of Isle Royale*. Fauna Nat. Parks of the U.S., fauna ser. 7, Washington.
———. 1970. *The wolf*. Nat. Hist. Press, New York.
Meinertzhagen, R. 1938. Some weights and measurements of large mammals. *Proc. zool. Soc. London* 108:433–39.
Mitchell, B., J. Shenton, and J. Uys. 1965. Predation on large mammals in the Kafue National Park, Zambia. *Zool. Africana* 1(2):297–318.
Moore, A. 1938. *Serengeti*. C. Scribner's Sons, London.
Moorehead, A. 1967. Survival in Serengeti. *The Sunday Times Magazine*, London. August 27, pp. 4–15.
Murie, A. 1944. *The wolves of Mount McKinley*. Fauna of the Natl. Parks of the U.S., fauna ser. 5, Washington.
Murray, M. 1967. The pathology of some diseases found in wild animals in East Africa. *E. Afr. Wildl. J.* 5:37–45.

Nellis, C. and L. Keith. 1968. Hunting activities and success of lynxes in Alberta. *J. Wildl. Mgmt.* 32(4):718–22.

Neseni, R. and G. Heidler. 1966. Ernährungsphysiologische Untersuchungen an Zootieren. *Zool. Garten.* 33(1–3):110–18.

Owen, J. n.d. Primary objects of management of the Serengeti National Park. Mimeographed report in the Tanzania National Parks files.

Ozoga, J. and E. Harger. 1966. Winter activities and feeding habits of northern Michigan coyotes. *J. Wildl. Mgmt.* 30(4): 809–18.

Palen, G. and G. Goddard. 1966. Catnip and oestrous behaviour in the cat. *Animal Behaviour* 14(2–3):372–77.

Pearsall, W. 1957. *Report on an ecological survey of the Serengeti National Park, Tanganyika.* Fauna Preservation Society, London.

Pearson, O. 1964. Carnivore-mouse predation: an example of its intensity and bioenergetics. *J. Mammal* 45(2):177–88.

Pease, A. 1914. *The book of the lion.* John Murray, London.

Percival, A. 1924. *A game ranger's note book.* Nisbet and Co., London.

Pienaar, U. de. 1961. A second outbreak of anthrax amongst game animals in the Kruger National Park, 5th June to 11th October, 1960. *Koedoe* 4:4–14.

———. 1966. An aerial census of elephant and buffalo in the Kruger National Park. *Koedoe* 9:40–107.

———. 1969. Predator-prey relations amongst the larger mammals of the Kruger National Park. *Koedoe* 12:108–76.

Pickering, R. 1968. *Ngorongoro's geological history.* Ngorongoro Conservation Area Booklets, no. 2, Arusha.

Pimlott, D. 1967. Wolf predation and ungulate populations. *Am. Zool.* 7:267–78.

Pimlott, D., J. Shannon, and G. Kolenosky. 1969. *The ecology of the timber wolf in Algonquin Provincial Park.* Dept. Lands and Forests, Ontario.

Pitelka, F. 1959. Numbers, breeding schedule, and territoriality in the pectoral sandpipers of Northern Alaska. *Condor.* 61: 233–64.

Rainsford, W. 1909. *The land of the lion.* Doubleday, New York.

Reuther, R. and J. Doherty. 1968. Birth seasons of mammals at San Francisco Zoo. In *Int. zoo yearbook*, vol. 8, ed. C. Jarvis, pp. 97–101. Zool. Soc. London, London.

Robinette, W. 1963. Weights of some of the larger mammals of N. Rhodesia. *Puku* 1: 207–15.

Robinette, W., J. Gashwiler, and O. Morris. 1961. Notes on cougar productivity and life history. *J. Mammal.* 42(2): 204–17.

Roosevelt, T. and E. Heller, 1922. *Life-histories of African game animals*, vol. 1. John Murray, London.

Rowell, T. 1967. Variability in the social organization of primates. In *Primate ethology*, ed. D. Morris, pp. 219–35. Aldine, Chicago.

Russo, J. 1964. *The Kaibab North deer herd—its history, problems and management.* State of Arizona Game and Fish Dept., bull. no. 7.

Sachs, R. 1967. Liveweights and body measurements of Serengeti game animals. *E. Afr. Wildl. J.* 5:24–36.

Sachs, R. 1969. Untersuchungen zur Artbestimmung und Differenzierung der Muskelfinnen ostafrikanischer Wildtiere. *Z. Tropenmedizin und Parasitologie* 20(1): 39–50.
Sachs, R. and C. Sachs. 1968. A survey of parasitic infestation of wild herbivores in the Serengeti region in Northern Tanzania and the Lake Rukwa region in Southern Tanzania. *Bull. epizoot. Dis. Afr.* 16:455–72.
Sachs, R., G. Schaller, and J. Baker. 1967. Isolation of trypanosomes of the *T. brucei* group from lion. *Acta Tropica* 24(2): 109–12.
Sachs, R., C. Staak, and C. Groocock. 1968. Serological investigation of brucellosis in game animals in Tanzania. *Bull. epizoot. dis. Afr.* 16:93–100.
Sadleir, R. 1966. Notes on reproduction in the larger Felidae. In *Int. zoo yearbook*, vol. 6, ed. C. Jarvis, pp. 184–87. Zool. Soc. London, London.
Saunders, J. 1963. Food habits of the lynx in Newfoundland. *J. Wildl. Mgmt.* 27(3):384–390.
Schaller, G. 1967. *The deer and the tiger*. University of Chicago Press, Chicago.
———. 1968. Hunting behaviour of the cheetah in the Serengeti National Park, Tanzania. *E. Afr. Wildl. J.* 6:95–100.
———. 1969. Life with the King of Beasts. *Natl. Geogr.* 135(4):494–519.
———. 1970. This gentle and elegant cat. *Nat. Hist.* 79(6):31–39.
Schaller, G. and G. Lowther. 1969. The relevance of carnivore behavior to the study of early hominids. *SW. J. of Anthro.* 25(4):307–41.
Schenkel, R. 1966a. Zum Problem der Territorialität und des Markierens bei Säugern—am Beispiel des Schwarzen Nashorns and des Löwens. *Z. Tierpsych.* 23(5):593–626.
———. 1966b. Play, exploration and territoriality in the wild lion. *Symp. Zool. Soc. Lond.* 18:11–22.
Schoener, T. 1968. Sizes of feeding territories among birds. *Ecology* 49(1):123–41.
———. 1969. Models of optimal size for solitary predators. *Am. Nat.* 103(931): 277–313.
Schultze-Westrum, T. 1965. Innerartliche Verständigung durch Düfte beim Gleitbeutler *Petaurus breviceps papuanus* Thomas (Marsupialia, Phalangeridae). *Z. vgl. Physiol.* 50:151–220.
Selous, F. 1908. *African nature notes and reminiscences*. Macmillan, London.
Shortridge, G. 1934. *The mammals of South West Africa*. W. Heinemann, London.
Simon, N. 1966. *Red Data Book*, vol. 1, Mammalia. IUCN, Morges, Switzerland.
Sinclair, A. 1970. Studies of the ecology of the East African buffalo. Ph.D. thesis. Oxford University, Oxford.
Sparks, J. 1967. Allogrooming in primates: a review. In *Primate Ethology*, ed. D. Morris, pp. 148–75. Aldine, Chicago.
Spinelli, P. and L. Spinelli. 1968. Second successful breeding of cheetahs in a private zoo. In *Int. zoo yearbook*, vol. 8, ed. C. Jarvis, pp. 76–78. Zool. Soc. London, London.
Stevenson-Hamilton, J. 1954. *Wild life in South Africa*. Cassell, London.

Steyn, T. 1951. The breeding of lions in captivity. *Fauna and Flora* (Pretoria). 2:37–55.

Sugiyama, Y. 1967. Social organization of Hanuman langurs. In *Social communication among primates*, ed. S. Altmann, pp. 221–36. University of Chicago Press, Chicago.

Swynnerton, G. 1958. Fauna of the Serengeti National Park. *Mammalia* 22:435–50.

Talbot, L. and D. Stewart. 1964. First wildlife census of the entire Serengeti-Mara Region, East Africa. *J. Wildl. Mgmt.* 28(1):815–27.

Talbot, L. and M. Talbot. 1963. *The wildebeest in Western Masailand, East Africa.* Wildl. Monogr. No. 12. The Wildlife Society.

Theberge, J. and J. Falls. 1967. Howling as a means of communication in timber wolves. *Am. Zool.* 7:331–38.

Thenius, E. 1966. Zur Stammesgeschichte der Hyänen (Carnivora, Mammalia). *Z. Säugetierk.* 31:293–300.

Thesiger, W. 1970. Wild dog at 5894 m (19,340 ft). *E. Afr. Wildl. J.* 8:202.

Thom, W. 1934. Tiger shooting in Burma. *J. Bombay Nat. Hist. Soc.* 37(3):571–603.

Tjader, R. 1911. *The big game of Africa.* D. Appleton, London.

Turnbull-Kemp, P. 1967. *The leopard.* Bailey Bros. and Swinfen, London.

Turner, M. 1965. Tanzania's Serengeti. *Africana* 2(5):11–18.

Ulmer, F. 1966. Voices of the felidae. In *Int. zoo yearbook*, vol. 6, ed. C. Jarvis, pp. 259–62. Zool. Soc. London, London.

Van de Werken, H. 1968. Cheetahs in captivity. *Zool. Garten.* 35(3):156–61.

Van Hooff, J. 1967. The facial displays of the Catarrhine monkeys and apes. In *Primate Ethology*, ed. D. Morris, pp. 7–68. Aldine, Chicago.

Verberne, G. 1970. Beobachtungen und Versuche über das Flehmen katzenartiger Raubtiere. *Z. Tierpsych.* 27:807–27.

Walther, F. 1968. *Verhalten der Gazellen.* Die neue Brehm Bücherei. A. Ziemsen Verlag, Wittenberg Lutherstadt.

———. 1969. Flight behaviour and avoidance of predators in Thomson's gazelle (*Gazella Thomsoni* Guenther 1884). *Behaviour* 34(3):184–221.

Watson, R. 1967. The population ecology of the wildebeest (*Connochaetes taurinus albojubatus* Thomas) in the Serengeti. Ph.D. thesis. Cambridge University, Cambridge.

———. 1969. Reproduction of wildebeest *Connochaetes taurinus* in the Serengeti region and its significance to conservation. *J. Reprod. Fert.*, suppl. 6:287–310.

Watson, R., A. Graham, and I. Parker. 1969. A census of the large mammals of Loliondo Controlled Area, northern Tanzania. *E. Afr. Wildl. J.* 7:43–59.

Watson, R. and O. Kerfoot. 1964. A short note on the grazing intensity of the Serengeti by plains game. *Z. Säugetierk.* 29(5):317–20.

Watson, R. and M. Turner. 1965. A count of the large mammals of the Lake Manyara National Park: results and discussion. *E. Afr. Wildl. J.* 3:95–98.

Weir, J. and E. Davison, 1965. Daily occurrence of African game animals at water holes during dry weather. *Zool. Africana* 1(2):353–68.

Wells, E. 1934. *Lions wild and friendly*. Viking Press, New York.

White, S. 1915. *The rediscovered country*. Doubleday, Page and Co., Garden City, New York.

Wilson, V. 1968. Weights of some mammals from Eastern Zambia. *Arnoldia* (Rhodesia) 3:1–19.

Woolpy, J. 1968. The social organization of wolves. *Nat. Hist.* 72:46–55.

Wright, B. 1960. Predation on big game in East Africa. *J. Wildl. Mgmt.* 24(1):1–15.

Wyman, J. 1967. The jackals of the Serengeti. *Animals* 10:79–83.

Wynne-Edwards, V. 1962. *Animal dispersion in relation to social behavior*. Oliver and Boyd, Edinburgh.

Yeoman, G. and J. Walker. 1967. *The ixodid ticks of Tanzania*. Commonwealth Inst. of Entomology, London.

Zuckerman, S. 1953. The breeding season of mammals in captivity. *Proc. Zool. Soc. London* 122(1):827–950.

Index

Note: Entries refer to the lion if not indicated otherwise. See entries on individual species for their characteristics.